Computer Modelling of Heat and Fluid Flow
in Materials Processing

Series in Materials Science and Engineering

Series Editors: **B Cantor**, University of York, UK
M J Goringe, School of Mechanical and Materials Engineering, University of Surrey, UK
E Ma, Department of Materials Science and Engineering, Johns Hopkins University, USA

Other titles in the series

Microelectronic Materials
C R M Grovenor
Department of Materials, University of Oxford, UK

Aerospace Materials
B Cantor, H Assender and P Grant (eds)
Department of Materials, University of Oxford, UK

Fundamentals of Ceramics
M Barsoum
Department of Materials Engineering, Drexel University, USA

Solidification and Casting
B Cantor and K O'Reilly (eds)
Department of Materials, University of Oxford, UK

Topics in the Theory of Solid Materials
J M Vail
Department of Physics and Astronomy, University of Manitoba, Canada

Physical Methods for Materials Characterization: Second Edition
P E J Flewitt, BNFL Magnox Generation and University of Bristol, UK, and R K Wild, University of Bristol, UK

Metal and Ceramic Composites
B Cantor, F P E Dunne and I C Stone (eds)
University of York, UK, and University of Oxford, UK

High Pressure Surface Science and Engineering
Y Gogotsi and V Domnich (eds)
Department of Materials Engineering, Drexel University, USA

Forthcoming titles in the series

Fundamentals of Fibre Reinforced Composite Materials
A R Bunsell and J Renard
Centre des Matériaux, Pierre-Marie Fourt, France

Series in Materials Science and Engineering

Computer Modelling of Heat and Fluid Flow in Materials Processing

Chun-Pyo Hong
Yonsei University, Korea

Institute of Physics Publishing
Bristol and Philadelphia

© IOP Publishing Ltd 2004

All rights reserved. No part of this publication may be reproduced, stored in a retrieval system or transmitted in any form or by any means, electronic, mechanical, photocopying, recording or otherwise, without the prior permission of the publisher. Multiple copying is permitted in accordance with the terms of licences issued by the Copyright Licensing Agency under the terms of its agreement with Universities UK (UUK).

British Library Cataloguing-in-Publication Data

A catalogue record for this book is available from the British Library.

ISBN 0 7503 0445 6

Library of Congress Cataloging-in-Publication Data are available

Series Editors: **B Cantor, M J Goringe and E Ma**

Commissioning Editor: Tom Spicer
Production Editor: Simon Laurenson
Production Control: Sarah Plenty and Leah Fielding
Cover Design: Victoria Le Billon
Marketing: Nicola Newey and Verity Cooke

Published by Institute of Physics Publishing, wholly owned by The Institute of Physics, London

Institute of Physics Publishing, Dirac House, Temple Back, Bristol BS1 6BE, UK

US Office: Institute of Physics Publishing, The Public Ledger Building, Suite 929, 150 South Independence Mall West, Philadelphia, PA 19106, USA

Typeset by Academic + Technical, Bristol
Printed in the UK by MPG Books Ltd, Bodmin, Cornwall

Contents

Preface	xi
1 Mechanisms of transport phenomena	**1**
1.1 Heat transfer	1
1.1.1 Conduction—Fourier's Law of Conduction	1
1.1.2 Convection	2
1.1.3 Radiation	5
1.2 Mass transfer	6
1.2.1 Diffusion—Fick's Law of Diffusion	6
1.2.2 Convective mass transfer	8
1.3 Momentum transfer	9
1.3.1 Viscous momentum transfer—Newton's Law of Viscosity	10
1.3.2 Convective momentum transfer	11
Reference	13
2 Governing equations for transport phenomena	**14**
2.1 Governing equations for mass transfer	14
2.1.1 Integral form of mass balance equation	14
2.1.2 Differential form of mass balance equation—equation of continuity	16
2.2 Governing equations for momentum transfer	19
2.2.1 Integral form of momentum balance equation	19
2.2.2 Differential form of momentum balance equation—equation of motion	20
2.2.3 Boundary conditions	22
2.3 Governing equations for energy transfer	23
2.3.1 Integral form of energy balance equation	23
2.3.2 Differential form of energy balance equation	24
2.3.3 Initial and boundary conditions	25
2.4 Governing equations for species transfer	26

vi   Contents

		2.4.1	Integral form of mass balance equation for species A	26
		2.4.2	Differential form of mass balance equation for species A	27
		2.4.3	Initial and boundary conditions	27
		References		29

3 Similarities among three types of transport phenomena — 30
- 3.1 Basic flux laws — 30
  - 3.1.1 Heat transfer (Fourier's law of conduction) — 31
  - 3.1.2 Mass transfer (Fick's law of diffusion) — 32
  - 3.1.3 Momentum transfer (Newton's law of viscosity) — 32
- 3.2 Convective transfer — 33
- 3.3 Governing equations — 33
- Further readings for chapters 1 through 3 — 36

4 Basics of finite difference methods — 37
- 4.1 Introduction — 37
- 4.2 Finite difference methods — 38
  - 4.2.1 Taylor-series formulation — 38
  - 4.2.2 Integral method — 41
  - 4.2.3 Finite volume method—control volume approach — 44
- References — 47

5 Steady state heat conduction — 48
- 5.1 Mathematical formulation — 48
  - 5.1.1 Governing equation — 48
  - 5.1.2 Boundary conditions — 48
- 5.2 Finite volume approach for steady state problems — 50
  - 5.2.1 Computational grids — 50
  - 5.2.2 Derivation of finite difference equations — 51
  - 5.2.3 Solution of linear algebraic equations — 55
- 5.3 One-dimensional cylindrical and spherical coordinates — 55
  - 5.3.1 Control volumes inside a domain — 56
  - 5.3.2 Control volumes on the outer boundary of a domain — 57
- 5.4 Multi-dimensional problems — 58
  - 5.4.1 Two-dimensional problems — 58
  - 5.4.2 Three-dimensional problems — 59
- 5.5 Worked examples — 60
  - 5.5.1 Example 5.1 — 60
  - 5.5.2 Example 5.2 — 62
  - 5.5.3 Example 5.3 — 64
- 5.6 Case study: one-dimensional steady state heat conduction problems — 67
  - 5.6.1 Description of the problem — 67
  - 5.6.2 Glossary of FORTRAN notation — 67

			Contents	vii

		5.6.3 Simulations	68
		5.6.4 Program list	69
6	**Transient heat conduction**	**73**
	6.1	Mathematical formulation	73
		6.1.1 Governing equation	73
		6.1.2 Initial and boundary conditions	73
	6.2	Finite volume approach for transient problems	74
		6.2.1 Computational grids	74
		6.2.2 Derivation of finite difference equations	74
	6.3	Solving schemes	75
		6.3.1 Fully explicit method	76
		6.3.2 Fully implicit method	78
		6.3.3 Crank–Nicolson method	79
	6.4	Stability analysis—von Neumann stability analysis	80
	6.5	Multi-dimensional problems	84
	6.6	Worked examples	85
		6.6.1 Example 6.1	86
		6.6.2 Example 6.2	87
		6.6.3 Example 6.3	90
	6.7	Case study: one-dimensional transient heat conduction problems	93
		6.7.1 Description of the problem	93
		6.7.2 Glossary of FORTRAN notation	94
		6.7.3 Simulations	95
		6.7.4 Program list	96
		References	100
7	**Phase change problems**	**101**
	7.1	Introduction	101
	7.2	Methods of solution for phase change	103
		7.2.1 Numerical methods	103
		7.2.2 Alloy solidification	105
	7.3	Case study: one-dimensional phase change problems	113
		7.3.1 Description of the problem	113
		7.3.2. Glossary of FORTRAN notation	114
		7.3.3 Simulation	115
		7.3.4 Program List	115
		References	121
8	**Discretization schemes for convection and diffusion terms**	**123**
	8.1	Introduction	123
	8.2	Steady one-dimensional convection and diffusion	124
		8.2.1 Governing equations	124

viii   Contents

		8.2.2 The analytical solution	124
		8.2.3 A control volume approach	126
		8.2.4 The central difference scheme	127
		8.2.5 The upwind difference scheme	129
		8.2.6 The hybrid difference scheme	130
		8.2.7 The power-law scheme	131
	8.3	Comparison among difference schemes	133
		References	134
9	**Solution algorithms for fluid flow analysis**		**135**
	9.1	Governing equations	135
	9.2	Solving schemes	136
		9.2.1 Vorticity-stream function approach	137
		9.2.2 Primitive variable approaches	138
	9.3	Summary	145
		References	147
10	**Fluid flow analysis using the SIMPLE method based on the Cartesian coordinate system**		**148**
	10.1	Governing equations	148
	10.2	Staggered and non-staggered grids	150
	10.3	Discretization method	152
		10.3.1 Discretization of the integral form of transport equation	152
		10.3.2 Discretization of momentum equations	156
		10.3.3 The SIMPLE algorithm	160
	10.4	Treatment of free surfaces	165
		10.4.1 The MAC method	166
		10.4.2 The VOF (volume of fluid) method	168
	10.5	Boundary conditions	171
	10.6	Turbulent flow	172
	10.7	Case studies	174
		10.7.1 Flow over a semi-circular core between two plates	174
		10.7.2 Internal flow in a U-tube	175
		References	177
11	**Fluid flow analysis using the SIMPLE method based on the body-fitted coordinate system**		**179**
	11.1	Introduction	179
	11.2	Transformation of coordinate systems	182
	11.3	Transformation of basic equations	183
	11.4	Discretization method	184
		11.4.1 Discretization of the integral form of transport equation	184

	11.4.2 Discretization of momentum equations	189
	11.4.3 The SIMPLE algorithm	191
11.5	The VOF method in the body-fitted coordinate system	195
11.6	Treatment of a surface cell	196
	11.6.1 Momentum equations	196
	11.6.2. Pressure at a surface cell	198
	11.6.3 Contravariant velocity	199
	11.6.4 Determination of the direction normal to a free surface	199
11.7	Case studies	200
	11.7.1 Flow over a semi-circular core between two plates	200
	11.7.2 Internal flow in a U-tube	201
	References	204

12 Modelling of mould filling — 205

12.1	Introduction	205
12.2	Numerical analysis of filling process	206
	12.2.1 Governing equations	206
	12.2.2 Free surface tracking in mould filling	207
	12.2.3 Algorithms for free surface tracking in the SIMPLE method	210
12.3	Examples of mould filling simulation	211
	12.3.1 Filling in a mould cavity with a straight and tapered gating system using the SIMPLE-VOF method	211
	12.3.2 Filling in a mould cavity with a curved gating system using the SIMPLE-BFC-VOF method	212
	12.3.3 Comparison of the standard SIMPLE-VOF and the SIMPLE-BFC-VOF methods	213
12.4	Case studies	215
	12.4.1 Filling in a rectangular cavity with a semicircular core on the bottom plate	215
	References	218

13 Modelling of microstructure evolution — 219

13.1	Introduction	219
13.2	Nucleation and growth kinetics	221
	13.2.1 Nucleation	221
	13.2.2 Growth kinetics	223
13.3	Classical cellular automaton models	225
	13.3.1 Model description	225
	13.3.2 Nucleation and growth algorithm implemented into CA	226

13.3.3 Coupling the macroscopic heat flow calculation
with CA models ... 227
13.3.4 Examples of classical CA simulation ... 231
13.4 Modified cellular automaton models ... 232
13.4.1 Model description ... 232
13.4.2 Growth ... 233
13.4.3 Coupling the continuum model with a modified CA
model ... 234
13.4.4 Examples of modified CA simulation ... 237
13.5 Case studies ... 244
13.5.1 Simulation of solidification grain structures by
classical cellular automaton models ... 244
13.5.2 Simulation of dendritic growth by modified cellular
automaton models ... 247
References ... 250

Index ... **253**

Preface

The importance of transport phenomena (heat, mass and momentum transfer) in materials processing has been recently re-evaluated. In the majority of materials processing, transport phenomena play an important role in that one or more types of transfer will be closely related to the processing of materials from one state to another. The understanding and analysis of transport phenomena related to materials processes is crucial to the design and optimization of the processes involved. Since 'transport phenomena' was first introduced into materials science and engineering curricula in the 1970s, the teaching of this subject has focused primarily on qualitative understanding of the physical phenomena involved in materials processes. The role of transport phenomena in materials science and engineering has been presented at a very scientific level, with little emphasis on practical application. Consequently, it has been very difficult for researchers and engineers who have this academic background in transport phenomena to put their knowledge into practice in real applications in materials processing. Recently, owing to greatly enhanced computer power, and development of numerical models and computational modeling software, transport phenomena coupled with computer simulations, i.e., 'Computational Transport Phenomena' can bridge this gap between 'Science' and 'Practice'.

The aims of this book are to enhance the capability of the reader (i) to utilize commercially available software, (ii) to develop tailored simulation software suitable for the processes of interest, and (iii) to apply the concepts of transport phenomena in materials science and engineering research. The book is not intended to include a comprehensive review of computational algorithms related to transport phenomena appearing in materials science and engineering problems and examples of applications in the field, rather it includes fundamentals of transport phenomena, basics of the finite-difference/finite-volume methods, algorithms of fluid flow simulations, and a few examples of applications.

The book is essentially intended to be self-contained. However, in order to maximize the benefit of this book the reader needs to have some knowledge of mathematics, especially integral and differential calculus, elementary vector/matrix algebra, and basic numerical methods. This book is intended for final-year undergraduate or graduate students, and also researchers and engineers who work in the field of materials processing and manufacturing.

The book consists of 13 chapters. The first part, consisting of chapters 1–3, provides basic concepts of transport phenomena, conservation laws for energy, mass and momentum, and derivation of governing integral and differential equations. The second part of this book, chapters 4–7, presents fundamentals of finite-difference/finite-volume methods, applications of finite volume methods to steady and transient potential flow problems, and heat transfer problems with phase change. The third part, chapters 8 and 9, deals with discretization schemes for convection and diffusion terms, and solution algorithms for solving fluid flow problems. The fourth part, chapters 10 and 11, describes basics of the SIMPLE methods for simulating fluid flow which include the discretization of governing equations and solution schemes, based on the Cartesian-coordinate and body-fitted-coordinate systems. The treatment of free surfaces in fluid flow is also included. The final part of this book, chapters 12 and 13, is devoted to applications of heat, mass and momentum transfer in materials processing, such as modelings of mould filling of molten metals in a die cavity and microstructure evolution in solidification of metals. The computer programs used in chapters 5–7 and 10–13 are available free online at www.bookmarkphysics.iop.org/.

Finally, the author would like to express his sincere thanks to many friends and colleagues who contributed to this textbook. First, I would like to thank Professor B Cantor, Professor T Umeda and Professor C S Choi for their kind suggestion and encouragement to write this book, and the team at Institute of Physics Publishing for their patience and support throughout the writing process. I especially thank my post-doc researchers, Dr S Y Lee, Dr J H Mok and Dr M F Zhu for their contributions to this book. I continue to be indebted to my students, W J Cho and H N Nam for their efforts in preparing the figures.

I would like to acknowledge my colleagues at Yonsei University, Professor C S Yoon, Professor T S Paik, Professor I W Paik, Professor J H Kim, and Professor C S Shin, for their continuous encouragement. I am also grateful to Dr M Itamura of Nano-Cast Corp. for his kind comments. Finally, I would like to thank my wife and two sons, Jung-Woo and Jin-Hyuk, for their patience and continued encouragement during the compilation of this book.

C P Hong
Yonsei University

'Your beginnings are humble, so prosperous will your future be' (Job 8:7)

Chapter 1

Mechanisms of transport phenomena

Transport phenomena in engineering fields involve three types of transfer: (1) energy or heat transport, (2) mass transport and (3) momentum transport or fluid dynamics. In this chapter, physical mechanisms of three types of transport phenomena will be briefly described.

1.1 Heat transfer

The irreversible phenomenon known as *heat transfer* occurs when there exists a temperature difference in a medium or between media. In order to understand the mechanisms of heat transfer, let us consider the cooling process of a heated carbon steel plate in a furnace, as shown in figure 1.1. Heat is transported from the steel plate to its surroundings by the three modes of heat transfer, which are generally recognized as *conduction, convection* and *radiation*.

1.1.1 Conduction—Fourier's Law of Conduction

The term *conduction* is used to refer to the transport of heat from high temperature to low temperature in a stationary medium, which may be a solid or a fluid, by the motion of molecules or electrons. In engineering applications, it is important to quantify heat transfer processes in terms of appropriate rate equations.

Consider the one-dimensional wall shown in figure 1.2, having a temperature distribution of $T(x)$. The temperature at $x = 0$ is higher than that at $x = L$, so heat is conducted from left to right, according to the rate equation known as *Fourier's law of conduction* expressed by

$$q_x = -\lambda \frac{dT}{dx}. \qquad (1.1.1)$$

The heat flux q_x, which is the heat transfer rate in the x direction per unit area, is proportional to the temperature gradient, dT/dx. The cgs and mks

1

Mechanisms of transport phenomena

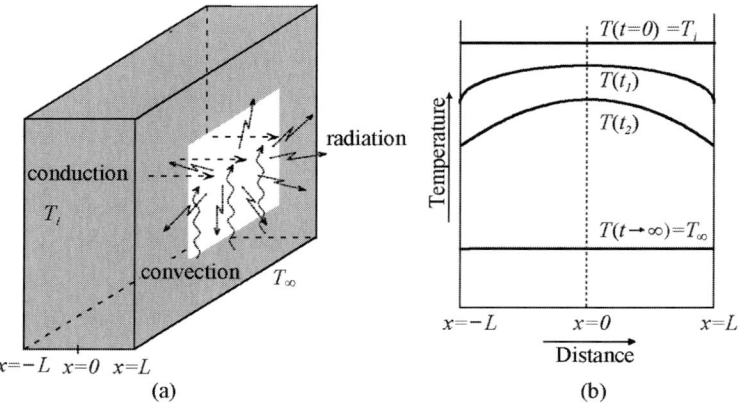

Figure 1.1. (a) Heat transfer mechanisms: conduction, convection and radiation, and (b) variation of temperature profiles with time.

units of the heat flux are $W \cdot cm^{-2}$ and $W \cdot m^{-2}$, respectively. Here the proportional constant λ is the *thermal conductivity* ($W \cdot cm^{-1} \cdot K^{-1}$ and $W \cdot m^{-1} \cdot K^{-1}$ in the cgs and mks units, respectively) and is a characteristic of the wall material. The negative sign in equation (1.1.1) indicates that heat is transferred in the direction of decreasing temperature.

1.1.2 Convection

Energy can be transported not only due to thermal gradient, but also due to bulk fluid flow. In order to estimate the heat transfer related to bulk fluid flow, let us consider a simple example, consisting of a fluid at a bulk temperature of T_∞ flowing with a bulk flow velocity of u_∞ through a circular channel

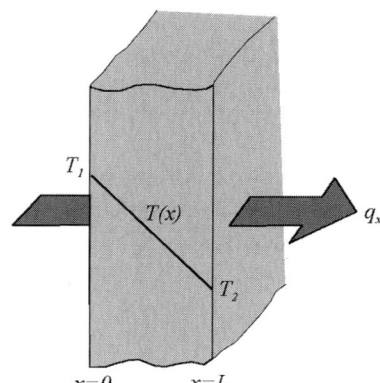

Figure 1.2. One-dimensional heat conduction through a plane wall.

Heat transfer 3

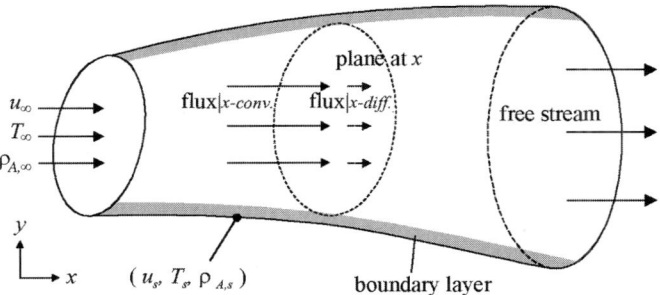

Figure 1.3. Bulk flow of fluids through a circular tube showing two components in heat, mass and momentum transport: conductive (diffusive or viscous) and convective flux terms.

whose inside surface is at temperature T_s, as shown in figure 1.3. If there is a temperature gradient in the fluid in the direction of bulk flow and T_∞ is different from T_s, two kinds of heat transfer can be considered in this system: one is the heat transfer in the direction of flow caused by the bulk fluid motion, and the other is the heat transfer which occurs between a fluid in motion and a bounding surface because of the temperature difference.

1.1.2.1 Energy flux by bulk flow

Consider first the heat transfer term in the direction of flow caused by the bulk fluid motion. This term consists of two components: the conductive (or diffusive) and convective components. The conductive heat flux per unit area across the plane at x is given by $q_{x_{\text{cond}}} = -\lambda (dT/dx)$, while the convective heat flux (or bulk heat flux) caused by bulk flow per unit area is given by $(\rho C_v T) u_\infty$, which is defined as the heat transferred across the plane at x resulting from the motion of the fluid itself. Here ρ is the density of the fluid and C_v is the specific heat capacity. Thus, the total heat flux $q_{x_{\text{total}}}$ is given by

$$q_{x_{\text{total}}} = -\lambda \frac{dT}{dx} + (\rho C_v T) u_\infty. \qquad (1.1.2)$$

Convection heat transfer can be classified according to the nature of the flow. If a fluid motion is induced externally by a pump or a fan, the heat transfer is said to be *forced convection*. If the fluid motion is set up by the buoyancy effect resulting from the density difference caused by the difference of temperature or solute concentration in the fluid, the heat transfer is said to be *free* (or *natural*) *convection*.

1.1.2.2 Thermal boundary layer

Let us now consider heat transfer, which occurs between a fluid in motion and a bounding surface because of the temperature difference. This

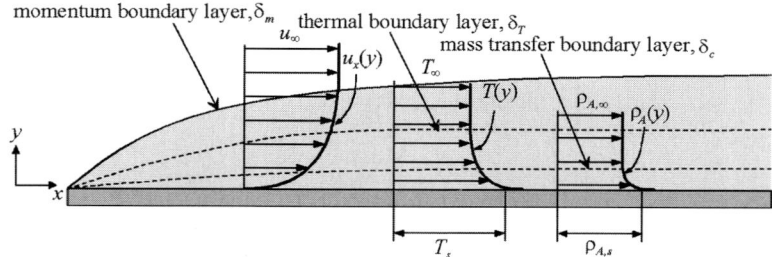

Figure 1.4. Development of momentum, thermal and mass transfer boundary layers in fluid flow over a solid surface.

mechanism of heat transfer, which is frequently encountered in materials processing, is also-called *convection*, since the motion of fluid plays an important role in determining the rate of heat transfer. As illustrated in figure 1.4, because of the interaction between a fluid and a solid surface and the effect of viscosity, there will be a region of fluid through which the fluid velocity varies from zero at the solid surface to u_∞ in the bulk flow. The velocity-affected region of flow is known as the *momentum* (or *velocity*) *boundary layer* δ_m, which is defined as the region where flow velocities are 99% or less of the bulk flow velocity u_∞, i.e., $(u_x(y) - u_s)/(u_\infty - u_s) \leq 0.99$. When u_s is equal to zero, $u_x(y)/u_\infty \leq 0.99$. Similarly, there will be a region of the fluid through which the temperature of the fluid varies from T_s at the solid surface to T_∞ in the free stream. The *thermal boundary layer* δ_T can also be defined as the distance from the solid surface at which the dimensionless temperature $(T(y) - T_s)/(T_\infty - T_s)$ reaches 0.99. As the flow rate increases, both the thicknesses of the velocity and thermal boundary layers decrease, resulting in the increase in both the velocity and temperature gradients.

The convection heat transfer between the fluid and the solid surface consists of two components: one is the contribution due to random molecular motion (diffusion or conduction) which dominates near the solid surface where the fluid velocity is zero, and the other is the contribution due to the bulk fluid motion within the boundary layer. It is, therefore, essential to understand boundary layer phenomena in treating convection heat transfer.

In engineering applications, in order to calculate the convective heat transfer between a fluid and a solid surface, the appropriate rate equation is considered as follow.

$$q_{y_{\text{conv}}} = h(T_\infty - T_s) \qquad (1.1.3)$$

where $q_{y_{\text{conv}}}$ is the convective heat flux between the solid surface and the fluid, which is proportional to the temperature difference between them, and the proportional constant h is referred to as the *convection heat transfer coefficient*. The cgs and mks units of h are $W \cdot cm^{-2} \cdot K^{-1}$ and $W \cdot m^{-2} \cdot K^{-1}$, respectively. This expression is known as *Newton's law of cooling*. In general, the determination of the convection heat transfer coefficient is very complicated since it depends on conditions in the boundary layer, such as the type of flow, the surface geometry, and the physical properties of the fluid.

1.1.3 Radiation

If the heated carbon steel plate, shown in figure 1.1, is located in a vacuum furnace, heat transfer between the steel plate and its surroundings takes place not by conduction or convection, but by some other mechanism. Every body at a finite temperature, which is higher than the absolute zero, emits energy due to its temperature, and the energy emitted is called *thermal radiation*. While the heat transfer by conduction or convection requires a medium, radiation does not since the radiation energy emitted by a body is transmitted in space in the form of electromagnetic waves or photons. In fact, radiation propagates most efficiently in a vacuum. Most gases transmit nearly all incident radiation, but liquids rapidly attenuate radiation. Most solids, except for glasses and transparent plastics, are completely opaque to radiation.

The maximum radiation flux ($W \cdot m^{-2}$) emitted by a body at a temperature T_s is given by the *Stefan–Boltzmann law*.

$$E_b = \sigma T_s^4 \tag{1.1.4}$$

where T_s is the absolute temperature (K) of the surface, σ is the *Stefan–Boltzmann constant* ($\sigma = 5.67 \times 10^{-8}\ W \cdot m^{-2} \cdot K^{-4}$), and E_b is called the blackbody emissive power. Only an ideal radiator (blackbody) can emit radiation flux according to equation (1.1.4). The heat flux emitted by a real surface at T_s is always less than that of the blackbody emissive power and is given by

$$q = \varepsilon E_b = \varepsilon \sigma T_s^4 \tag{1.1.5}$$

where ε is the emissivity of a real surface, whose value is in the range $0 < \varepsilon < 1$.

The analysis of radiation exchange between two or more surfaces is a complicated algebraic procedure. Consider a simple case that occurs frequently in practice where a small surface with a temperature of T_s and a constant emissivity ε is completely surrounded by a large surface at temperature, T_{sur}. When the small surface and the large surroundings are separated by a gas that has no effect on the radiation transfer, the net rate of radiation heat exchange between the surface and the surroundings is given by

$$q_{rad} = \varepsilon \sigma (T_{sur}^4 - T_s^4). \tag{1.1.6}$$

1.2 Mass transfer

Analogous to heat transfer, if there is a difference in the concentration of some chemical species in a mixture, mass transfer, which consists of two modes, *diffusive* and *convective mass transfer*, occurs. While a temperature gradient stands for the driving potential for heat transfer, a species concentration gradient in a mixture provides the driving potential for the mass transfer of that species.

Consider the situation illustrated in figure 1.5, in which CO_2 is flowing over a carbon steel plate at a high temperature. Then, carbon dioxide will be transported to the carbon steel surface by convective mass transfer, where it will react with carbon at the steel surface as follows:

$$C + CO_2 = 2CO.$$
$$(s) \quad (g) \quad \quad (g)$$

This will result in a depletion of carbon concentration at the surface, leading to a difference in carbon concentration within the carbon steel plate. Thus, carbon is transported from the center to the surface of the plate as a result of the diffusive mass transfer. The diffusive and convective mass transfer of carbon will continue until its activity (or chemical potential) in the carbon steel is equal to its activity in the surrounding atmosphere.

1.2.1 Diffusion—Fick's Law of Diffusion

Similar to conduction in heat transfer, the term *diffusion* is used to refer to the transport of a species from a high concentration to a low concentration in a mixture, which originates from molecular activity.

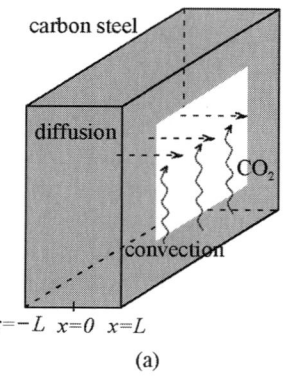
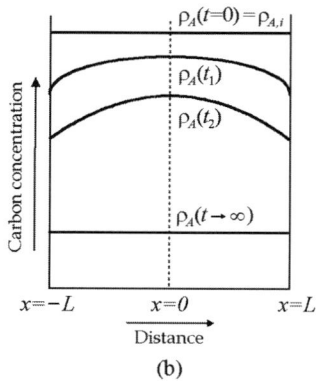

Figure 1.5. (a) Mass transfer mechanisms: diffusive and convective mass transfer, and (b) variation of concentration profiles with time.

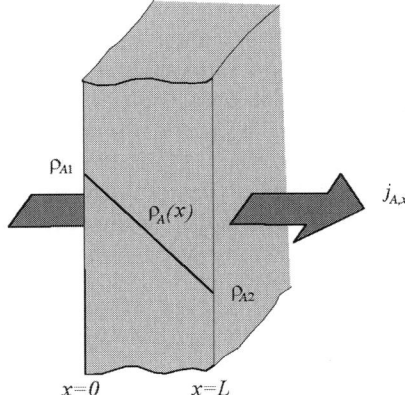

Figure 1.6. One-dimensional diffusion through a plane wall.

Consider the one-dimensional diffusion of species A through a planar medium of A and B, as shown in figure 1.6, having a concentration distribution, $\rho_A(x)$ or $C_A(x)$.

The concentration of species A at $x = 0$ is higher than that at $x = L$, so the transport of species A takes place from left to right. By analogy with the equation for heat conduction, the rate equation for the transfer of species A in a binary mixture of A and B known as *Fick's law of diffusion* is expressed as follows.

In the case of a mass concentration basis

$$j_{A,x} = -\rho D_{AB} \frac{d\omega_A}{dx}. \tag{1.2.1a}$$

Here ρ is the total mass concentration or the density of the binary mixture ($\rho = \rho_A + \rho_B$) and ω_A is the mass fraction of species A, defined by $\omega_A = \rho_A/\rho$ and $\omega_A + \omega_B = 1$. For constant ρ, equation (1.2.1a) simplifies to

$$j_{A,x} = -D_{AB} \frac{d\rho_A}{dx}. \tag{1.2.1b}$$

In the case of a molar concentration basis

$$J_{A,x} = -CD_{AB} \frac{dx_A}{dx}. \tag{1.2.2a}$$

For constant C, equation (1.2.2a) simplifies to

$$J_{A,x} = -D_{AB} \frac{dC_A}{dx}. \tag{1.2.2b}$$

8   *Mechanisms of transport phenomena*

Here C is the molar concentration of the binary mixture ($C = C_A + C_B$) and x_A is the mole fraction of species A, defined by $x_A = C_A/C$ and $x_A + x_B = 1$.

The quantity $j_{A,x}$ (kg·m^{-2}·s^{-1}), which is defined as the mass flux of species A in the x direction per unit area, is proportional to the concentration gradient of species A, $d\rho_A/dx$. The proportional constant D_{AB} is the *mass diffusivity* or *diffusion coefficient* (cm^2·s^{-1} in cgs and m^2·s^{-1} in mks units, respectively). For a molar basis, the quantity $J_{A,x}$ (kmol·m^{-2}·s^{-1}) is the molar flux of species A. The negative signs in equations (1.2.1) and (1.2.2) indicate that mass transfer occurs in the direction of decreasing concentration.

1.2.2 Convective mass transfer

In order to treat the mass transfer related to bulk fluid flow, consider a simple example, consisting of a fluid with the bulk concentration of $\rho_{A,\infty}$ (or $C_{A,\infty}$) flowing through a circular channel whose wall is made of a soluble material with the surface concentration of $\rho_{A,s}$ (or $C_{A,s}$), as illustrated in figure 1.3. Then the solid would dissolve and diffuse away from the solid surface. Analogous to the convective heat transfer, two kinds of mass transfer can be considered in a system where a fluid flows over a solid body or inside a channel. One is the mass transfer in the direction of flow caused by the bulk fluid motion and the other is the mass transfer, which occurs between a fluid in motion and a bounding surface because of a concentration difference.

1.2.2.1 *Mass flux by bulk flow*

Consider first the mass transfer term in the direction of flow caused by the bulk fluid motion. This term consists of two components: diffusive and convective components. The diffusive mass flux per unit area across the plane at x is given by $j_{A,x} = -D_{AB}(d\rho_A/dx)$. The convective component per unit area is given by $\rho_{A,\infty}u_\infty$ (or $C_{A,\infty}u_\infty$), which is defined as the mass flux of A transferred across the plane at x resulting from the motion of the fluid itself. In general, the convective mass flux (bulk mass flux) will dominate while only diffusive mass transfer takes place when there is no bulk motion of fluid. So, the total mass flux of species A is given by

$$n_{A,x} = -D_{AB}(d\rho_A/dx) + \rho_{A,\infty}u_\infty. \tag{1.2.3}$$

The molar flux of species A is given by

$$N_{A,x} = -D_{AB}(dC_A/dx) + C_{A,\infty}u_\infty. \tag{1.2.4}$$

where $n_{A,x}$ and $N_{A,x}$ are the total mass and molar fluxes of species A which consist of two components, diffusive and bulk flux terms.

1.2.2.2 *Mass transfer (or concentration) boundary layer*

Let us now consider the mass transfer between a solid surface and its surrounding fluid, which is frequently encountered in practical problems. As a result of diffusion, the fluid concentration $\rho_A(y)$ in the region near the solid surface is affected by the soluble solid, varying from $\rho_{A,s}$ at the solid surface to $\rho_{A,\infty}$ in the bulk fluid. Analogous to the thermal boundary layer in convection heat transfer, the *mass transfer* or *concentration boundary layer*, δ_C, can also be defined as the distance from the solid surface at which the dimensionless concentration $(\rho_A(y) - \rho_{A,s})/(\rho_{A,\infty} - \rho_{A,s})$ reaches 0.99, as illustrated in figure 1.4.

In order to evaluate the convective mass transfer between a fluid and a solid surface, the mass transfer coefficient K_C (m · s^{-1}) can also be defined as follows.

In the form of mass flux

$$n_{A,y} = K_C(\rho_{A,\infty} - \rho_{A,s}). \tag{1.2.5}$$

In the form of molar flux

$$N_{A,y} = K_C(C_{A,\infty} - C_{A,s}). \tag{1.2.6}$$

As in the case of the convective heat transfer coefficient, the value of the mass transfer coefficient K_C varies with the flow conditions, the fluid properties and the system geometry, and is experimentally determined.

1.3 Momentum transfer

Since the fundamental aspects of heat and mass transport have been discussed in the previous two sections, it is simple for us now to understand the more complex subject of momentum transport. Since the momentum of a body is the product of its mass and velocity, the velocity of a fluid at a given point can be considered as its momentum per unit mass. Therefore, variations in the velocity of a fluid can result in momentum transport, just as variations in temperature and concentration result in heat and mass transport.

In practical fluid flow systems, two different types of fluid flow are considered: laminar and turbulent flow. In laminar flow, the motion of fluid particles shows an orderly manner parallel to the direction of flow without mixing with each other. This type of fluid flow occurs when the flow velocity is small. When the flow velocity is increased, the motion of the fluid particles becomes irregular and is accompanied by fluctuations in velocity, which is called turbulent flow.

The transition from laminar to turbulent flow was studied experimentally by Reynolds in 1883. Based on his observation, Reynolds found that the nature of flow is affected by the four variables, fluid velocity (u), pipe diameter (D), fluid density (ρ) and fluid viscosity (μ), and defined a dimensionless parameter:

$$Re = \frac{\rho u D}{\mu}. \tag{1.3.1}$$

This dimensionless parameter Re is called the *Reynolds number*. The critical value of Re at which transition from laminar to turbulent flow occurs is approximately 2100 for fluid flow in a circular tube.

1.3.1 Viscous momentum transfer—Newton's Law of Viscosity

In order to treat the momentum transport, let us consider a fluid contained between two parallel plates, as shown in figure 1.7. At time $t = 0$ the lower plate is set in motion with a velocity u_s by applying a force F_x in the x direction while the upper plate is stationary. As the lower plate moves, the fluid adjacent to it moves at the same velocity. This is known as the *no-slip boundary condition*. The velocity of the fluid adjacent to the upper plate must be zero. As time proceeds, the movement of the lower plate leads to an increase in the velocity of the fluid in the x direction, from zero to some positive value, and finally to a steady-state value at $t \to \infty$, as shown in figure 1.7. This indicates that x-momentum is transported in the y direction from the lower to the upper region due to the velocity gradient in the fluid.

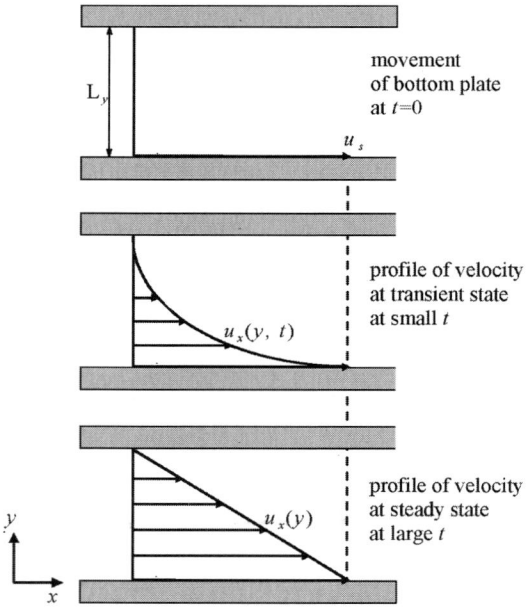

Figure 1.7. One-dimensional flow of a fluid between two parallel plates.

In order to keep the lower plate in motion with a constant velocity u_s, a force F_x must be maintained, which is expressed by

$$\frac{F_x}{A_y} = \mu \frac{u_s}{L_y}, \quad (1.3.2)$$

where L_y is the distance between the two plates and μ is the constant of proportionality. The force per unit area (F_x/A_y) is the shear stress since the force acts tangentially to the area of the plate A_y. At steady-state the velocity profile will be linear as illustrated in figure 1.7. Thus, equation (1.3.2) can be expressed as

$$\tau_{yx} = -\mu \frac{du_x}{dy} \quad (1.3.3)$$

where the proportional constant μ is the *viscosity* of the fluid. The first subscript of the shear stress refers to the area normal to the y axis over which it acts (i.e. the direction of momentum transport) and the second subscript indicates the direction in which the shear stress acts. Similar to the sign convention for heat or mass transfer, the minus sign in equation (1.3.3) represents the fact that momentum is transferred from the lower layers of fluid to the upper layers, in the direction of decreasing velocity. This empirical relationship is known as *Newton's law of viscosity*, and fluids that obey this law are called *Newtonian fluids*.

The cgs unit of stress is $\text{dyn} \cdot \text{cm}^{-2}$ or $\text{g} \cdot \text{cm}^{-1} \cdot \text{s}^{-2}$. Therefore, referring to equation (1.3.3), viscosity can be expressed in $\text{g} \cdot \text{cm}^{-1} \cdot \text{s}^{-1}$, known as a poise (P) or in units of 0.01 P, known as a centipoise (cP). In the mks unit viscosity is expressed in pascal-seconds (Pa·s), where $1\,\text{Pa}\cdot\text{s} = 10\,\text{P} = 10^3\,\text{cP} = 1\,\text{kg}\cdot\text{m}^{-1}\cdot\text{s}^{-1}$.

In many engineering problems, the kinematic viscosity ν, which is defined as the value of viscosity of a fluid divided by its density, is generally expressed as

$$\nu = \frac{\mu}{\rho}. \quad (1.3.4)$$

The kinematic viscosity is commonly expressed in $\text{cm}^2 \cdot \text{s}^{-1}$ and $\text{m}^2 \cdot \text{s}^{-1}$ in the cgs and mks units. Analogous to mass diffusivity, it is called *momentum diffusivity*.

1.3.2 Convective momentum transfer

In order to estimate the momentum transfer related to bulk flow, let us now consider a simple example, consisting of a fluid of uniform velocity u_∞ flowing through a circular channel, as shown in figure 1.3. If there is a velocity gradient in the fluid in the direction of bulk flow (in the x direction), two types of momentum transfer can be considered in this system: one is the momentum transfer in the direction of bulk flow and the other is the momentum transfer which occurs between a fluid in motion and a stationary channel surface because of a velocity difference.

1.3.2.1 Bulk flux of x-momentum

Similar to the cases of the heat and mass transfer due to the bulk flow, the bulk flux of x-momentum in the x direction consists of two components: viscous and bulk flux terms. The viscous momentum flux is given by $\tau_{xx} = -\mu \, du_x/dx$. The bulk flux of x-momentum is given by $(\rho u_x) \cdot u_x$, which is the product of the x-momentum concentration and the bulk flow velocity in the x direction. Thus, the total x-momentum flux π_{xx} in the x direction is given by

$$\pi_{xx} = -\mu(du_x/dx) + (\rho u_x) \cdot u_x. \tag{1.3.5}$$

Since momentum is a vector (whereas temperature and concentration are scalars), it is also possible to have a bulk flux of y-momentum in the x direction, which would be given by the product of the y-momentum concentration (ρu_y) and the bulk flow velocity in the x direction (u_x).

1.3.2.2 Momentum boundary layer

According to the concept of the momentum boundary layer which was first proposed by Prandtl [1], a bulk fluid which flows over a flat plate or in a tube can be divided into two regions: (i) a thin layer adjacent to the solid surface, the *boundary layer*, in which the velocity of the fluid is reduced by the solid because of the effect of viscosity and (ii) the rest of the bulk fluid, the *free stream*, in which the velocity is uniform and the effect of viscosity is no longer important.

Consider the momentum transfer in the momentum boundary layer. At the solid/fluid interface momentum transfer occurs only by viscous flux term since there is no fluid motion. Therefore, the momentum flux across the solid/fluid interface is given by

$$\tau_{yx}|_{y=0} = -\mu \left. \frac{\partial u_x}{\partial y} \right|_{y=0}. \tag{1.3.6}$$

However, the velocity gradient at the interface is not known. Similar to the convective heat and mass transfer at the solid/fluid interface the *momentum transfer coefficient* can be defined as

$$C'_f = \frac{\tau_{yx}|_{y=0}}{u_\infty - u_x|_{y=0}} = \frac{-\mu(\partial u_x/\partial y)|_{y=0}}{u_\infty - u_s}. \tag{1.3.7}$$

The convective momentum transfer between a fluid and a solid surface is then given by

$$\tau_{yx}|_{y=0} = C'_f(u_\infty - u_s). \qquad (1.3.8)$$

This expression is similar to *Newton's law of cooling*.

Reference

[1] Prandtl L 1904 *Proceedings of the 3rd International Congress on Mathematics*, Heidelberg.

Chapter 2

Governing equations for transport phenomena

In materials engineering problems the subject of transport phenomena is mainly concerned with the prediction of temperature, concentration and fluid velocity fields within a melt or a solid medium. In order to model the problems mathematically, we need to set up a series of governing equations and appropriate initial and boundary conditions, which describe the physical phenomena in materials processing. The governing equations can be derived based on the two sets of equations: (1) balance equations (conservation laws) and (2) rate equations (flux laws). There are three fundamental physical laws, which are related to these problems: (i) the law of conservation of mass, (ii) the law of conservation of momentum (Newton's second law of motion), and (iii) the law of conservation of energy (the first law of thermodynamics). The general balance equation for a system can be expressed as

$$\left\{\begin{array}{c} \text{rate of} \\ \text{accumulation} \end{array}\right\} = \left\{\begin{array}{c} \text{rate of} \\ \text{in} \end{array}\right\} - \left\{\begin{array}{c} \text{rate of} \\ \text{out} \end{array}\right\} + \left\{\begin{array}{c} \text{rate of} \\ \text{generation} \end{array}\right\}. \quad (2.0.1)$$

2.1 Governing equations for mass transfer

2.1.1 Integral form of mass balance equation

Let us consider a hypothetical volume element Ω located in a fluid flow field, as illustrated in figure 2.1. We call this volume element a *control volume* and its boundary the *control-volume surface* Γ. With respect to the control volume, the law of conservation of mass for a homogeneous fluid having no chemical reactions may be stated as

$$\underbrace{\left\{\begin{array}{c} \text{rate of mass} \\ \text{accumulation} \end{array}\right\}}_{(1)} = \left\{\begin{array}{c} \text{rate of} \\ \text{mass in} \end{array}\right\} - \left\{\begin{array}{c} \text{rate of} \\ \text{mass out} \end{array}\right\} = \underbrace{\left\{\begin{array}{c} \text{net rate of} \\ \text{mass in} \end{array}\right\}}_{(2)}. \quad (2.1.1)$$

Now the terms in equation (2.1.1) can be evaluated as follows.

Governing equations for mass transfer

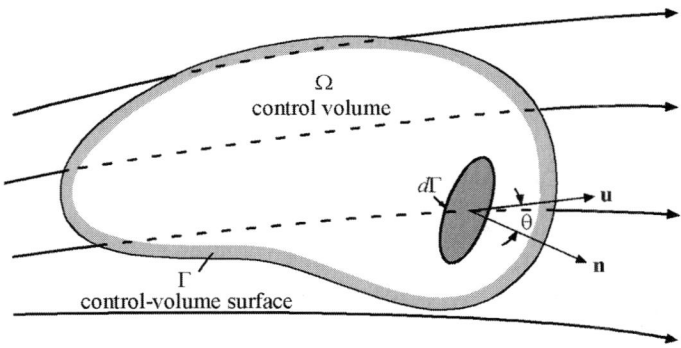

Figure 2.1. Control volume in a fluid flow field.

The rate of mass accumulation in the infinitesimal volume element $d\Omega$ is $\rho\,d\Omega$, which can be integrated over Ω to obtain the total rate of mass accumulation in the control volume.

$$\frac{\partial}{\partial t}\iiint_\Omega \rho\,d\Omega. \tag{2.1.2a}$$

For the infinitesimal surface $d\Gamma$ on the control-volume surface, the rate of mass efflux is equal to $(\rho u)(d\Gamma \cos\theta)$, where θ is the angle between the velocity vector $\mathbf{u}$ and the outward unit normal vector $\mathbf{n}$ to $d\Gamma$, as shown in figure 2.1. From vector algebra we have the relation

$$(\rho u)(d\Gamma \cos\theta) = \rho\,d\Gamma |\mathbf{u}|\,|\mathbf{n}|\cos\theta$$
$$= \rho \mathbf{u}\cdot\mathbf{n}\,d\Gamma.$$

Since $\mathbf{n}$ is the outward normal vector, the rate of mass influx through $d\Gamma$ is $-\rho\mathbf{u}\cdot\mathbf{n}\,d\Gamma$, which can be integrated over the control-volume surface Γ to obtain the total net rate of mass influx into the control volume Ω as follow.

$$-\iint_\Gamma \rho\mathbf{u}\cdot\mathbf{n}\,d\Gamma. \tag{2.1.2b}$$

Substitution of equations (2.1.2a) and (2.1.2b) into equation (2.1.1) gives the mass balance equation

$$\underbrace{\frac{\partial}{\partial t}\iiint_\Omega \rho\,d\Omega}_{(1)} = -\underbrace{\iint_\Gamma \rho\mathbf{u}\cdot\mathbf{n}\,d\Gamma}_{(2)} \tag{2.1.3a}$$

or in the form

$$\frac{\partial}{\partial t}\iiint_\Omega \rho\,d\Omega + \iint_\Gamma \rho\mathbf{u}\cdot\mathbf{n}\,d\Gamma = 0. \tag{2.1.3b}$$

This equation is called the control-volume form or the *integral form of mass balance equation*. The terms (1) and (2) in the above equation involve the physical meanings of the terms (1) and (2) in equation (2.1.1). It is to be noted that equation (2.1.3) is available for any shapes of control volume.

2.1.2 Differential form of mass balance equation—equation of continuity

In the previous section we derived the integral form of mass balance equation based on the concept of the conservation of mass for a general control volume. Let us now consider the derivation of the conservation law in partial differential form.

The differential form of mass balance equation can be derived in two different ways: one is a purely mathematical procedure, in which the integral form of mass balance equation can be transferred into the differential form of mass balance equation with the aid of the *Gauss divergence theorem* [1], and the other is based on a procedure in which the concept of mass conservation can be directly applied to the three-dimensional differential control volumes, as defined in figure 2.2. In the latter method, the differential equations for rectangular, cylindrical and spherical coordinates can be easily derived using the differential control volume appropriate for each coordinate system [2].

First, let us consider the derivation of the differential form of mass balance equation using the integral form of mass balance equation with the aid of the Gauss divergence theorem. The surface integral in equation (2.1.3) can be transformed into a volume integral as

$$\iint_\Gamma \rho \mathbf{u} \cdot \mathbf{n} \, d\Gamma = \iiint_\Omega \nabla \cdot (\rho \mathbf{u}) \, d\Omega. \tag{2.1.4}$$

Substituting equation (2.1.4) into equation (2.1.3) gives

$$\iiint_\Omega \left\{ \frac{\partial \rho}{\partial t} + \nabla \cdot (\rho \mathbf{u}) \right\} d\Omega = 0. \tag{2.1.5}$$

Since this integral must vanish for an arbitrary region Ω and the integrand is a continuous function, it follows that the integrand must be equal to zero. Therefore, we have the differential form of mass balance equation

$$\frac{\partial \rho}{\partial t} + \nabla \cdot (\rho \mathbf{u}) = 0. \tag{2.1.6}$$

This equation is called the *equation of continuity*.

For an incompressible fluid, this equation reduces to

$$\nabla \cdot \mathbf{u} = 0. \tag{2.1.7}$$

The equations of continuity for rectangular, cylindrical and spherical coordinate systems are given in table 2.1.

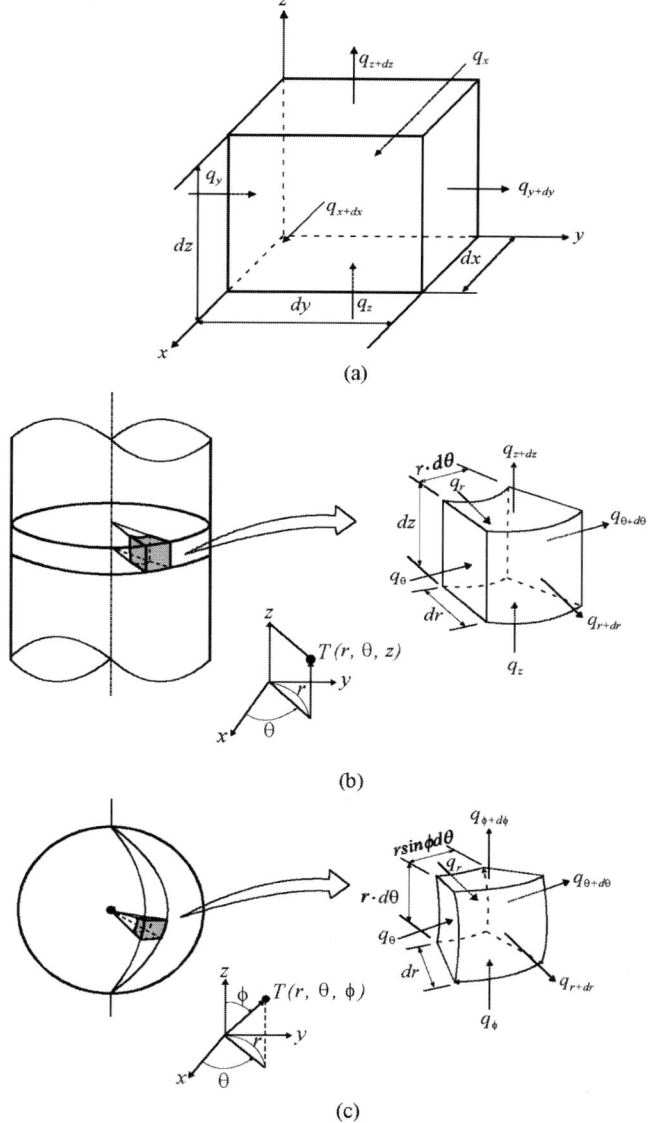

Figure 2.2. Differential control volumes in three types of coordinate system: (a) Cartesian coordinates (x, y, z), (b) cylindrical coordinates (r, θ, z), and spherical coordinates (r, θ, ϕ).

Secondly, let us derive the equation of continuity by applying the concept of mass conservation to the three-dimensional differential control volumes shown in figure 2.2. We now consider the case of rectangular coordinate. The left-hand side of equation (2.1.1), which indicates the rate

Table 2.1. Equation of continuity.

Rectangular coordinates (x, y, z):

$$\frac{\partial \rho}{\partial t} + \frac{\partial}{\partial x}(\rho u_x) + \frac{\partial}{\partial y}(\rho u_y) + \frac{\partial}{\partial z}(\rho u_z) = 0$$

Cylindrical coordinates (r, θ, z):

$$\frac{\partial \rho}{\partial t} + \frac{1}{r}\frac{\partial}{\partial r}(\rho r u_r) + \frac{1}{r}\frac{\partial}{\partial \theta}(\rho u_\theta) + \frac{\partial}{\partial z}(\rho u_z) = 0$$

Spherical coordinates (r, θ, ϕ):

$$\frac{\partial \rho}{\partial t} + \frac{1}{r^2}\frac{\partial}{\partial r}(\rho r^2 u_r) + \frac{1}{r \sin\theta}\frac{\partial}{\partial \theta}(\rho u_\theta \sin\theta) + \frac{1}{r \sin\theta}\frac{\partial}{\partial \phi}(\rho u_\phi) = 0$$

---

of mass accumulation within the differential control volume $(\Delta x \, \Delta y \, \Delta z)$, can be expressed as

$$\frac{\partial}{\partial t}(\rho \, \Delta x \, \Delta y \, \Delta z). \tag{2.1.8a}$$

The right-hand side of equation (2.1.1) consists of three parts, in the x, y and z directions as follows. The net mass influx into the control volume in the x direction is

$$(\rho u_x|_x - \rho u_x|_{x+\Delta x})(\Delta y \, \Delta z), \tag{2.1.8b}$$

in the y direction it is

$$(\rho u_y|_y - \rho u_y|_{y+\Delta y})(\Delta x \, \Delta z), \tag{2.1.8c}$$

and in the z direction it is

$$(\rho u_z|_z - \rho u_z|_{z+\Delta z})(\Delta x \, \Delta y). \tag{2.1.8d}$$

Thus, the total net rate of mass influx is the sum of equations (2.1.8b) through (2.1.8d). Substituting equations (2.1.8a) through (2.1.8d) into equation (2.1.1) yields

$$\frac{\partial}{\partial t}(\rho \, \Delta x \, \Delta y \, \Delta z) = (\rho u_x|_x - \rho u_x|_{x+\Delta x})(\Delta y \, \Delta z)$$
$$+ (\rho u_y|_y - \rho u_y|_{y+\Delta y})(\Delta x \, \Delta z)$$
$$+ (\rho u_z|_z - \rho u_z|_{z+\Delta z})(\Delta x \, \Delta y). \tag{2.1.9}$$

By dividing both sides of equation (2.1.9) by $(\Delta x \, \Delta y \, \Delta z)$ and taking the limit as Δx, Δy and Δz approach zero, we obtain

$$\frac{\partial \rho}{\partial t} + \frac{\partial}{\partial x}(\rho u_x) + \frac{\partial}{\partial y}(\rho u_y) + \frac{\partial}{\partial z}(\rho u_z) = 0 \tag{2.1.10}$$

or in a vector form as

$$\frac{\partial \rho}{\partial t} + \nabla \cdot (\rho \mathbf{u}) = 0. \tag{2.1.11}$$

2.2 Governing equations for momentum transfer

2.2.1 Integral form of momentum balance equation

The concept of momentum balance can be expressed with respect to a control volume as

$$\left\{ \begin{array}{c} \text{rate of} \\ \text{momentum} \\ \text{accumulation} \end{array} \right\} = \left\{ \begin{array}{c} \text{rate of} \\ \text{momentum} \\ \text{in} \end{array} \right\} - \left\{ \begin{array}{c} \text{rate of} \\ \text{momentum} \\ \text{out} \end{array} \right\} + \left\{ \begin{array}{c} \text{sum of} \\ \text{forces acting} \\ \text{on system} \end{array} \right\}. \tag{2.2.1a}$$

The first and second terms on the right-hand side of the above equation consist of two components, convective and viscous momentum transfer. Thus, equation (2.2.1a) can be expressed in a different form as

$$\underbrace{\left\{ \begin{array}{c} \text{rate of} \\ \text{momentum} \\ \text{accumulation} \end{array} \right\}}_{(1)} = \underbrace{\left\{ \begin{array}{c} \text{net rate of} \\ \text{convective} \\ \text{momentum in} \end{array} \right\}}_{(2)} + \underbrace{\left\{ \begin{array}{c} \text{net rate of} \\ \text{viscous} \\ \text{momentum in} \end{array} \right\}}_{(3)} + \underbrace{\left\{ \begin{array}{c} \text{sum of} \\ \text{forces acting} \\ \text{on system} \end{array} \right\}}_{(4)}. \tag{2.2.1b}$$

Similarly, the following integral form of momentum balance equation can be obtained by applying equation (2.2.1) to the control volume Ω shown in figure 2.3 where $\mathbf{f}_\Phi = \mathbf{\tau}$.

$$\underbrace{\frac{\partial}{\partial t} \iiint_\Omega \rho \mathbf{u} \, d\Omega}_{(1)} = \underbrace{-\iint_\Gamma \rho \mathbf{uu} \cdot \mathbf{n} \, d\Gamma}_{(2)} \underbrace{-\iint_\Gamma \mathbf{\tau} \cdot \mathbf{n} \, d\Gamma}_{(3)} \underbrace{-\iint_\Gamma p\mathbf{n} \, d\Gamma}_{(4)} + \underbrace{\iiint_\Omega \mathbf{f}_b \, d\Omega}_{(5)} \tag{2.2.2}$$

where p is the pressure acting on a fluid per unit area and $\mathbf{f}_b$ is the body force per unit volume, which may involve gravitational force, centrifugal force, Coriolis force or electromagnetic force. Term (1) in equation (2.2.2) is the rate of momentum accumulation. Terms (2) and (3) indicate the net rate of convective momentum *in* and the net rate of viscous momentum *in*. Terms (4) and (5) indicate the pressure force acting on the control-volume surface Γ and the body force acting on the control volume, Ω, respectively.

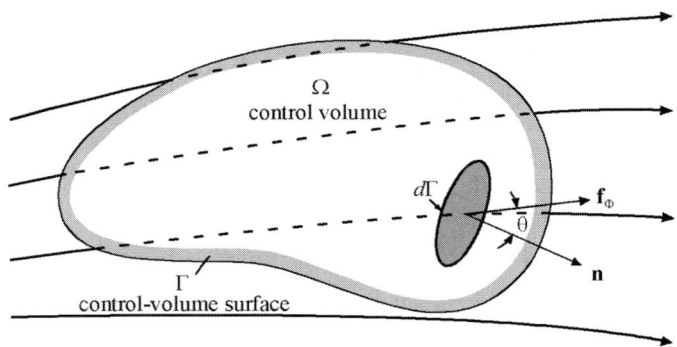

Figure 2.3. Control volume in a fluid field with heat and species transfer. Here, $f_\Phi = \tau$, q and j_A for fluid flow, heat and species transfer, respectively.

By applying the Gauss divergence theorem to term (4) in equation (2.2.2), we obtain the following expression in which terms (1) through (4) represent the integral forms of terms (1) through (4) in equation (2.2.1b).

$$\underbrace{\frac{\partial}{\partial t}\iiint_\Omega \rho \mathbf{u}\, d\Omega}_{(1)} = \underbrace{-\iint_\Gamma \rho \mathbf{uu}\cdot \mathbf{n}\, d\Gamma}_{(2)} - \underbrace{\iint_\Gamma \tau\cdot \mathbf{n}\, d\Gamma}_{(3)} + \underbrace{\iiint_\Omega (\mathbf{f_b} - \nabla p)\, d\Omega}_{(4)}. \quad (2.2.3)$$

2.2.2 Differential form of momentum balance equation—equation of motion

According to the similar procedure used in the derivation of the differential form of mass balance equation, the differential form of momentum balance equation, i.e., the equation of motion is given by

$$\frac{\partial(\rho \mathbf{u})}{\partial t} = -\nabla\cdot(\rho \mathbf{uu}) - \nabla\cdot\tau - \nabla p + \mathbf{f_b}. \quad (2.2.4)$$

By substituting Newton's law of viscosity into the above equation and assuming constant ρ and μ, we obtain

$$\rho\frac{\partial \mathbf{u}}{\partial t} + \rho \mathbf{u}\cdot\nabla \mathbf{u} = \mu\nabla^2\mathbf{u} - \nabla p + \mathbf{f_b}. \quad (2.2.5)$$

If the gravitational force is the only body force involved, then $\mathbf{f_b} = \rho \mathbf{g}$. This equation is the *Navier–Stokes equation* for an incompressible fluid flow. The Navier–Stokes equation can be expressed in terms of the substantial derivative notation, D/Dt, which is defined as

$$\frac{D}{Dt} = \frac{\partial}{\partial t} + u_x\frac{\partial}{\partial x} + u_y\frac{\partial}{\partial y} + u_z\frac{\partial}{\partial z}. \quad (2.2.6)$$

Table 2.2. Equation of motion in rectangular coordinates (x, y, z) (for a Newtonian fluid with constant ρ and μ).

---

x-component:

$$\rho\left\{\frac{\partial u_x}{\partial t} + u_x\frac{\partial u_x}{\partial x} + u_y\frac{\partial u_x}{\partial y} + u_z\frac{\partial u_x}{\partial z}\right\} = \mu\left\{\frac{\partial^2 u_x}{\partial x^2} + \frac{\partial^2 u_x}{\partial y^2} + \frac{\partial^2 u_x}{\partial z^2}\right\} - \frac{\partial p}{\partial x} + \rho g_x$$

y-component:

$$\rho\left\{\frac{\partial u_y}{\partial t} + u_x\frac{\partial u_y}{\partial x} + u_y\frac{\partial u_y}{\partial y} + u_z\frac{\partial u_y}{\partial z}\right\} = \mu\left\{\frac{\partial^2 u_y}{\partial x^2} + \frac{\partial^2 u_y}{\partial y^2} + \frac{\partial^2 u_y}{\partial z^2}\right\} - \frac{\partial p}{\partial y} + \rho g_y$$

z-component:

$$\rho\left\{\frac{\partial u_z}{\partial t} + u_x\frac{\partial u_z}{\partial x} + u_y\frac{\partial u_z}{\partial y} + u_z\frac{\partial u_z}{\partial z}\right\} = \mu\left\{\frac{\partial^2 u_z}{\partial x^2} + \frac{\partial^2 u_z}{\partial y^2} + \frac{\partial^2 u_z}{\partial z^2}\right\} - \frac{\partial p}{\partial z} + \rho g_z$$

Then equation (2.2.5) becomes

$$\rho\frac{D\mathbf{u}}{Dt} = \mu\nabla^2\mathbf{u} - \nabla p + \mathbf{f_b}. \qquad (2.2.7)$$

Equation (2.2.7) is a statement of Newton's second law of motion in the differential form. The left-hand side of equation (2.2.7) is mass(ρ) × acceleration($D\mathbf{u}/Dt$), which is equal to the sum of the viscous force ($\mu\nabla^2\mathbf{u}$), the pressure force ($-\nabla p$) and the gravitational force ($\mathbf{f_b}$). The Navier–Stokes equations are summarized for rectangular, cylindrical and spherical coordinates in tables 2.2–2.4.

Table 2.3. Equation of motion in cylindrical coordinates (r, θ, z) (for a Newtonian fluid with constant ρ and μ).

---

r-component:

$$\rho\left\{\frac{\partial u_r}{\partial t} + u_r\frac{\partial u_r}{\partial r} + \frac{u_\theta}{r}\frac{\partial u_r}{\partial \theta} - \frac{u_\theta^2}{r} + u_z\frac{\partial u_r}{\partial z}\right\}$$

$$= \mu\left[\frac{\partial}{\partial r}\left\{\frac{1}{r}\frac{\partial}{\partial r}(ru_r)\right\} + \frac{1}{r^2}\frac{\partial^2 u_r}{\partial \theta^2} - \frac{2}{r^2}\frac{\partial u_\theta}{\partial \theta} + \frac{\partial^2 u_r}{\partial z^2}\right] - \frac{\partial p}{\partial r} + \rho g_r$$

θ-component:

$$\rho\left\{\frac{\partial u_\theta}{\partial t} + u_r\frac{\partial u_\theta}{\partial r} + \frac{u_\theta}{r}\frac{\partial u_\theta}{\partial \theta} + \frac{u_r u_\theta}{r} + u_z\frac{\partial u_\theta}{\partial z}\right\}$$

$$= \mu\left[\frac{\partial}{\partial r}\left\{\frac{1}{r}\frac{\partial}{\partial r}(ru_\theta)\right\} + \frac{1}{r^2}\frac{\partial^2 u_\theta}{\partial \theta^2} + \frac{2}{r^2}\frac{\partial u_r}{\partial \theta} + \frac{\partial^2 u_\theta}{\partial z^2}\right] - \frac{1}{r}\frac{\partial p}{\partial \theta} + \rho g_\theta$$

z-component:

$$\rho\left\{\frac{\partial u_z}{\partial t} + u_r\frac{\partial u_z}{\partial r} + \frac{u_\theta}{r}\frac{\partial u_z}{\partial \theta} + u_z\frac{\partial u_z}{\partial z}\right\} = \mu\left[\frac{1}{r}\frac{\partial}{\partial r}\left(r\frac{\partial u_z}{\partial r}\right) + \frac{1}{r^2}\frac{\partial^2 u_z}{\partial \theta^2} + \frac{\partial^2 u_z}{\partial z^2}\right] - \frac{\partial p}{\partial z} + \rho g_z$$

Table 2.4. Equation of motion in spherical coordinates (r, θ, ϕ) (for a Newtonian fluid with constant ρ and μ).

r-component:

$$\rho\left\{\frac{\partial u_r}{\partial t} + u_r\frac{\partial u_r}{\partial r} + \frac{u_\theta}{r}\frac{\partial u_r}{\partial \theta} + \frac{u_\phi}{r\sin\theta}\frac{\partial u_r}{\partial \phi} - \frac{u_\theta^2 + u_\phi^2}{r}\right\}$$

$$= \mu\left[\nabla^2 u_r - \frac{2}{r^2}u_r - \frac{2}{r^2}\frac{\partial u_\theta}{\partial \theta} - \frac{2}{r^2}u_\theta\cot\theta - \frac{2}{r^2\sin\theta}\frac{\partial u_\phi}{\partial \phi}\right] - \frac{\partial p}{\partial r} + \rho g_r$$

θ-component:

$$\rho\left\{\frac{\partial u_\theta}{\partial t} + u_r\frac{\partial u_\theta}{\partial r} + \frac{u_\theta}{r}\frac{\partial u_\theta}{\partial \theta} + \frac{u_\phi}{r\sin\theta}\frac{\partial u_\theta}{\partial \phi} + \frac{u_r u_\theta}{r} - \frac{u_\phi^2\cot\theta}{r}\right\}$$

$$= \mu\left[\nabla^2 u_\theta + \frac{2}{r^2}\frac{\partial u_r}{\partial \theta} - \frac{u_\theta}{r^2\sin^2\theta} - \frac{2\cos\theta}{r^2\sin^2\theta}\frac{\partial u_\phi}{\partial \phi}\right] - \frac{1}{r}\frac{\partial p}{\partial \theta} + \rho g_\theta$$

ϕ-component:

$$\rho\left\{\frac{\partial u_\phi}{\partial t} + u_r\frac{\partial u_\phi}{\partial r} + \frac{u_\theta}{r}\frac{\partial u_\phi}{\partial \theta} + \frac{u_\phi}{r\sin\theta}\frac{\partial u_\phi}{\partial \phi} + \frac{u_r u_\phi}{r} + \frac{u_\theta u_\phi}{r}\cot\theta\right\}$$

$$= \mu\left[\nabla^2 u_\phi - \frac{u_\phi}{r^2\sin^2\theta} + \frac{2}{r^2\sin\theta}\frac{\partial u_r}{\partial \phi} + \frac{2\cos\theta}{r^2\sin^2\theta}\frac{\partial u_\theta}{\partial \phi}\right] - \frac{1}{r\sin\theta}\frac{\partial p}{\partial \phi} + \rho g_\phi$$

2.2.3 Boundary conditions

One of the governing differential equations of momentum transfer can be solved using the appropriate initial or boundary conditions, or both, to describe the fluid flow phenomena. Initial conditions refer to the value of **u** at the start of the time interval, for non-steady-state problems. Boundary conditions refer to the values of **u** existing at specific positions on the boundary.

Boundary conditions frequently encountered for momentum transfer include:

1. Prescribed inlet or outlet condition. A fluid flows into a system through a nozzle or flows out of a system, with a prescribed flow velocity.
2. Free-slip condition. The free-slip condition indicates that the momentum flux (hence the velocity gradient) in the liquid is nearly zero and can be assumed to be zero under such conditions as follows: (i) at a plane or axis of symmetry and (ii) at a liquid/gas interface or at the free surface of a liquid.
3. No-slip condition. At a solid/fluid interface, the fluid is assumed to cling to the solid surface with which it is in contact. Thus, at a stationary solid wall the fluid velocity is assumed to be zero. At a moving solid wall, on the other hand, the fluid velocity is equal to the velocity of the solid wall.
4. At a liquid/liquid interface, the momentum flux perpendicular to the interface and the velocity are both continuous across the interface.

2.3 Governing equations for energy transfer

2.3.1 Integral form of energy balance equation

The conservation law of energy is originated from the first law of thermodynamics. The law of energy conservation can be expressed for an open system under an unsteady condition as

$$\begin{Bmatrix} \text{rate of energy} \\ \text{accumulation} \end{Bmatrix} = \begin{Bmatrix} \text{rate of} \\ \text{energy in} \end{Bmatrix} - \begin{Bmatrix} \text{rate of} \\ \text{energy out} \end{Bmatrix} - \begin{Bmatrix} \text{rate of} \\ \text{work done} \end{Bmatrix} + \begin{Bmatrix} \text{rate of heat} \\ \text{generation} \end{Bmatrix}.$$
$$\quad (1) \qquad\qquad\qquad (2) \qquad\qquad\qquad (3) \qquad\qquad\qquad (4) \qquad\qquad\qquad (5)$$
$$(2.3.1)$$

The energy in terms (1) through (3) in the above equation includes the thermal, kinetic, and potential energies per unit volume of the fluid, given by

$$E = \rho(C_v T + u^2/2 + \text{potential energy}).$$

Term (4), indicating the rate of work done by the fluid on the surroundings, includes the pressure, viscous and shaft works. Term (5) indicates the rate of heat generation caused by Joule heating, phase transformations, or chemical reactions.

However, in most problems of materials processing, the kinetic and potential energies in terms (1) through (3) are negligible as compared to the thermal energy. In addition, the pressure, viscous and shaft works included in term (4) are also negligible in most heat transfer problems. When the thermal energy is only considered, equation (2.3.1) simplifies to

$$\begin{Bmatrix} \text{rate of thermal} \\ \text{energy accumulation} \end{Bmatrix} = \begin{Bmatrix} \text{rate of thermal} \\ \text{energy in} \end{Bmatrix} - \begin{Bmatrix} \text{rate of thermal} \\ \text{energy out} \end{Bmatrix}$$
$$+ \begin{Bmatrix} \text{rate of heat} \\ \text{generation} \end{Bmatrix}. \qquad (2.3.2a)$$

The first and second terms on the right-hand side of the above equation can be expressed in a different form as

$$\begin{Bmatrix} \text{rate of thermal} \\ \text{energy} \\ \text{accumulation} \end{Bmatrix} = \begin{Bmatrix} \text{net rate of} \\ \text{thermal energy in} \\ \text{by convection} \end{Bmatrix} + \begin{Bmatrix} \text{net rate of} \\ \text{thermal energy in} \\ \text{by conduction} \end{Bmatrix}$$
$$\quad (1) \qquad\qquad\qquad\qquad (2) \qquad\qquad\qquad\qquad (3)$$
$$+ \begin{Bmatrix} \text{rate of} \\ \text{heat} \\ \text{generation} \end{Bmatrix}. \qquad (2.3.2b)$$
$$\qquad\qquad (4)$$

Similarly, the integral form of energy balance equation can be derived by applying equation (2.3.2b) to the control volume Ω shown in figure 2.3 where $\mathbf{f}_\Phi = \mathbf{q}$.

$$\underbrace{\frac{\partial}{\partial t} \iiint_\Omega (\rho C_v T) \, \mathrm{d}\Omega}_{(1)} = \underbrace{- \iint_\Gamma (\rho C_v T)(\mathbf{u} \cdot \mathbf{n}) \, \mathrm{d}\Gamma}_{(2)} \underbrace{- \iint_\Gamma (\mathbf{q} \cdot \mathbf{n}) \, \mathrm{d}\Gamma}_{(3)} + \underbrace{\iiint_\Omega \dot{g} \, \mathrm{d}\Omega}_{(4)}.$$

(2.3.3)

Term (1) in equation (2.3.3) is the rate of thermal energy accumulation. Terms (2) and (3), which are called *surface phenomena* related to the phenomena across the control-volume surface Γ, indicate the net influxes of thermal energy due to *convection* (bulk fluid flow) and due to *conduction*. Term (4) indicates the rate of heat generation within the control volume Ω where $\dot{g}$ is the rate of heat generation per unit volume. So, terms (1) through (4) in equation (2.3.3) represent the integral forms of terms (1) through (4) in equation (2.3.2b).

2.3.2 Differential form of energy balance equation

The differential energy balance equation can also be derived in two different ways, which were described in section 2.1.2 for the derivation of the differential form of mass balance equation. Then, we have the following equation.

$$\frac{\partial}{\partial t}(\rho C_v T) = -\nabla \cdot (\rho C_v T \mathbf{u}) - \nabla \cdot \mathbf{q} + \dot{g} + \Psi. \qquad (2.3.4a)$$

Here Ψ indicates the viscous dissipation term, which is a function of fluid viscosity and shear–strain rates. The effect of the dissipation function becomes significant when the velocity gradients and viscosity are very high as in supersonic boundary layers. However, in most incompressible fluids its effect is negligible, and we have

$$\frac{\partial}{\partial t}(\rho C_v T) = -\nabla \cdot (\rho C_v T \mathbf{u}) - \nabla \cdot \mathbf{q} + \dot{g}. \qquad (2.3.4b)$$

By substituting the Fourier law of conduction, equation (1.1.1), into the above equation, the differential energy balance equation can be expressed for incompressible fluids as follows.

$$\rho C_v \frac{\partial T}{\partial t} = -\rho \mathbf{u} \cdot \nabla(C_v T) + \nabla \cdot (\lambda \nabla T) + \dot{g}. \qquad (2.3.5a)$$

Assuming constant λ and C_v

$$\rho C_v \left\{ \frac{\partial T}{\partial t} + \mathbf{u} \cdot \nabla T \right\} = \lambda \nabla^2 T + \dot{g}. \qquad (2.3.5b)$$

Table 2.5. Equation of energy conservation for constant C_v and λ.

Rectangular coordinates (x, y, z):

$$\rho C_v \left\{ \frac{\partial T}{\partial t} + u_x \frac{\partial T}{\partial x} + u_y \frac{\partial T}{\partial y} + u_z \frac{\partial T}{\partial z} \right\} = \lambda \left\{ \frac{\partial^2 T}{\partial x^2} + \frac{\partial^2 T}{\partial y^2} + \frac{\partial^2 T}{\partial z^2} \right\} + \dot{g}$$

Cylindrical coordinates (r, θ, z):

$$\rho C_v \left\{ \frac{\partial T}{\partial t} + u_r \frac{\partial T}{\partial r} + u_\theta \frac{1}{r} \frac{\partial T}{\partial \theta} + u_z \frac{\partial T}{\partial z} \right\} = \lambda \left[\frac{1}{r} \frac{\partial}{\partial r} \left(r \frac{\partial T}{\partial r} \right) + \frac{1}{r^2} \frac{\partial^2 T}{\partial \theta^2} + \frac{\partial^2 T}{\partial z^2} \right] + \dot{g}$$

Spherical coordinates (r, θ, ϕ):

$$\rho C_v \left\{ \frac{\partial T}{\partial t} + u_r \frac{\partial T}{\partial r} + u_\theta \frac{1}{r} \frac{\partial T}{\partial \theta} + u_\phi \frac{1}{r \sin \theta} \frac{\partial T}{\partial \phi} \right\}$$
$$= \lambda \left[\frac{1}{r^2} \frac{\partial}{\partial r} \left(r^2 \frac{\partial T}{\partial r} \right) + \frac{1}{r^2 \sin \theta} \frac{\partial}{\partial \theta} \left(\sin \theta \frac{\partial T}{\partial \theta} \right) + \frac{1}{r^2 \sin^2 \theta} \frac{\partial^2 T}{\partial \phi^2} \right] + \dot{g}$$

The energy equation can also be expressed in terms of the substantial derivative notation, D/Dt.

$$\rho C_v \frac{DT}{Dt} = \lambda \nabla^2 T + \dot{g}. \tag{2.3.6}$$

The differential energy equations are summarized for rectangular, cylindrical and spherical coordinates in table 2.5.

2.3.3 Initial and boundary conditions

In order to solve the differential governing equations of heat transfer, we need appropriate initial and boundary conditions.

The initial condition in heat transfer processes is the temperature distributions in the domain at the start of the time interval of interest. The initial temperature distribution may be simply equal to a constant or expressed as a function of space variables.

$$T = T_0 \quad \text{or} \quad T = T_0(x, y, z) \quad \text{at } t = 0. \tag{2.3.7}$$

The boundary conditions frequently encountered in heat transfer are mainly of four kinds as follows:

2.4.1.1 Prescribed temperature

$$T = \bar{T}_b \tag{2.3.8}$$

where $\bar{T}_b$ is the prescribed temperature at a boundary $(x = x_b)$.

26   *Governing equations for transport phenomena*

2.4.1.2   Prescribed heat flux (or adiabatic)

$$-\lambda \frac{\partial T}{\partial x}\bigg|_{x=x_b} = \bar{q}_b. \tag{2.3.9}$$

When a plane is symmetrical thermally and geometrically, then $\bar{q}_b = 0$ at this plane.

2.4.1.3   Convection (h_∞ and T_∞ are prescribed values)

$$-\lambda \frac{\partial T}{\partial x}\bigg|_{x=x_b} = h(T_\infty - T_s) \tag{2.3.10}$$

where h_∞ and T_∞ are the convective heat transfer coefficient and the bulk temperature of the fluid, and T_s is the surface temperature of the solid.

2.4.1.4   Radiation (ε and T_{sur} are prescribed values)

$$-\lambda \frac{\partial T}{\partial x}\bigg|_{x=x_b} = h_r(T_{sur} - T_s) \tag{2.3.11}$$

where T_{sur} is the temperature of the surrounding and h_r is called the *radiation heat transfer coefficient*, given by

$$h_r \equiv \varepsilon\sigma(T_{sur} + T_s)(T_{sur}^2 + T_s^2). \tag{2.3.12}$$

2.4   Governing equations for species transfer

2.4.1   Integral form of mass balance equation for species A

The integral form of mass balance equation for a single component medium has been derived in section 2.1.1. Let us now apply the law of species conservation to a fixed control volume Ω bounded by a control-volume surface Γ, through which a fluid containing species A in a binary solution is flowing, as shown in figure 2.3. If we consider the conservation of a given species A for a binary system, this relation should include a term accounting for the generation or disappearance of A by chemical reaction in the control volume.

Then, the law of conservation of species A can be stated as

$$\left\{\begin{array}{c}\text{rate of}\\ \text{species } A\\ \text{accumulation}\end{array}\right\} = \left\{\begin{array}{c}\text{net rate of}\\ \text{species } A \text{ in}\\ \text{by convection}\end{array}\right\} + \left\{\begin{array}{c}\text{net rate of}\\ \text{species } A \text{ in}\\ \text{by diffusion}\end{array}\right\} + \left\{\begin{array}{c}\text{rate of}\\ \text{species } A\\ \text{generation}\end{array}\right\}.$$

$$\quad(1) \qquad\qquad\qquad (2) \qquad\qquad\qquad (3) \qquad\qquad\qquad (4)$$

$$\tag{2.4.1}$$

Governing equations for species transfer 27

Analogous to the derivation of the integral form of energy balance equation, we have the following integral form of species balance equation by applying equation (2.4.1) to the control volume Ω shown in figure 2.3 where $\mathbf{f}_\phi = \mathbf{j}_A$.

$$\frac{\partial}{\partial t}\iiint_\Omega \rho_A \, d\Omega = -\iint_\Gamma \rho_A \mathbf{u}\cdot\mathbf{n}\, d\Gamma - \iint_\Gamma \mathbf{j}_A\cdot\mathbf{n}\, d\Gamma + \iiint_\Omega r_A\, d\Omega \quad (2.4.2a)$$

(1) \qquad (2) \qquad (3) \qquad (4)

or in the form

$$\frac{\partial}{\partial t}\iiint_\Omega \rho\omega_A \, d\Omega = -\iint_\Gamma \rho\omega_A \mathbf{u}\cdot\mathbf{n}\, d\Gamma - \iint_\Gamma \mathbf{j}_A\cdot\mathbf{n}\, d\Gamma + \iiint_\Omega r_A\, d\Omega. \quad (2.4.2b)$$

Terms (1) through (4) in the above equation represent the integral forms of terms (1) through (4) in equation (2.4.1). This equation is called the *integral form of species balance equation*.

2.4.2 Differential form of mass balance equation for species A

Analogous to the case of heat transfer, the differential equation for species conservation can be given for the rectangular coordinate system as

$$\frac{\partial \rho_A}{\partial t} = -\nabla\cdot(\rho_A \mathbf{u}) - \nabla\cdot\mathbf{j}_A + r_A \quad (2.4.3a)$$

or in the form

$$\frac{\partial(\rho\omega_A)}{\partial t} = -\nabla\cdot(\rho\omega_A \mathbf{u}) - \nabla\cdot\mathbf{j}_A + r_A. \quad (2.4.3b)$$

Substituting Fick's law of diffusion into equation (2.4.3) and assuming constant ρ and D_{AB} yields the equation of continuity for species A

$$\frac{\partial \rho_A}{\partial t} + \mathbf{u}\cdot\nabla\rho_A = D_{AB}\nabla^2\rho_A + r_A \quad (2.4.4)$$

where D_{AB} is the mass diffusivity or diffusion coefficient for species A diffusing through the binary mixture. Equation (2.4.4) is the equation of continuity for species A or the *differential form of species balance equation*.

The above equation can also be expressed in terms of the substantial derivative notation, D/Dt.

$$\frac{D\rho_A}{Dt} = D_{AB}\nabla^2\rho_A + r_A. \quad (2.4.5)$$

The differential forms of species balance equations for rectangular, cylindrical and spherical coordinate systems are given in table 2.6.

2.4.3 Initial and boundary conditions

The initial condition in mass transfer processes is the concentration of the species at the start of the time interval of interest. The initial concentration

Table 2.6. Equation of species conservation for constant ρ and D_{AB}.

Rectangular coordinates (x, y, z):

$$\left\{\frac{\partial \rho_A}{\partial t} + u_x \frac{\partial \rho_A}{\partial x} + u_y \frac{\partial \rho_A}{\partial y} + u_z \frac{\partial \rho_A}{\partial z}\right\} = D_{AB}\left\{\frac{\partial^2 \rho_A}{\partial x^2} + \frac{\partial^2 \rho_A}{\partial y^2} + \frac{\partial^2 \rho_A}{\partial z^2}\right\} + r_A$$

Cylindrical coordinates (r, θ, z):

$$\left\{\frac{\partial \rho_A}{\partial t} + u_r \frac{\partial \rho_A}{\partial r} + u_\theta \frac{1}{r}\frac{\partial \rho_A}{\partial \theta} + u_z \frac{\partial \rho_A}{\partial z}\right\} = D_{AB}\left[\frac{1}{r}\frac{\partial}{\partial r}\left(r\frac{\partial \rho_A}{\partial r}\right) + \frac{1}{r^2}\frac{\partial^2 \rho_A}{\partial \theta^2} + \frac{\partial^2 \rho_A}{\partial z^2}\right] + r_A$$

Spherical coordinates (r, θ, ϕ):

$$\left\{\frac{\partial \rho_A}{\partial t} + u_r \frac{\partial \rho_A}{\partial r} + u_\theta \frac{1}{r}\frac{\partial \rho_A}{\partial \theta} + u_\phi \frac{1}{r \sin\theta}\frac{\partial \rho_A}{\partial \phi}\right\}$$

$$= D_{AB}\left[\frac{1}{r^2}\frac{\partial}{\partial r}\left(r^2 \frac{\partial \rho_A}{\partial r}\right) + \frac{1}{r^2 \sin\theta}\frac{\partial}{\partial \theta}\left(\sin\theta \frac{\partial \rho_A}{\partial \theta}\right) + \frac{1}{r^2 \sin^2\theta}\frac{\partial^2 \rho_A}{\partial \phi^2}\right] + r_A$$

may be simply equal to a constant or expressed as a function of space variables.

$$\rho_A = \rho_{A0} \quad \text{or} \quad \rho_A = \rho_{A0}(x, y, z) \quad \text{at} \quad t = 0. \quad (2.4.6)$$

The boundary conditions frequently encountered in mass transfer are as follows.

2.4.3.1 Prescribed concentration

$$\rho_A = \bar{\rho}_{A,b} \quad \text{at} \quad x = x_b \quad (2.4.7)$$

where $\bar{\rho}_{A,b}$ is the prescribed concentration of species A at a boundary $(x = x_b)$.

2.4.3.2 Prescribed species flux

$$-D_{AB}\frac{\partial \rho_A}{\partial x}\bigg|_{x=x_b} = \bar{j}_{A,b} \quad (2.4.8)$$

where $\bar{j}_{A,b}$ is the prescribed mass flux of species A at a boundary $(x = x_b)$.

2.4.3.3 Convective species transfer (K_C and $\rho_{A,\infty}$ are prescribed values)

$$-D_{AB}\frac{\partial \rho_A}{\partial x}\bigg|_{x=x_b} = K_C(\rho_{A,\infty} - \rho_{A,s}) \quad (2.4.9)$$

where K_C and $\rho_{A,\infty}$ are the convection mass transfer coefficient and the concentration of species A in the bulk fluid, and $\rho_{A,s}$ is the concentration of species A in the fluid adjacent to the surface.

References

[1] Sokolnikoff I S and Redheffer R M 1966 *Mathematics of Physics and Modern Engineering* 2nd edition (McGraw-Hill) p. 425
[2] Bird R B, Stewart W E and Lightfoot E N 1960 *Transport Phenomena* (New York: Wiley) p. 50

Chapter 3

Similarities among three types of transport phenomena

In chapters 1 and 2, the mechanisms of transfer and the governing equations for the three types of transport phenomena were briefly described. It was also found that there are similarities among the transfer mechanisms and the governing equations describing the processes of heat, mass and momentum transport. As a consequence, understanding of one type of transport phenomenon may lead to understanding of other types of phenomenon. It is, therefore, very important to understand the similarities and the differences among the three types of transport phenomena, so that we can solve the practical problems of transport phenomena in materials processing. In this chapter, we will consider the similarities among the three types of transport phenomena with regards to flux laws, convective transfer, and governing equations, which are important in modelling transport phenomena.

3.1 Basic flux laws

Fundamental aspects on heat, mass and momentum transport have been studied to understand how these transport phenomena be related in materials engineering problems, and be treated in numerical simulation.

In order to understand the similarities among the flux laws on heat, mass and momentum transport, consider figure 3.1, which illustrates three types of transfer. The transfer of heat, species A and x-momentum occurs in the direction of decreasing T, ρ_A (or C_A) and u_x. The basic flux laws are summarized as follows.

(i) Heat transfer (Fourier's law of conduction)

$$q_y = -\lambda \frac{dT}{dy}. \tag{3.1.1}$$

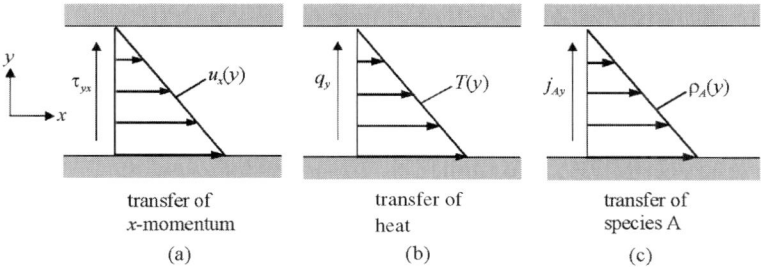

transfer of x-momentum
(a)

transfer of heat
(b)

transfer of species A
(c)

Figure 3.1. Three types of transfer: (a) x-momentum, (b) heat, and (c) species A. The vertical arrows indicate the direction of transport.

(ii) Mass transfer (Fick's law of diffusion)

$$j_{A,y} = -D_{AB} \frac{d\rho_A}{dy} \quad \text{(mass concentration)} \qquad (3.1.2a)$$

$$J_{A,y} = -D_{AB} \frac{dC_A}{dy} \quad \text{(molar concentration)}. \qquad (3.1.2b)$$

(iii) Momentum transfer (Newton's law of viscosity)

$$\tau_{yx} = -\mu \frac{du_x}{dy} \qquad (3.1.3)$$

where λ, D_{AB} and μ indicate the transport properties, and T, ρ_A (or C_A) and u_x the potentials. If the units of the transport properties are unified, these equations can be completely analogous. As mentioned previously, the unit of mass diffusivity is $\text{cm}^2 \cdot \text{s}^{-1}$ in cgs unit (or $\text{m}^2 \cdot \text{s}^{-1}$ in mks unit). Thus we can define the thermal diffusivity as

$$\alpha \equiv \frac{\lambda}{\rho C_v} \qquad (3.1.4a)$$

where α is the thermal diffusivity ($\text{cm}^2 \cdot \text{s}^{-1}$ or $\text{m}^2 \cdot \text{s}^{-1}$). Similarly, the momentum diffusivity is also defined as

$$\nu \equiv \frac{\mu}{\rho}. \qquad (3.1.4b)$$

The flux laws for heat, mass and momentum transport can then be rewritten as follows.

3.1.1 Heat transfer (Fourier's law of conduction)

$$q_y = -\alpha \frac{d(\rho C_v T)}{dy}. \qquad (3.1.5)$$

3.1.2 Mass transfer (Fick's law of diffusion)

$$j_{A,y} = -D_{AB}\frac{d(\rho\omega_A)}{dy} \quad \text{(in mass concentration)} \tag{3.1.6a}$$

$$J_{A,y} = -D_{AB}\frac{d(Cx_A)}{dy} \quad \text{(in molar concentration)}. \tag{3.1.6b}$$

3.1.3 Momentum transfer (Newton's law of viscosity)

$$\tau_{yx} = -\nu\frac{d(\rho u_x)}{dy}. \tag{3.1.7}$$

The terms $\rho C_v T$, $\rho\omega_A$ (or Cx_A) and ρu_x refer to the concentrations of energy, mass and momentum. The flux laws can be expressed simply as

$$\text{(flux)} = -\text{(diffusivity)} \times \text{(concentration gradient)}. \tag{3.1.8}$$

There is an analogy among heat, mass and momentum transport phenomena in many engineering problems. By interchanging analogous transport properties (or diffusivities) and potentials (or concentration), a known solution in heat transport problems can be used to obtain the solution in mass or momentum transport problems or vice versa. It is convenient to apply the similarity among these transport phenomena in computer modelling of materials processing. Table 3.1 summarizes the analogous terms in heat, mass and momentum transport.

Table 3.1. Analogous terms in heat, mass and momentum transfer.

	Energy	Species	Momentum
Flux	q_y	$j_{A,y}$ $J_{A,y}$	τ_{yx}
Transport property	λ	D_{AB}	μ
Potential	T	ρ_A C_A	u_x
Potential gradient	$\partial T/\partial y$	$\partial\rho_A/\partial y$ $\partial C_A/\partial y$	$\partial u_x/\partial y$
Diffusivity	$\alpha \equiv \lambda/(\rho C_v)$	D_{AB}	$\nu \equiv \mu/\rho$
Concentration	$\rho C_v T$	$\rho\cdot\omega_A$ $C\cdot x_A$	ρu_x
Concentration gradient	$\partial(\rho C_v T)/\partial y$	$\partial(\rho\cdot\omega_A)/\partial y$ $\partial(C\cdot x_A)/\partial y$	$\partial(\rho u_x)/\partial y$
Flux law	$q_y = -\alpha\dfrac{\partial(\rho C_v T)}{\partial y}$	$j_{A,y} = -D_{AB}\dfrac{\partial(\rho\cdot\omega_A)}{\partial y}$ $J_{A,y} = -D_{AB}\dfrac{\partial(C\cdot x_A)}{\partial y}$	$\tau_{yx} = -\nu\dfrac{\partial(\rho u_x)}{\partial y}$

3.2 Convective transfer

In materials processing, we frequently need to estimate the transport of heat, mass or momentum to or from a solid surface or a liquid/solid interface, i.e., a boundary between phases. In heat, mass and momentum transport of materials processing, convective transfer caused by bulk fluid motion is concomitant with diffusive transfer governed by the basic flux law. In order to model heat, mass and momentum transfer in materials processing, we also need to understand the similarities not only on the basic flux laws, but also on the convective transfer among the three types of transport phenomena. In this section, we will consider similarities in regards to convective transfer.

The rate equations for convective heat, mass and momentum transfer are summarized as follows.

(1) Convective heat transfer

$$q_{y|y=0}(\equiv -\lambda(\partial T/\partial y)|_{y=0}) = h(T_\infty - T_s). \tag{3.2.1}$$

(2) Convective mass transfer

$$n_{A,y|y=0}(\equiv -D_{AB}(\partial \rho_A/\partial y)|_{y=0}) = K_C(\rho_{A,\infty} - \rho_{A,s}). \tag{3.2.2}$$

(3) Convective momentum transfer

$$\tau_{yx|y=0}(\equiv -\mu(\partial u_x/\partial x) = C'_f(u_\infty - u_s) \tag{3.2.3}$$

where h, K_C and C'_f are the convective heat, mass and momentum transfer coefficients between a fluid and a bounding solid surface, and the terms in the parentheses on the right-hand side of the above equations indicate the potential differences between the bulk fluid and the fluid adjacent to the solid surface. Thus, equations (3.2.1), (3.2.2) and (3.2.3) can be expressed simply as

(flux at the solid surface) = −(transfer coefficient) × (potential difference). (3.2.4)

Similarities among the three boundary layers, thermal, concentration and momentum boundary layers can also be considered in a similar way.

3.3 Governing equations

As discussed on the similarities among the flux laws in heat, mass and momentum transport, we may consider the similarities among the governing equations for three different types of transport phenomena in a similar way.

First, let us consider the integral or control volume form of balance equations. The integral forms of governing equations for mass, momentum, heat and species transfer, equations (2.1.3), (2.2.3), (2.3.3) and (2.4.2), are summarized as follows.

(mass transfer)

$$\frac{\partial}{\partial t}\iiint_\Omega \rho\, d\Omega = -\iint_\Gamma \rho \mathbf{u}\cdot\mathbf{n}\, d\Gamma + \{0\} + \{0\} \qquad (3.3.1)$$

(momentum transfer)

$$\frac{\partial}{\partial t}\iiint_\Omega \rho\mathbf{u}\, d\Omega = -\iint_\Gamma \rho\mathbf{u}\mathbf{u}\cdot\mathbf{n}\, d\Gamma - \iint_\Gamma \tau\cdot\mathbf{n}\, d\Gamma + \iiint_\Omega (\mathbf{f_b} - \nabla p)\, d\Omega \quad (3.3.2)$$

(heat transfer)

$$\frac{\partial}{\partial t}\iiint_\Omega \rho C_v T\, d\Omega = -\iint_\Gamma \rho C_v T\mathbf{u}\cdot\mathbf{n}\, d\Gamma - \iint_\Gamma \mathbf{q}\cdot\mathbf{n}\, d\Gamma + \iiint_\Omega \dot{g}\, d\Omega \quad (3.3.3)$$

(species transfer)

$$\frac{\partial}{\partial t}\iiint_\Omega \rho\omega_A\, d\Omega = -\iint_\Gamma \rho\omega_A\mathbf{u}\cdot\mathbf{n}\, d\Gamma - \iint_\Gamma \mathbf{j_A}\cdot\mathbf{n}\, d\Gamma + \iiint_\Omega r_A\, d\Omega. \quad (3.3.4)$$

Equations (3.3.2) through (3.3.4) have the similar form as

$$\left\{\begin{array}{c}\text{rate of}\\ \text{accumulation}\end{array}\right\} = \left\{\begin{array}{c}\text{net rate of in}\\ \text{by convective}\\ \text{transfer}\end{array}\right\} + \left\{\begin{array}{c}\text{net rate of in}\\ \text{by diffusive or}\\ \text{viscous transfer}\end{array}\right\} + \left\{\begin{array}{c}\text{rate of}\\ \text{generation}\end{array}\right\}.$$

$$(1) \qquad\qquad (2) \qquad\qquad\qquad (3) \qquad\qquad\qquad (4)$$

$$(3.3.5)$$

If we consider the case in which the second and third terms on the right-hand side of equation (3.3.5), which represent the net rate in by diffusion and the rate of generation, to be zero, the general relation, equation (3.3.5), is simplified to equation (3.3.1).

This relation can be expressed as the following general form of integral equation.

$$\frac{\partial}{\partial t}\iiint_\Omega \rho\Phi\, d\Omega = -\iint_\Gamma (\rho\Phi\mathbf{u})\cdot\mathbf{n}\, d\Gamma - \iint_\Gamma \mathbf{f}_\Phi\cdot\mathbf{n}\, d\Gamma + \iiint_\Omega S\, d\Omega. \quad (3.3.6)$$

$$(1) \qquad\qquad\quad (2) \qquad\qquad\quad (3) \qquad\qquad (4)$$

The definitions of the parameters Φ, $\mathbf{f}_\Phi$ and S are given in table 3.2. Terms (1) through (4) in equation (3.3.6) represent the integral forms of terms (1) through (4) in equation (3.3.5).

In the above equation the diffusion flux term due to the gradient of the general transport property $\mathbf{f}_\Phi$ represents viscous stress, heat flux or species

Governing equations 35

Table 3.2. Definitions of Φ, $\mathbf{f}_\Phi$ and S in equations (3.3.6) and (3.3.12) for mass, momentum, heat and species transfer.

Mode of transfer	Φ	$\mathbf{f}_\Phi$	S
Mass transfer	1	0	0
Momentum transfer	$\mathbf{u}$	$\boldsymbol{\tau}$	$\mathbf{f_b} - \nabla p$
Heat transfer	$C_v T$	$\mathbf{q}$	$\dot{g}$
Species transfer	ω_A	$\mathbf{j_A}$	r_A

flux, and is expressed by $-K\nabla\Phi$. Then, the above equation can be expressed as

$$\frac{\partial}{\partial t}\iiint_\Omega \rho\Phi \, d\Omega = -\iint_\Gamma (\rho\Phi\mathbf{u})\cdot\mathbf{n}\,d\Gamma + \iint_\Gamma K\nabla\Phi\cdot\mathbf{n}\,d\Gamma + \iiint_\Omega S\,d\Omega \quad (3.3.7)$$

where K is the diffusion coefficient.

Let us now consider the differential forms of governing equations for mass, momentum, heat and species transfer. Equations (2.1.11), (2.2.4), (2.3.4) and (2.4.3) can be rearranged for the comparison as follows.

(mass transfer)

$$\frac{\partial(\rho)}{\partial t} = -\nabla\cdot(\rho\mathbf{u}) + \{0\} + \{0\} \quad (3.3.8)$$

(momentum transfer)

$$\frac{\partial(\rho\mathbf{u})}{\partial t} = -\nabla\cdot(\rho\mathbf{u}\mathbf{u}) - \nabla\cdot\boldsymbol{\tau} + (\mathbf{f_b} - \nabla p) \quad (3.3.9)$$

(heat transfer)

$$\frac{\partial(\rho C_v T)}{\partial t} = -\nabla\cdot(\rho C_v T\mathbf{u}) - \nabla\cdot\mathbf{q} + \dot{g} \quad (3.3.10)$$

(species transfer)

$$\frac{\partial(\rho\omega_A)}{\partial t} = -\nabla\cdot(\rho\omega_A\mathbf{u}) - \nabla\cdot\mathbf{j_A} + r_A. \quad (3.3.11)$$

Similar to the integral form of balance equation, the above equations can also be expressed as the following general form of differential equation.

$$\frac{\partial(\rho\Phi)}{\partial t} = -\nabla\cdot(\rho\Phi\mathbf{u}) - \nabla\cdot\mathbf{f}_\Phi + S. \quad (3.3.12)$$

The definitions of the parameters Φ, $\mathbf{f}_\Phi$ and S are given in table 3.2. Similarly, the above equation can also be expressed by

$$\frac{\partial(\rho\Phi)}{\partial t} = -\nabla\cdot(\rho\Phi\mathbf{u}) + K\nabla^2\Phi + S. \quad (3.3.13)$$

Further readings for chapters 1 through 3

Bird R B, Stewart W E and Lightfood E N 1960 *Transport Phenomena* (New York: Wiley)
Fahien R W 1983 *Fundamentals of Transport Phenomena* (New York: McGraw-Hill)
Gaskell D R 1992 *An Introduction to Transport Phenomena in Materials Engineering* (London: Macmillan)
Guthrie R I L 1989 *Engineering in Process Metallurgy* (Oxford: Clarendon Press)
Kou S 1996 *Transport Phenomena and Materials Processing* (New York: Wiley)
Poirier D R and Geiger G H 1994 *Transport Phenomena in Materials Processing* (Pennsylvania: TMS)
Welty J R, Wicks C E and Wilson R E 1984 *Fundamentals of Momentum, Heat and Mass Transfer* 3rd edition (New York: Wiley)

Chapter 4

Basics of finite difference methods

In chapters 1 through 3, we have studied the fundamentals of heat, mass and momentum transfer, and derived the integral and differential forms of balance equations, which govern transport phenomena in materials processing. In this chapter, we will briefly describe the fundamentals of finite difference methods, which will be used throughout this textbook for numerical modelling of transport phenomena in materials processing.

4.1 Introduction

In many materials processing problems we frequently encounter problems in which the distributions of temperature and species concentration, and fluid flow velocity fields are to be predicted for optimizing the processes. In order to solve these problems we need to set up a series of partial differential equations and boundary conditions, which must be satisfied in a domain and on its boundary. This mathematical procedure is referred to as *mathematical formulation*. When a mathematical formulation, which characterizes *physical phenomena*, is made, the set of partial differential equations need to be solved. There are two main types of solution procedure: analytical and numerical methods.

Analytical methods for solving partial differential equations have been extensively developed in heat and mass transfer, including the textbooks by Carslaw and Jaeger [1], and by Crank [2]. The analytical solutions, obtained by separation of variables, reflection and superposition of solutions, Laplace transform, are generally expressed as algebraic formulae, power series and transcendental functions, which satisfy the differential equations at every point in the continuous problem domain. However, analytical methods are available with the limitation that the equations should be linear, i.e., thermal and physical properties of materials involved must be regarded as independent of temperature or position. The solution becomes much more complicated when an attempt is made to take phase change problems into consideration.

38   *Basics of finite difference methods*

On the other hand, numerical methods can cope with non-linearity problems. Numerical methods are considered to have much more potential than analytical methods in practical problems. There are three typical numerical methods being used for solving the partial differential equations: (i) finite difference method, (ii) finite element method and (iii) boundary element method.

The finite difference method (FDM) based on Taylor-series formulation, pioneered by Dusinberre [3], offers advantages in regard to numerical formulation, data preparation and computing times, but it may not be appropriate for complex geometries because of its restrictions on the element shape. The finite element method (FEM), originally developed for solving complex stress analysis problems, has been applied to heat transfer problems [4] and problems including phase change [5]. Several studies on the applications of the boundary element method (BEM), which is based on a combination of integral equations and weighted residual methods, have been reported on heat transfer [6] and solidification problems [7]. Recently, the finite volume (or control volume) method (FVM), which is another way of using finite difference approximations, has been extensively applied to analyze heat and fluid flow problems in materials processing. In the finite volume method, finite difference equations are developed by constraining the integral form of governing equations to a finite control volume and conserving the specific physical quantity such as energy, mass, species or momentum over the control volume [8].

In this chapter, the finite difference methods, which are mathematically much simpler than the finite element and boundary element methods, will be briefly described.

4.2 Finite difference methods

Several procedures commonly used to develop finite difference equations include (i) Taylor-series formulation, (ii) integral method and (iii) finite volume (control volume) method.

4.2.1 Taylor-series formulation

4.2.1.1 Expansion of Taylor series

We now demonstrate how to derive finite difference equations using Taylor-series expansion. Figure 4.1 indicates one-dimensional grid points for the Taylor-series expansion.

The definition of the derivative of the function $\Phi(x)$ at $x = x_0$ given by equation (4.2.1) is the basic idea of finite-difference representation of a derivative.

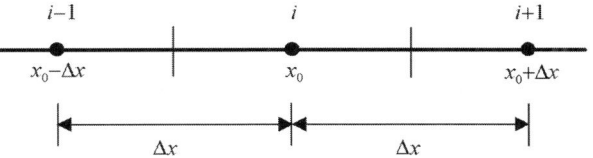

Figure 4.1. One-dimensional grid points used for the Taylor-series expansion.

$$\frac{\partial \Phi}{\partial x} = \lim_{\Delta x \to 0} \frac{\Phi(x_0 + \Delta x) - \Phi(x_0)}{\Delta x}. \quad (4.2.1)$$

If Δx is sufficiently small but finite, the right-hand side of the above equation can be a reasonable approximation to the left-hand side, $\partial \Phi / \partial x$.

Taylor-series expansion of a function $\Phi(x)$ about a point x_0 can be given both in the forward (i.e., positive x) and backward (i.e., negative x) directions as follows.

$$\Phi(x_0 + \Delta x) = \Phi(x_0) + \left.\frac{\partial \Phi}{\partial x}\right|_{x_0} \Delta x + \left.\frac{\partial^2 \Phi}{\partial x^2}\right|_{x_0} \frac{(\Delta x)^2}{2!} + \left.\frac{\partial^3 \Phi}{\partial x^3}\right|_{x_0} \frac{(\Delta x)^3}{3!} + \cdots \quad (4.2.2)$$

$$\Phi(x_0 - \Delta x) = \Phi(x_0) - \left.\frac{\partial \Phi}{\partial x}\right|_{x_0} \Delta x + \left.\frac{\partial^2 \Phi}{\partial x^2}\right|_{x_0} \frac{(\Delta x)^2}{2!} - \left.\frac{\partial^3 \Phi}{\partial x^3}\right|_{x_0} \frac{(\Delta x)^3}{3!} + \cdots \quad (4.2.3)$$

These two expressions form the basis for developing finite difference approximations for the first and the second derivatives, $\partial \Phi / \partial x$ and $\partial^2 \Phi / \partial x^2$.

4.2.1.2 Finite difference approximation for the first derivatives

By rearranging the above equations, we can obtain the forward and the backward finite difference approximations for the first derivative as follows.

$$\left.\frac{\partial \Phi}{\partial x}\right|_{x_0} = \frac{\Phi(x_0 + \Delta x) - \Phi(x_0)}{\Delta x} + O(\Delta x) \quad \text{(forward)} \quad (4.2.4)$$

$$\left.\frac{\partial \Phi}{\partial x}\right|_{x_0} = \frac{\Phi(x_0) - \Phi(x_0 - \Delta x)}{\Delta x} + O(\Delta x) \quad \text{(backward)} \quad (4.2.5)$$

where the order of notation $O(\Delta x)$ characterizes the truncation error associated with the finite difference approximation. It represents the difference between the partial derivative and its finite-difference representation.

By subtracting equation (4.2.3) from equation (4.2.2), we obtain

$$\left.\frac{\partial \Phi}{\partial x}\right|_{x_0} = \frac{\Phi(x_0 + \Delta x) - \Phi(x_0 - \Delta x)}{2\Delta x} + O(\Delta x^2) \quad \text{(central).} \quad (4.2.6)$$

This representation is known as the central difference approximation of order $(\Delta x)^2$, hence is a more accurate approximation than the forward and backward differences of order (Δx).

In order to simplify the above finite difference approximations, let i be the grid point at x_0. The notations $(i+1)$ and $(i-1)$ refer to the grid points at $x_0 + \Delta x$ and $x_0 - \Delta x$, respectively. Applying this notation and neglecting the error terms yields the following finite difference approximations for the first derivative.

$$\left.\frac{\partial \Phi}{\partial x}\right|_i = \frac{\Phi_{i+1} - \Phi_i}{\Delta x} \quad \text{(forward)} \quad (4.2.7)$$

$$\left.\frac{\partial \Phi}{\partial x}\right|_i = \frac{\Phi_i - \Phi_{i-1}}{\Delta x} \quad \text{(backward)} \quad (4.2.8)$$

$$\left.\frac{\partial \Phi}{\partial x}\right|_i = \frac{\Phi_{i+1} - \Phi_{i-1}}{2\Delta x} \quad \text{(central)} \quad (4.2.9)$$

4.2.1.3 Finite difference approximation for the second derivatives

The Taylor series expansions given by equations (4.2.2) and (4.2.3) can also be used to develop finite difference approximations for the second derivatives.

In the case of the second derivative the central finite difference approximation is commonly employed. By adding equations (4.2.2) and (4.2.3), we obtain the central finite difference approximation as

$$\left.\frac{\partial^2 \Phi}{\partial x^2}\right|_{x_0} = \frac{\Phi(x_0 - \Delta x) - 2\Phi(x_0) + \Phi(x_0 + \Delta x)}{\Delta x^2} + O(\Delta x^2) \quad (4.2.10a)$$

or in a simple form as

$$\left.\frac{\partial^2 \Phi}{\partial x^2}\right|_i = \frac{\Phi_{i-1} - 2\Phi_i + \Phi_{i+1}}{\Delta x^2} \quad \text{(central)}. \quad (4.2.10b)$$

4.2.1.4 Application of the finite difference approximation

As a simple example, let us now derive finite difference equations of the following one-dimensional, transient convection–diffusion equation. For the sake of simplicity, the flow velocity u is assumed to be constant.

$$\underbrace{\frac{\partial \Phi}{\partial t}}_{(1)} = \underbrace{-u \frac{\partial \Phi}{\partial x}}_{(2)} + \underbrace{\lambda \frac{\partial^2 \Phi}{\partial x^2}}_{(3)} \quad (4.2.11)$$

Term (1) indicates the transient term containing the time derivative, and terms (2) and (3) are the convection and diffusion terms, respectively.

Finite difference methods

By applying the forward difference scheme for the time derivative and the central difference scheme for the space derivatives (convection and diffusion terms), we obtain the following finite difference equation.

$$\frac{\Phi_i^{t+\Delta t} - \Phi_i^t}{\Delta t} = -u\frac{\Phi_{i+1}^t - \Phi_{i-1}^t}{2\Delta x} + \lambda \frac{\Phi_{i-1}^t - 2\Phi_i^t + \Phi_{i+1}^t}{\Delta x^2}. \tag{4.2.12}$$

Here, the superscripts t and $t + \Delta t$ represent the present and future time levels, respectively. We can solve equation (4.2.12) explicitly for obtaining $\Phi_i^{t+\Delta t}$ in terms of the known values at the present time level. However, the use of the central difference scheme causes a stability constraint on the time step Δt when applied to discretize the convection term, i.e., term (2) in equation (4.2.11). In this case the upwind scheme can be used to discretize the convection term, which will be described in chapter 8.

4.2.2 Integral method

Let us now consider the derivation of finite difference approximations for the one-dimensional convection–diffusion equation, equation (4.2.11), by the integral method.

$$\frac{\partial \Phi}{\partial t} = -u\frac{\partial \Phi}{\partial x} + \lambda \frac{\partial^2 \Phi}{\partial x^2}. \tag{4.2.11}$$

By integrating the partial differential equation with respect to the independent variables, such as x and t over the local neighborhood of grid point (t, x) as shown in figure 4.2, we can develop the finite difference equations.

If we integrate both sides of the above equation over the intervals t to $t + \Delta t$ and $\Gamma_{i-1/2}$ to $\Gamma_{i+1/2}$, we have

$$\int_\Omega \left\{ \int_t^{t+\Delta t} \frac{\partial \Phi}{\partial t} dt \right\} d\Omega = -u \int_\Omega \left\{ \int_t^{t+\Delta t} \frac{\partial \Phi}{\partial x} dt \right\} d\Omega$$
$$+ \lambda \int_\Omega \left\{ \int_t^{t+\Delta t} \frac{\partial^2 \Phi}{\partial x^2} dt \right\} d\Omega. \tag{4.2.13}$$

In the case of one-dimensional rectangular coordinate system, equation (4.2.13) can be simplified as

$$\int_{\Gamma_{i-1/2}}^{\Gamma_{i+1/2}} \left\{ \int_t^{t+\Delta t} \frac{\partial \Phi}{\partial t} dt \right\} dx = -u \int_{\Gamma_{i-1/2}}^{\Gamma_{i+1/2}} \left\{ \int_t^{t+\Delta t} \frac{\partial \Phi}{\partial x} dt \right\} dx$$
$$+ \lambda \int_{\Gamma_{i-1/2}}^{\Gamma_{i+1/2}} \left\{ \int_t^{t+\Delta t} \frac{\partial^2 \Phi}{\partial x^2} dt \right\} dx. \tag{4.2.14}$$

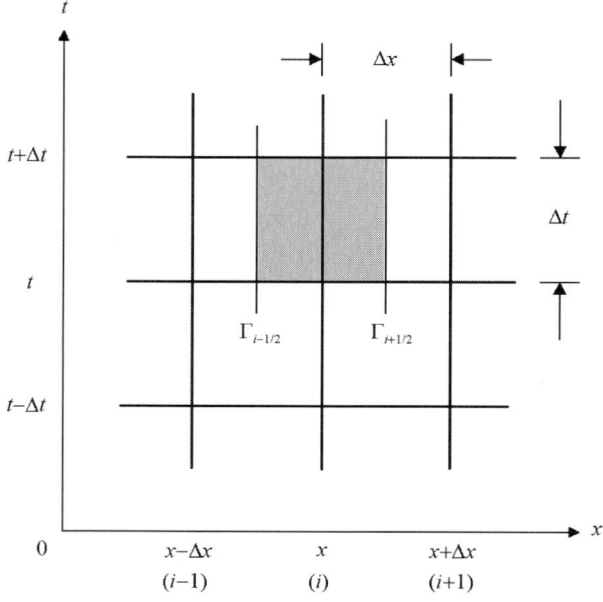

Figure 4.2. One-dimensional grid points (t, x) used for the integral method.

Let us now evaluate the terms in equation (4.2.14). Firstly, the integration of the term on the left-hand side of equation (4.2.14) can be done as follows.

$$\int_{\Gamma_{i-1/2}}^{\Gamma_{i+1/2}} \left\{ \int_{t}^{t+\Delta t} \frac{\partial \Phi}{\partial t} dt \right\} dx = \int_{\Gamma_{i-1/2}}^{\Gamma_{i+1/2}} (\Phi_x^{t+\Delta t} - \Phi_x^{t}) dx = (\Phi_i^{t+\Delta t} - \Phi_i^{t}) \Delta x. \quad (4.2.15)$$

The integral for the first term on the right-hand side of equation (4.2.14) can be represented as

$$-u \int_{\Gamma_{i-1/2}}^{\Gamma_{i+1/2}} \left\{ \int_{t}^{t+\Delta t} \frac{\partial \Phi}{\partial x} dt \right\} dx = -u \int_{t}^{t+\Delta t} \left\{ \int_{\Gamma_{i-1/2}}^{\Gamma_{i+1/2}} \frac{\partial \Phi}{\partial x} dx \right\} dt$$

$$= -u \int_{t}^{t+\Delta t} (\Phi|_{\Gamma_{i+1/2}} - \Phi|_{\Gamma_{i-1/2}}) dt. \quad (4.2.16a)$$

For the next level of integration, let us now consider the mean-value theorem for integrals, which is defined as

$$\int_{x}^{x+\Delta x} f(\Phi) dx \cong f(\bar{\Phi}) \Delta x \qquad \text{where } \bar{\Phi} \in [x, x + \Delta x].$$

Finite difference methods 43

Then, the integration of the right-hand side of equation (4.2.16a) can be given by

$$-u \int_t^{t+\Delta t} (\Phi|_{\Gamma_{i+1/2}} - \Phi|_{\Gamma_{i-1/2}}) \, dt = -u(\Phi|^t_{\Gamma_{i+1/2}} - \Phi|^t_{\Gamma_{i-1/2}}) \Delta t. \quad (4.2.16b)$$

If we take the simple arithmetic mean between the adjacent nodal values, the terms $\Phi|^t_{\Gamma_{i\pm 1/2}}$ on the control-volume faces at $x \pm \Delta x/2$ in equation (4.2.16b) can be represented by

$$\Phi|^t_{\Gamma_{i\pm 1/2}} = \tfrac{1}{2}(\Phi^t_{x\pm\Delta x} + \Phi^t_x) = \tfrac{1}{2}(\Phi^t_{i\pm 1} + \Phi^t_i). \quad (4.2.16c)$$

By substituting equation (4.2.16c) into equation (4.2.16b), we obtain the integral of the first term on the right-hand side of equation (4.2.14).

$$-u \int_{\Gamma_{i-1/2}}^{\Gamma_{i+1/2}} \left\{ \int_t^{t+\Delta t} \frac{\partial \Phi}{\partial x} \, dt \right\} dx = -\frac{\Delta t}{2} u(\Phi^t_{i+1} - \Phi^t_{i-1}). \quad (4.2.17)$$

Finally, the integral of the second term on the right-hand side of equation (4.2.14) can be obtained as follows.

$$\lambda \int_{\Gamma_{i-1/2}}^{\Gamma_{i+1/2}} \left\{ \int_t^{t+\Delta t} \frac{\partial^2 \Phi}{\partial x^2} \, dt \right\} dx = \lambda \int_t^{t+\Delta t} \left\{ \int_{\Gamma_{i-1/2}}^{\Gamma_{i+1/2}} \frac{\partial^2 \Phi}{\partial x^2} \, dx \right\} dt$$

$$= \lambda \int_t^{t+\Delta t} \left\{ \frac{\partial \Phi}{\partial x} \bigg|_{\Gamma_{i+1/2}} - \frac{\partial \Phi}{\partial x} \bigg|_{\Gamma_{i-1/2}} \right\} dt$$

$$= \lambda \left\{ \frac{\partial \Phi}{\partial x} \bigg|^t_{\Gamma_{i+1/2}} - \frac{\partial \Phi}{\partial x} \bigg|^t_{\Gamma_{i-1/2}} \right\} \Delta t. \quad (4.2.18)$$

If we apply the central difference scheme for the spacial derivatives in equation (4.2.18), we obtain

$$\frac{\partial \Phi}{\partial x} \bigg|^t_{\Gamma_{i-1/2}} = \frac{\Phi^t_i - \Phi^t_{i-1}}{\Delta x} \quad (4.2.19a)$$

and

$$\frac{\partial \Phi}{\partial x} \bigg|^t_{\Gamma_{i+1/2}} = \frac{\Phi^t_{i+1} - \Phi^t_i}{\Delta x}. \quad (4.2.19b)$$

Thus, the integral of the second term on the right-hand side of equation (4.2.14) is given by

$$\lambda \int_{\Gamma_{i-1/2}}^{\Gamma_{i+1/2}} \left\{ \int_t^{t+\Delta t} \frac{\partial^2 \Phi}{\partial x^2} \, dt \right\} dx = \lambda \left\{ \frac{\Phi^t_{i-1} - 2\Phi^t_i + \Phi^t_{i+1}}{\Delta x^2} \right\} \Delta t. \quad (4.2.20)$$

By substituting equations (4.2.15), (4.2.17) and (4.2.20) into equation (4.2.14) and rearranging it, we obtain the finite difference equation for one-dimensional, transient convection–diffusion equation as

$$\frac{\Phi_i^{t+\Delta t} - \Phi_i^t}{\Delta t} = -u\frac{\Phi_{i+1}^t - \Phi_{i-1}^t}{2\Delta x} + \lambda\frac{\Phi_{i-1}^t - 2\Phi_i^t + \Phi_{i+1}^t}{\Delta x^2}. \quad (4.2.21)$$

4.2.3 Finite volume method—control volume approach

In the Taylor-series approach and the integral method, the differential form of governing equations was employed as the correct and appropriate form of the conservation law governing the problems in considered, and a purely mathematical procedure was used to develop the finite difference approximations to derivatives. However, in the control volume or finite volume approach, the physical law represented by the partial differential equations is not considered, instead the conservation of a specific physical quantity such as energy, mass or momentum over a finite control volume is considered. Thus, the control volume approach to derive finite difference equations has the more physical meaning compared to the use of Taylor-series expansion.

As an example, consider one-dimensional convection–diffusion problems, and define a finite control volume in space about the location (i) as depicted in figure 4.3. The one-dimensional convection–diffusion equation, equation (4.2.11), can be represented as the following integral form of balance equation.

$$\underbrace{\frac{\partial}{\partial t}\iiint_\Omega \Phi \, d\Omega}_{(1)} = \underbrace{-\iint_\Gamma \mathbf{u}\Phi \cdot \mathbf{n}\, d\Gamma}_{(2)} + \underbrace{\iint_\Gamma \lambda \nabla \Phi \cdot \mathbf{n}\, d\Gamma}_{(3)} \quad (4.2.22a)$$

or in the general form of balance equation as

$$\left\{\begin{array}{c}\text{rate of}\\ \text{potential }\Phi\\ \text{accumulation in}\\ \text{a control volume}\end{array}\right\} = \left\{\begin{array}{c}\text{net flux of }\Phi\\ \text{into a control}\\ \text{volume by}\\ \text{convection}\end{array}\right\} + \left\{\begin{array}{c}\text{net flux of }\Phi\\ \text{into a control}\\ \text{volume by}\\ \text{diffusion}\end{array}\right\}.$$

$$(4.2.22b)$$

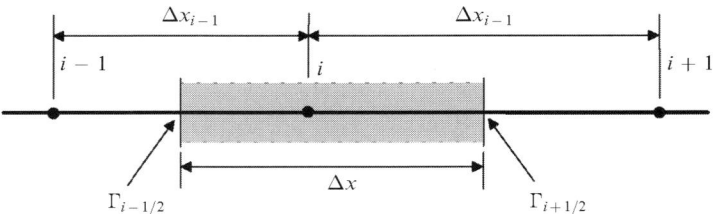

Figure 4.3. Control volumes for one-dimensional problems.

Term (1) in equation (4.2.22a), an integral over the control volume (Ω), represents the time rate of increase of Φ in the control volume, and terms (2) and (3), the integrals over the boundaries of the control volume (control-volume face Γ), indicate the net fluxes of Φ into the control volume by convection and diffusion through its control-volume faces. It is noted that equation (4.2.22) is valid for control volumes of any shapes.

By applying the conservation principles, either in the integral form of balance equation, equation (4.2.22a), or in the general form of balance equation, equation (4.2.22b), to a fixed region in space known as a control volume, we can develop finite difference equations for one-dimensional convection–diffusion problems. The methods based on equation (4.2.22a) and equation (4.2.22b) are referred to as the finite volume method [9] and the direct finite difference method [10], respectively. However, principally these two methods are identical to each other.

Let us now consider the derivation of finite difference equations by the finite volume method. In transient problems it is also necessary to integrate equation (4.2.22a) with respect to time over a small time interval Δt from t to $t + \Delta t$. Then, equation (4.2.22a) can be expressed as

$$\underbrace{\frac{1}{\Delta t} \int_t^{t+\Delta t} \frac{\partial}{\partial t} \left\{ \iiint_\Omega \Phi \, d\Omega \right\} dt}_{(1)} = \underbrace{-\frac{1}{\Delta t} \int_t^{t+\Delta t} \left\{ \iint_\Gamma \mathbf{u}\Phi \cdot \mathbf{n} \, d\Gamma \right\} dt}_{(2)}$$

$$+ \underbrace{\frac{1}{\Delta t} \int_t^{t+\Delta t} \left\{ \iint_\Gamma \lambda \nabla \Phi \cdot \mathbf{n} \, d\Gamma \right\} dt}_{(3)}. \quad (4.2.23)$$

Let us evaluate the terms in equation (4.2.23) by applying this equation into the control volume in figure 4.3. Firstly, term (1), which represents the total increase of Φ in the control volume per unit time, can be evaluated by assuming that the potential Φ at a point x represents the mean value for the control volume (i).

$$\frac{1}{\Delta t} \int_t^{t+\Delta t} \frac{\partial}{\partial t} \left\{ \iiint_\Omega \Phi \, d\Omega \right\} dt = \frac{1}{\Delta t} \int_{\Gamma_{i-1/2}}^{\Gamma_{i+1/2}} \left\{ \int_t^{t+\Delta t} \frac{\partial \Phi}{\partial t} dt \right\} dx$$

$$= \frac{1}{\Delta t} \int_{\Gamma_{i-1/2}}^{\Gamma_{i+1/2}} (\Phi^{t+\Delta t} - \Phi^t) \, dx$$

$$= \frac{(\Phi_i^{t+\Delta t} - \Phi_i^t)}{\Delta t} V_i. \quad (4.2.24)$$

Here, the value Φ_i at a nodal point (i) refers to the average value over the control volume (i) and V_i is its volume.

Term (2) in equation (4.2.23) represents the net influx of Φ into the control volume through its control-volume faces by convection. If we take the fully explicit scheme for the time integration and assume that the flow velocity u is constant, the integral is given by

$$-\frac{1}{\Delta t}\int_t^{t+\Delta t}\left\{\iint_\Gamma \mathbf{u}\Phi\cdot\mathbf{n}\,d\Gamma\right\}dt = -u(\Phi|_{\Gamma_{i+1/2}}^t \times A_{\Gamma_{i+1/2}} - \Phi|_{\Gamma_{i-1/2}}^t \times A_{\Gamma_{i-1/2}}).$$
(4.2.25)

Here, $A_{\Gamma_{i-1/2}}$ and $A_{\Gamma_{i+1/2}}$ are the areas of the control-volume faces $\Gamma_{i-1/2}$ and $\Gamma_{i+1/2}$. The values $\Phi|_{\Gamma_{i-1/2}}$ and $\Phi|_{\Gamma_{i+1/2}}$ refer to the potential values on the control-volume faces $\Gamma_{i-1/2}$ and $\Gamma_{i+1/2}$.

Term (3) in equation (4.2.23), indicating the net influx of Φ into the control volume through its control-volume faces by diffusion, can be represented by

$$\frac{1}{\Delta t}\int_t^{t+\Delta t}\left\{\iint_\Gamma \lambda\nabla\Phi\cdot\mathbf{n}\,d\Gamma\right\}dt = \left\{\lambda\frac{\partial\Phi}{\partial x}\bigg|_{\Gamma_{i+1/2}}^t \times A_{\Gamma_{i+1/2}} - \lambda\frac{\partial\Phi}{\partial x}\bigg|_{\Gamma_{i-1/2}}^t \times A_{\Gamma_{i-1/2}}\right\}.$$
(4.2.26)

By substituting equations (4.2.24), (4.2.25) and (4.2.26) into equation (4.2.23), we obtain the following equation.

$$\frac{(\Phi_i^{t+\Delta t}-\Phi_i^t)}{\Delta t}V_i = -u(\Phi|_{\Gamma_{i+1/2}}^t \times A_{\Gamma_{i+1/2}} - \Phi|_{\Gamma_{i-1/2}}^t \times A_{\Gamma_{i-1/2}})$$

$$+\left\{\lambda\frac{\partial\Phi}{\partial x}\bigg|_{\Gamma_{i+1/2}}^t \times A_{\Gamma_{i+1/2}} - \lambda\frac{\partial\Phi}{\partial x}\bigg|_{\Gamma_{i-1/2}}^t \times A_{\Gamma_{i-1/2}}\right\}. \quad (4.2.27)$$

In a one-dimensional rectangular coordinate system, $V_i = 1\times\Delta x$ and $A_{\Gamma_{i-1/2}} = A_{\Gamma_{i+1/2}} = 1$. Now, we need to evaluate the potential values $\Phi|_{\Gamma_{i-1/2}}^t$ and $\Phi|_{\Gamma_{i+1/2}}^t$, and their gradients on the control-volume faces $\Gamma_{i-1/2}$ and $\Gamma_{i+1/2}$. Let us first consider the evaluation of the potential values $\Phi|_{\Gamma_{i-1/2}}^t$ and $\Phi|_{\Gamma_{i+1/2}}^t$. If we take the simple arithmetic mean between the adjacent nodal values, the terms $\Phi|_{\Gamma_{i-1/2}}^t$ and $\Phi|_{\Gamma_{i+1/2}}^t$ in equation (4.2.27) can be represented by

$$\Phi|_{\Gamma_{i-1/2}}^t = \tfrac{1}{2}(\Phi_{i-1}^t + \Phi_i^t) \quad (4.2.28a)$$

$$\Phi|_{\Gamma_{i+1/2}}^t = \tfrac{1}{2}(\Phi_{i+1}^t + \Phi_i^t) \quad (4.2.28b)$$

Let us now consider the evaluation of the flux rates by diffusion in equation (4.2.27). If we apply the central difference scheme for the spatial derivatives, we have

$$\frac{\partial\Phi}{\partial x}\bigg|_{\Gamma_{i-1/2}}^t = \frac{\Phi_i^t - \Phi_{i-1}^t}{\Delta x_{i-1}} \quad (4.2.29a)$$

$$\frac{\partial\Phi}{\partial x}\bigg|_{\Gamma_{i+1/2}}^t = \frac{\Phi_{i+1}^t - \Phi_i^t}{\Delta x_{i+1}}. \quad (4.2.29b)$$

For the sake of simplicity, all the distances between two nodes are assumed to be equal, i.e., $\Delta x_{i-1} = \Delta x_{i+1} = \Delta x$. Then, by substituting equations (4.2.28) and (4.2.29) into equation (4.2.27) and rearranging it, we obtain the finite difference equation for one-dimensional convection–diffusion problems.

$$\frac{\Phi_i^{t+\Delta t} - \Phi_i^t}{\Delta t} = -u\frac{\Phi_{i+1}^t - \Phi_{i-1}^t}{2\Delta x} + \lambda\frac{\Phi_{i-1}^t - 2\Phi_i^t + \Phi_{i+1}^t}{\Delta x^2}. \qquad (4.2.30)$$

It is to be noted that the resultant finite difference equations, equations (4.2.12), (4.2.21) and (4.2.30) derived by three different discretization methods, are identical to each other.

References

[1] Carslaw H S and Jaeger J C 1959 *Conduction of Heat in Solids* 2nd edition (Oxford: Oxford University Press)
[2] Crank J 1956 *The Mathematics of Diffusion* 1st edition (Oxford: Oxford University Press)
[3] Dusinberre G M 1961 *Heat Transfer Calculations by Finite Differences* (Scranton, PA: International Textbook Co)
[4] Zienkiewicz O C and Cheung Y K 1965 *The Engineer* **220** 507
[5] Comini G and Del Guidice S 1974 *Int. J. Numer. Methods Engng.* **8** 612
[6] Brebbia C A and Wrobel L C 1980 *Recent Advances in Numerical Methods in Fluids* vol. 1 (ed. C Taylor *et al*) (Swansea: Pineridge Press)
[7] Hong C P, Umeda T and Kimura Y 1984 *Metall. Trans.* **15B** 91, 101
[8] Roache P J 1976 *Computational Fluid Dynamics* (New Mexico: Hermora Publishers)
[9] Tannehill J C, Anderson D A and Pletcher R H 1984 *Computational Fluid Mechanics and Heat Transfer* (Washington: Hemisphere Publishing Co)
[10] Ohnaka I 1979 *Tetsu-to-Hagane* **65** 1737

Chapter 5

Steady state heat conduction

In the previous chapter we studied the basics of finite difference methods including three kinds of discretization schemes. In this chapter we will examine steady-state potential flow problems by the control-volume based finite volume method, which will be used throughout this textbook. Because of the similarities between heat and species transfer, it might be enough only to consider heat conduction problems here.

5.1 Mathematical formulation

5.1.1 Governing equation

As we described in chapter 2, the integral form of governing equation for steady-state heat conduction with constant thermal and physical properties is given by

$$-\iint_\Gamma (\mathbf{q} \cdot \mathbf{n})\, d\Gamma + \iiint_\Omega \dot{g}\, d\Omega = 0. \tag{5.1.1}$$

As illustrated in chapter 4, the integral form of the conservation law, equation (5.1.1), is the starting point for the derivation of finite difference equations in the finite volume method.

5.1.2 Boundary conditions

The boundary conditions frequently encountered in heat transfer have already been mentioned in chapter 2. The boundary conditions are generally considered to describe the heat flux rate around the outer boundary of the calculation domain. In the finite volume method each control volume has several control-volume faces around it, and heat flow occurs across these control-volume faces whether the control-volume faces are positioned inside the domain or on the outer boundary of the domain.

In this chapter we will consider boundary conditions both for the outer boundary of the domain and for the inner boundary between the control

Mathematical formulation

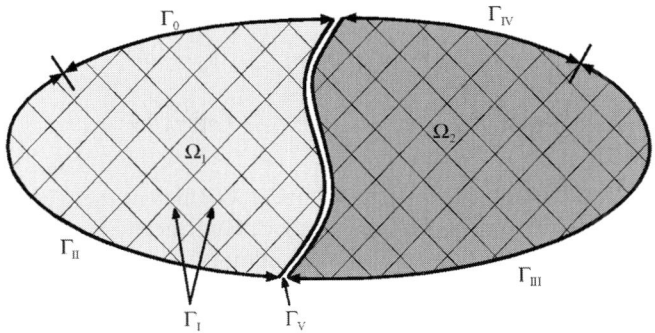

Figure 5.1. A computational domain with two sub-domains.

volumes. The boundary condition between two different solid regions (two sub-domains) will also be considered. Figure 5.1 describes a computational domain with two sub-domains and associated notations for different types of boundaries.

1. Prescribed temperature (Γ_0-type boundary)

$$T = \bar{T}_{\Gamma_0}. \qquad (5.1.2a)$$

2. Conduction between two control volumes across an inner boundary (Γ_I-type boundary)

$$q_{\Gamma_I} = -\lambda \frac{\partial T}{\partial x}. \qquad (5.1.2b)$$

3. Prescribed heat flux or insulated boundary (Γ_{II}-type boundary)

$$q_{\Gamma_{II}} \left(\equiv -\lambda \frac{\partial T}{\partial x} \bigg|_{\Gamma_{II}} \right) = \bar{q}_{\Gamma_{II}}. \qquad (5.1.2c)$$

4. Convection: h_∞ and T_∞ are prescribed (Γ_{III}-type boundary)

$$q_{\Gamma_{III}} \left(\equiv -\lambda \frac{\partial T}{\partial x} \bigg|_{\Gamma_{III}} \right) = h_\infty (T_\infty - T). \qquad (5.1.2d)$$

5. Radiation: ε and T_{sur} are prescribed (Γ_{IV}-type boundary)

$$q_{\Gamma_{VI}} \left(\equiv -\lambda \frac{\partial T}{\partial x} \bigg|_{\Gamma_{IV}} \right) = h_r (T_{sur} - T) \qquad (5.1.2e)$$

where $h_r \equiv \varepsilon \sigma (T_{sur} + T)(T_{sur}^2 + T^2)$ is called the *radiation heat transfer coefficient*, T_{sur} is the temperature of the surrounding surface and T is the surface temperature of the solid.

6. Heat transfer occurs between two different solids with thermal resistance (Γ_V-type boundary)}

$$q_{\Gamma_V}\left(\equiv -\lambda \left.\frac{\partial T}{\partial x}\right|_{\Gamma_V}\right) = h_{int}(T_{\Omega_1} - T_{\Omega_2}) \qquad (5.1.2f)$$

where h_{int} is the interfacial heat transfer coefficient between two sub-domains (solids), and T_{Ω_1} and T_{Ω_2} are the surface temperatures of the adjacent control volumes in the sub-domains Ω_1 and Ω_2.

5.2 Finite volume approach for steady state problems

5.2.1 Computational grids

Let us consider the steady-state heat conduction in the one-dimensional calculation domain shown in figure 5.2, in which the control volumes for the internal and boundary points are defined. The first step for computer simulation of this problem by the finite volume method is to divide the computational domain into discrete control volumes. A number of nodes (nodal points) are defined in the domain surrounded by the outer boundaries (Γ_A and Γ_B) of the computational domain.

Figure 5.3 indicates a schematic representation of control volumes and associated notations in a one-dimensional domain. In a one-dimensional domain, each node (nodal point i) in the internal region has two neighboring nodes ($i-1$ and $i+1$) and two control-volume faces ($\Gamma_{i-1/2}$ and $\Gamma_{i+1/2}$). The control-volume faces are positioned at the center between two neighboring nodes. However, in the case of control volumes on the outer boundary of the computational domain, nodes are positioned on the outer boundary and each node has one neighboring node. The distances between the nodes ($i-1$)

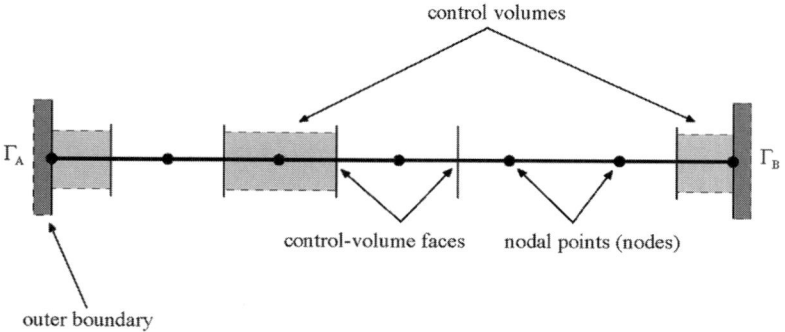

Figure 5.2. One-dimensional computational domain used in the finite volume method.

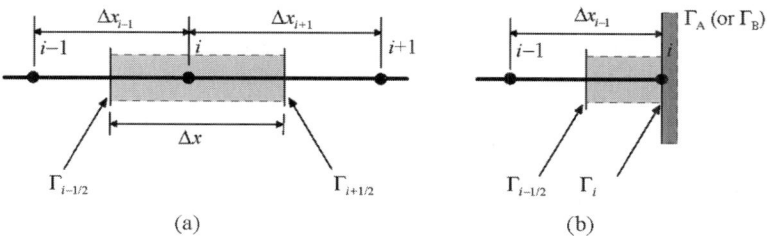

Figure 5.3. A schematic representation of control volumes: (a) a control volume in the internal region and (b) a control volume on the outer boundary.

and (i), and between the nodes $(i+1)$ and (i), are identified by Δx_{i-1} and Δx_{i+1}, respectively. The control volume width is Δx.

5.2.2 Derivation of finite difference equations

As mentioned in chapter 4, the first step for the derivation of finite difference equations in the finite volume method is to apply the integral form of the energy balance equation to control volumes.

Now, let us apply the integral form of energy balance equation (5.1.1) into a control volume (i) in figure 5.3(a).

$$-\iint_\Gamma (\mathbf{q}\cdot\mathbf{n})\,d\Gamma + \iiint_\Omega \dot{g}\,d\Omega = 0. \qquad (5.1.1)$$

5.2.2.1 A conventional scheme for discretizing the diffusion term

First, consider the conventional finite volume method. For the sake of simplicity, consider the one-dimensional rectangular coordinate system. In the conventional finite volume method described in chapter 4, the first term in equation (5.1.1) can be evaluated as follow.

$$\begin{aligned}
-\iint_\Gamma (\mathbf{q}\cdot\mathbf{n})\,d\Gamma &= -\{(q_{\Gamma_{i+1/2}}A_{\Gamma_{i+1/2}}) - (q_{\Gamma_{i-1/2}}A_{\Gamma_{i-1/2}})\} \\
&= (q_{\Gamma_{i-1/2}}A_{\Gamma_{i-1/2}}) - (q_{\Gamma_{i+1/2}}A_{\Gamma_{i+1/2}}) \\
&= \left(-\lambda A_{\Gamma_{i-1/2}} \frac{T_i - T_{i-1}}{\Delta x_{i-1}}\right) - \left(-\lambda A_{\Gamma_{i+1/2}} \frac{T_{i+1} - T_i}{\Delta x_{i+1}}\right)
\end{aligned}$$
$$(5.2.1)$$

where $A_{\Gamma_{i-1/2}} = A_{\Gamma_{i-1/2}} = 1.0$ and $\Delta x_{i-1} = \Delta x_{i+1} = \Delta x$. The term on the right-hand side of equation (5.2.1) indicates that the net heat influx by conduction in the x direction is evaluated based on the assumption that heat conduction occurs from left to right in the $(+x)$ direction, as described in figure 5.4(a).

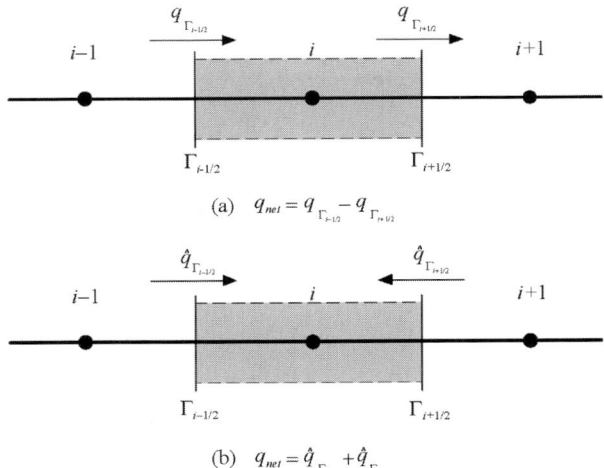

Figure 5.4. Direction of heat flow in a control volume through its control-volume face: (a) a conventional scheme and (b) a modified scheme.

The second term in equation (5.1.1) is simply given by

$$\iiint_\Omega \dot{g}\, d\Omega = V_i \dot{g}_i \tag{5.2.2}$$

where $V_i = \Delta x$.

Then, by substituting equations (5.2.1) and (5.2.2) into equation (5.1.1) we obtain the following discretized equation.

$$T_{i-1} - 2T_i + T_{i+1} = \Delta x^2 \dot{g}_i / \lambda. \tag{5.2.3}$$

Equation (5.2.3) can only be applied for the inner nodes of the computational domain. The discretized equations for the nodes on the outer boundary shown in figure 5.3(b) must be derived according to the boundary conditions.

5.2.2.2 A modified discretization scheme

Let us now consider a modified scheme for discretizing the diffusion term. The first term in equation (5.1.1), which stands for the *surface phenomena* mentioned in chapter 2, indicates the total net heat influx into a control volume (*i*) through all its control-volume faces. As mentioned in the previous section, a control-volume face may be positioned inside the domain between two inner control volumes or on the outer boundary of the domain. It is therefore reasonable to consider that the *surface phenomena* represented by the first term may include not only conductive heat transfer by equation

(5.1.2b), but also other types of heat transfer such as convection and radiation, given by equations (5.1.2c)–(5.1.2f).

Based on the concept of *surface phenomena*, we can express the first term on the left-hand side of equation (5.1.1) as follows.

$$-\iint_\Gamma (\mathbf{q} \cdot \mathbf{n})\, d\Gamma = (\hat{q}_{\Gamma_{i-1/2}} \times A_{\Gamma_{i-1/2}}) + (\hat{q}_{\Gamma_{i+1/2}} \times A_{\Gamma_{i+1/2}}). \quad (5.2.4)$$

Here $\hat{q}_{\Gamma_{i-1/2}}$ and $\hat{q}_{\Gamma_{i+1/2}}$ indicate the net heat flux rates into the control volume (i) across the control-volume faces $\Gamma_{i-1/2}$ and $\Gamma_{i+1/2}$, as described in figure 5.4(b). It is to be noted that the meaning of $\hat{q}$ in equation (5.2.4) is different from that of q in equation (5.2.1) in several viewpoints: (i) Equation (5.2.1) can only be applied to evaluate the heat flux rate by conduction. However, the integration of equation (5.2.4) can be applied to evaluate the net heat flux rates into a control volume through its control-volume faces by any types of heat transfer modes, given by equations (5.1.2b)–(5.1.2f). (ii) Equation (5.2.4) is available both in structured and unstructured grids, but equation (5.2.1) is only available for the structured grids.

Equation (5.2.4) can now be rewritten in a modified form as

$$-\iint_\Gamma (\mathbf{q} \cdot \mathbf{n})\, d\Gamma = \sum_{m=1}^{nf} (\hat{q}_{\Gamma_m} A_{\Gamma_m}). \quad (5.2.5)$$

Here, nf is the number of control-volume faces of the control volume (i) (for example, $nf = 2$, 4 and 6 in one-, two- and three-dimensional problems, respectively). The number m indicates the mth control-volume face (Γ_m) of the control volume (i), $\hat{q}_{\Gamma_m}$ the net heat flux through the mth control-volume face into a control volume (i), and A_{Γ_m} the cross-sectional area of the mth control-volume face. It is also to be noted that the control-volume face might be a part of the inner boundary between a control volume (i) and its neighboring control volume or a part of the outer boundary of the computational domain (Ω). It is clear that equation (5.2.5) is available for evaluating the net heat influx across the control-volume faces having any types of boundary conditions given by equation (5.1.2). However, it should be noted that $\hat{q}_{\Gamma_m}$ does not include the heat flux term by fluid flow.

By substituting equations (5.2.2) and (5.2.5) into equation (5.1.1), we have the general form of finite difference equation for steady-state heat conduction problems as follows.

$$\sum_{m=1}^{nf} (\hat{q}_{\Gamma_m} A_{\Gamma_m}) + V_i \dot{g}_i = 0. \quad (5.2.6)$$

Equation (5.2.6) can be applied to any type of coordinate, such as rectangular, cylindrical and spherical coordinate systems. Further, equation (5.2.6) allows the use of unstructured grids, for example, triangles or pentagons in two-dimensional problems.

Let us now consider how to evaluate the term $\sum_{m=1}^{nf}(\hat{q}_{\Gamma_m} A_{\Gamma_m})$ in equation (5.2.6), which indicates the total net heat flux into the control volume (i) through its control-volume faces. As mentioned above, the control-volume faces might be a part of the inner boundary between control volumes inside the domain or the outer boundary of the domain (Ω).

The boundary conditions, equations (5.1.2a)–(5.1.2f), described in section 5.1.2, indicate the mechanism of heat flow across the control-volume faces both on the inner and the outer boundaries of the computational domain. Thus, the first term in equation (5.2.6), representing the net heat flux rate across the control-volume faces of a control volume, can be evaluated as follows.

(1) Temperature is prescribed on the outer boundary (Γ_0-type boundary)

$$T_i = \bar{T}_{\Gamma_0}. \qquad (5.2.7a)$$

In this case, there is no need to evaluate the temperature of a node (i) on the outer boundary of the computational domain (Ω).

(2) Conduction in a homogeneous medium (Γ_I-type boundary)

Heat flow occurs by conduction across the mth control-volume face, i.e. the inner boundary between a control volume (i) and its neighboring control volume (nb) in a homogeneous medium. The net heat flux into the control volume (i) through the mth control-volume face of Γ_I-type boundary is given by

$$\hat{q}_{\Gamma_I} = \lambda \frac{T_{nb} - T_i}{\Delta x_{nb}} \qquad (5.2.7b)$$

where T_{nb} is the temperature of the neighboring node (nb) and Δx_{nb} is the distance between the node (i) and its neighboring node (nb).

(3) Prescribed heat flux boundary condition (Γ_{II}-type boundary)

If the mth control-volume face of a control volume is a part of the Γ_{II}-type outer boundary of the domain (Ω), the net heat flux into the control volume (i) through the mth control-volume face is prescribed as

$$\hat{q}_{\Gamma_{II}} = \bar{q}_{\Gamma_{II}} \qquad (5.2.7c)$$

where $\bar{q}_{\Gamma_{II}}$ is the prescribed heat flux across the mth control-volume face. The prescribed heat flux is $\bar{q}_{\Gamma_{II}} = 0$ on the control-volume face, which is thermally or geometrically symmetrical, such as on the centerline of a solid cylinder or a sphere.

(4) Convection heat transfer boundary condition (Γ_{III}-type boundary)

If the mth control-volume face of a control volume (i) is a part of the Γ_{III}-type outer boundary of the domain (Ω), convection heat transfer occurs across the mth control-volume face. The net heat flux into the control volume (i) is given by

$$\hat{q}_{\Gamma_{III}} = h_\infty(T_\infty - T_i) \qquad (5.2.7d)$$

where h_∞ and T_∞ are the convection heat transfer coefficient and the temperature of the fluid adjacent to the control volume (i).

(5) Radiation heat transfer boundary condition (Γ_{IV}-type boundary)

If the mth control-volume face of a control volume (i) is a part of the Γ_{IV}-type outer boundary of the domain (Ω), radiation heat transfer occurs across the mth control-volume face. The net heat flux into the control volume (i) is given by

$$\hat{q}_{\Gamma_{IV}} = h_r(T_{sur} - T_i) \qquad (5.2.7e)$$

where $h_r = \sigma\varepsilon(T_{sur} + T_i)(T_{sur}^2 + T_i^2)$ and T_{sur} is the temperature of the surrounding surface. This boundary can be treated in a similar way to the Γ_{III}-type boundary.

(6) Heat transfer occurs between two different solids with thermal resistance (Γ_V-type boundary)

The net heat flux into the control volume (i) is given by

$$\hat{q}_{\Gamma_V} = h_{int}(T_{nb} - T_i). \qquad (5.2.7f)$$

Here, h_{int} is the interfacial heat transfer coefficient between two sub-domains (solids), and T_i and T_{nb} are the temperatures of the node (i) and its neighboring node (nb) at the interface between the two sub-domains, Ω_1 and Ω_2.

5.2.3 Solution of linear algebraic equations

The general discretized equation (5.2.6) can be applied to all nodal points in a computational domain including the nodes on the outer boundary. The resulting system of linear algebraic equations can be solved by the standard Gauss elimination method to obtain the temperature distribution at nodal points. In heat conduction problems, the coefficient matrix in the linear equation has a tridiagonal system, in which all the non-zero elements in the coefficient matrix must be on the main diagonal or on the two diagonals just above and below the main diagonal. The simultaneous solution can be achieved by a particular form of Gauss elimination, known as the Thomas Algorithm or Tri-Diagonal Matrix Algorithm (TDMA).

5.3 One-dimensional cylindrical and spherical coordinates

The general form of finite difference equation, equation (5.2.6), can also be applied to one-dimensional problems for cylindrical and spherical coordinate systems.

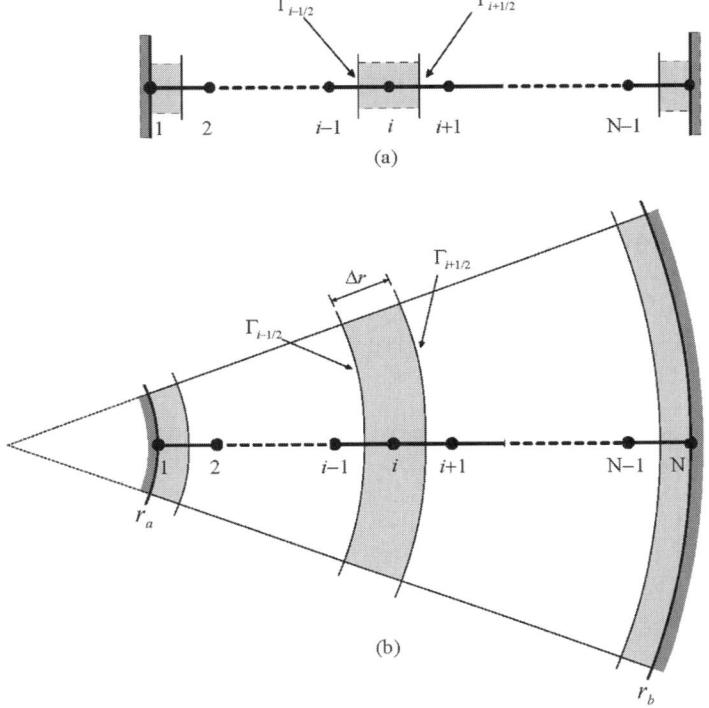

Figure 5.5. Grids used for one-dimensional problems: (a) rectangular and (b) cylindrical or spherical coordinate systems.

Figure 5.5 indicates the grids used for one-dimensional heat conduction: (a) rectangular and (b) cylindrical or spherical coordinate systems. The physical domain is divided into N control volumes, having N nodes. The values V_i and A_{Γ_m} in equation (5.2.6) can be evaluated according to the coordinate systems. For the sake of simplicity, the control volume width is assumed to be constant, Δr.

5.3.1 Control volumes inside a domain

For the inner control volumes $(2, 3, \ldots N - 1)$, which have two neighboring control volumes, we have the following relations.

(1) In the case of the rectangular coordinate system,

$$\left. \begin{array}{l} A_{\Gamma_{i-1/2}} = A_{\Gamma_{i+1/2}} = 1.0 \\ V_i = 1 \cdot \Delta x = \Delta x \end{array} \right\} \quad (5.3.1a)$$

where L is the thickness of the plate and $\Delta x = L/(N-1)$.

(2) In the case of the cylindrical coordinate system,
$$\left.\begin{array}{l} A_{\Gamma_{i-1/2}} = 2\pi\{r_a + (i-1)\cdot \Delta r - \Delta r/2\} \\ A_{\Gamma_{i+1/2}} = 2\pi\{r_a + (i-1)\cdot \Delta r + \Delta r/2\} \\ V_i = \pi[\{r_a + (i-1)\cdot \Delta r + \Delta r/2\}^2 - \{r_a + (i-1)\cdot \Delta r - \Delta r/2\}^2] \end{array}\right\}.$$
(5.3.1b)

(3) In the case of the spherical coordinate system,
$$\left.\begin{array}{l} A_{\Gamma_{i-1/2}} = 4\pi\{r_a + (i-1)\cdot \Delta r - \Delta r/2\}^2 \\ A_{\Gamma_{i+1/2}} = 4\pi\{r_a + (i-1)\cdot \Delta r + \Delta r/2\}^2 \\ V_i = 4\pi/3[\{r_a + (i-1)\cdot \Delta r + \Delta r/2\}^3 - \{r_a + (i-1)\cdot \Delta r - \Delta r/2\}^3] \end{array}\right\}.$$
(5.3.1c)

5.3.2 Control volumes on the outer boundary of a domain

For the control volumes on the outer boundary of the domain, i.e., control volume 1 or N, we have the following relations.

(1) In the case of the rectangular coordinate system,
$$\left.\begin{array}{l} A_{\Gamma_{i-1/2}} = A_{\Gamma_{i+1/2}} = 1.0 \\ V_1 = V_N = \Delta x/2 \end{array}\right\}.$$
(5.3.2a)

(2) In the case of the cylindrical coordinate system:
For node 1,
$$\left.\begin{array}{l} A_{\Gamma_{i-1/2}} = 2\pi r_a, \qquad A_{\Gamma_{i+1/2}} = 2\pi(r_a + \Delta r/2) \\ V_1 = \pi\{(r_a + \Delta r/2)^2 - r_a^2\} \end{array}\right\}.$$
(5.3.2b)

For node N,
$$\left.\begin{array}{l} A_{\Gamma_{i-1/2}} = 2\pi(r_b - \Delta r/2), \qquad A_{\Gamma_{i+1/2}} = 2\pi r_b \\ V_N = \pi\{r_b^2 - (r_b - \Delta r/2)^2\} \end{array}\right\}.$$
(5.3.2c)

(3) In the case of the spherical coordinate system:
For node 1,
$$\left.\begin{array}{l} A_{\Gamma_{i-1/2}} = 4\pi r_a^2, \qquad A_{\Gamma_{i+1/2}} = 4\pi(r_a + \Delta r/2)^2 \\ V_1 = 4\pi/3\{(r_a + \Delta r/2)^3 - (r_a)^3\} \end{array}\right\}.$$
(5.3.2d)

For node N,
$$\left.\begin{array}{l} A_{\Gamma_{i-1/2}} = 4\pi(r_b - \Delta r/2)^2, \qquad A_{\Gamma_{i+1/2}} = 4\pi r_b^2 \\ V_N = 4\pi/3\{r_b^3 - (r_b - \Delta r/2)^3\} \end{array}\right\}.$$
(5.3.2e)

where r_a and r_b are the inner and outer diameters of a hollow cylinder or a hollow sphere, and Δr is the distance between the two nodal points which is given by $\Delta r = (r_b - r_a)/(N - 1)$. In the cases of a solid cylinder or a solid sphere, $r_a = A_{\Gamma_{i-1/2}} = 0$, and the heat flux across the control-volume face $A_{\Gamma_{i-1/2}}$, i.e., the centerline, becomes zero because of the geometrical symmetry.

5.4 Multi-dimensional problems

5.4.1 Two-dimensional problems

As mentioned before, the general form of finite difference equation for steady-state heat conduction problems, equation (5.2.6), is available for one-, two- and three-dimensional problems. In case of two-dimensional problems, the number of control-volume faces for a control volume can be 3, 4 or 5 depending upon the grid system used. For the sake of simplicity, let us consider a two-dimensional structured grid in which each control volume has 4 control-volume faces, as shown in figure 5.6.

Then, equation (5.2.6) is expressed by

$$\sum_{m=1}^{nf} (\hat{q}_{\Gamma_m} \times A_{\Gamma_m}) + V_{i,j}\dot{g}_{i,j} = \hat{q}_{\Gamma_{i-1/2}} A_{\Gamma_{i-1/2}} + \hat{q}_{\Gamma_{i+1/2}} A_{\Gamma_{i+1/2}}$$
$$+ \hat{q}_{\Gamma_{j-1/2}} A_{\Gamma_{j-1/2}} + \hat{q}_{\Gamma_{j+1/2}} A_{\Gamma_{j+1/2}}$$
$$+ V_{i,j}\dot{g}_{i,j} = 0. \qquad (5.4.1)$$

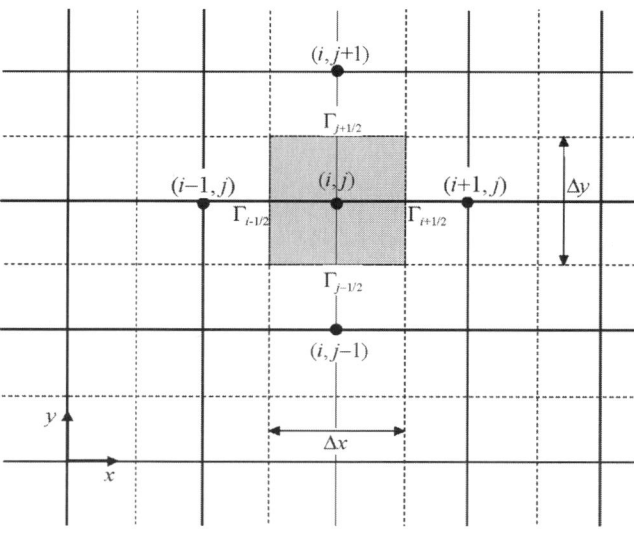

Figure 5.6. A part of two-dimensional structured grids.

Here, $\Gamma_{i\pm1/2}$ and $\Gamma_{j\pm1/2}$ indicate the control-volume faces $\Gamma_{i\pm1/2,j}$ and $\Gamma_{i,j\pm1/2}$, respectively, as shown in figure 5.6. $\hat{q}_{\Gamma_{i\pm1/2}}$ and $\hat{q}_{\Gamma_{j\pm1/2}}$ indicate the net heat flux rates into a control volume (i,j) through its control-volume faces $\Gamma_{i\pm1/2,j}$ and $\Gamma_{i,j\pm1/2}$. $A_{\Gamma_{i\pm1/2}}$ and $A_{\Gamma_{j\pm1/2}}$ are their cross-sectional areas: $A_{\Gamma_{i-1}} = A_{\Gamma_{i+1}} = \Delta y$ and $A_{\Gamma_{j-1}} = A_{\Gamma_{j+1}} = \Delta x$. $V_{i,j}(= \Delta x \Delta y)$ and $\dot{g}_{i,j}$ are the volume and the rate of heat generation of the control volume (i,j). The values of $\hat{q}_{\Gamma_{i\pm1/2}}$ and $\hat{q}_{\Gamma_{j\pm1/2}}$ can be evaluated using equations (5.2.7b) through (5.2.7f) according to the boundary conditions of the control-volume faces.

5.4.2 Three-dimensional problems

Similarly, let us consider a three-dimensional structured grid in which each control volume has 6 control-volume faces, as shown in figure 5.7.

In this case each control volume has two more control-volume faces, $\Gamma_{k-1/2}$ and $\Gamma_{k+1/2}$. Then, equation (5.2.6) is expressed by

$$\sum_{m=1}^{nf} (\hat{q}_{\Gamma_m} \times A_{\Gamma_m}) + V_{i,j,k}\dot{g}_{i,j,k} = \hat{q}_{\Gamma_{i-1/2}}A_{\Gamma_{i-1/2}} + \hat{q}_{\Gamma_{i+1/2}}A_{\Gamma_{i+1/2}}$$
$$+ \hat{q}_{\Gamma_{j-1/2}}A_{\Gamma_{j-1/2}} + \hat{q}_{\Gamma_{j+1/2}}A_{\Gamma_{j+1/2}} + \hat{q}_{\Gamma_{k-1/2}}A_{\Gamma_{k-1/2}}$$
$$+ \hat{q}_{\Gamma_{k+1/2}}A_{\Gamma_{k+1/2}} + V_{i,j,k}\dot{g}_{i,j,k} = 0. \quad (5.4.2)$$

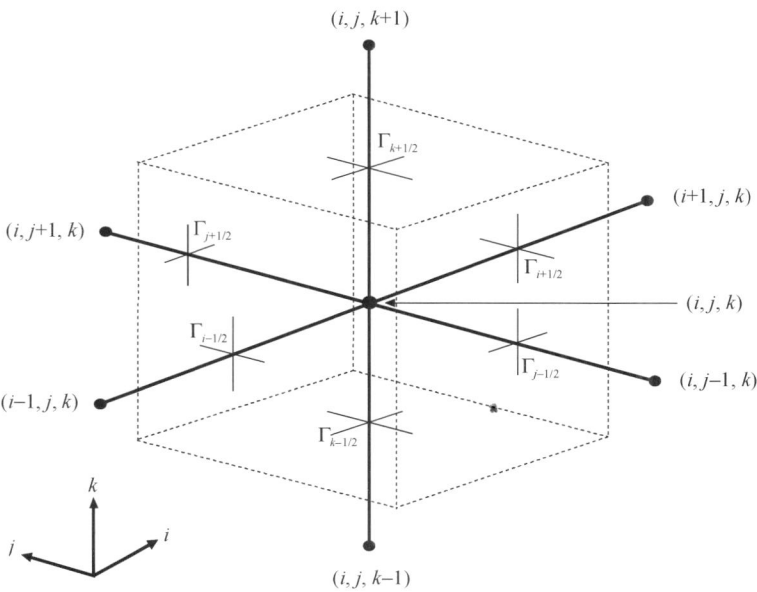

Figure 5.7. A three-dimensional structured grid.

Here, $\hat{q}_{\Gamma_m}$ indicates the net heat flux rates into a control volume (i,j,k) through its mth control-volume face. Similarly, $\Gamma_{i\pm 1/2}$, $\Gamma_{j\pm 1/2}$, and $\Gamma_{k\pm 1/2}$ indicate the control-volume faces $\Gamma_{i\pm 1/2,j,k}$ and $\Gamma_{i,j\pm 1/2,k}$, and $\Gamma_{i,j,k\pm 1/2}$, respectively. The cross-sectional areas are given by $A_{\Gamma_{i-1/2}} = A_{\Gamma_{i+1/2}} = \Delta y\, \Delta z$, $A_{\Gamma_{j-1/2}} = A_{\Gamma_{j+1/2}} = \Delta x\, \Delta z$, and $A_{\Gamma_{k-1/2}} = A_{\Gamma_{k+1/2}} = \Delta x\, \Delta y$. The volume of the control volume (i,j,k) is $V_{i,j,k} = \Delta x\, \Delta y\, \Delta z$. The value $\hat{q}_{\Gamma_m}$ can also be evaluated using equations (5.2.7b) through (5.2.7f) according to the boundary conditions of the control-volume faces.

5.5 Worked examples

In this section we will consider the application of the finite volume method to one-dimensional steady-state heat conduction problems having various types of boundary conditions. In one-dimensional problems, each control volume has two control-volume faces, $\Gamma_{i-1/2}$ and $\Gamma_{i+1/2}$. Thus, equation (5.2.6) can be expressed by

$$\sum_{m=1}^{nf} (\hat{q}_{\Gamma_m} \times A_{\Gamma_m}) + V_i \dot{g}_i = \{\hat{q}_{\Gamma_{i-1/2}} \times A_{\Gamma_{i-1/2}} + \hat{q}_{\Gamma_{i+1/2}} \times A_{\Gamma_{i+1/2}}\} + V_i \dot{g}_i = 0. \tag{5.5.1}$$

In the case of the rectangular coordinate system, the cross-sectional areas, $A_{\Gamma_{i-1/2}}$ and $A_{\Gamma_{i+1/2}}$, are equal to unity, and the volume of a control volume, V_i, is equal to the control volume width, Δx_i. For the sake of simplicity, the control volume width is assumed to be constant ($\Delta x_i = \Delta x$). Then, equation (5.5.1) is simplified to

$$(\hat{q}_{\Gamma_{i-1/2}} + \hat{q}_{\Gamma_{i+1/2}}) + \Delta x\, \dot{g}_i = 0 \tag{5.5.2}$$

The values of $\hat{q}_{\Gamma_{i-1/2}}$ and $\hat{q}_{\Gamma_{i-1/2}}$ can be evaluated using equations (5.2.7b) through (5.2.7f) according to the boundary conditions of the control-volume faces $\Gamma_{i-1/2}$ and $\Gamma_{i+1/2}$.

5.5.1 Example 5.1

Consider one-dimensional steady-state heat conduction in an insulated steel rod, as indicated in figure 5.8. The steel rod is heated electrically by the passage of electric current, which generates energy at a rate of $\dot{g} = 10^5$ W/m^3. Thermal conductivity, λ, is 50 W/m · K. The length of the rod is 0.2 m. One end of the rod is maintained at a constant temperature of $\bar{T}_{\Gamma_A} = 100\,°C$ (Γ_0-type boundary), and the other end is insulated, i.e., zero heat flux, $\bar{q}_{\Gamma_B} = 0$ (Γ_{II}-type boundary).

Develop the finite difference equations for this problem and calculate the temperature distribution in the steel rod.

Solution

Let us divide the length of the rod into five equal control volumes having six nodes with $\Delta x = 0.04$ m, as shown in figure 5.8.

By applying the boundary conditions given by equations (5.2.7b)–(5.2.7f) into equation (5.5.2), we have the finite difference equations for nodes 1 through 5 as follows.

For node i ($i = 1$–4), each control volume has two control-volume faces of Γ_1-type boundary (equation (5.2.7b)), thus the finite difference equation is

$$\lambda \frac{T_{i-1} - T_i}{\Delta x} + \lambda \frac{T_{i+1} - T_i}{\Delta x} + \Delta x \dot{g} = 0 \tag{1a}$$

where $T_0 = \bar{T}_{\Gamma_A}$.

For node 5, $\bar{q}_{\Gamma_B} = 0$.

$$\lambda \frac{T_4 - T_5}{\Delta x} + 0 + \Delta x \dot{g} = 0. \tag{1b}$$

By summarizing equations (1a) and (1b) for nodes 1 through 5, we obtain

$$\begin{aligned}
-2T_1 + T_2 &= -(\bar{T}_{\Gamma_A} + \Delta x^2 \cdot \dot{g}/\lambda) \\
T_1 - 2T_2 + T_3 &= -\Delta x^2 \cdot \dot{g}/\lambda \\
T_2 - 2T_3 + T_4 &= -\Delta x^2 \cdot \dot{g}/\lambda \\
T_3 - 2T_4 + T_5 &= -\Delta x^2 \cdot \dot{g}/\lambda \\
T_4 - T_5 &= -\Delta x^2 \cdot \dot{g}/2\lambda
\end{aligned} \tag{2}$$

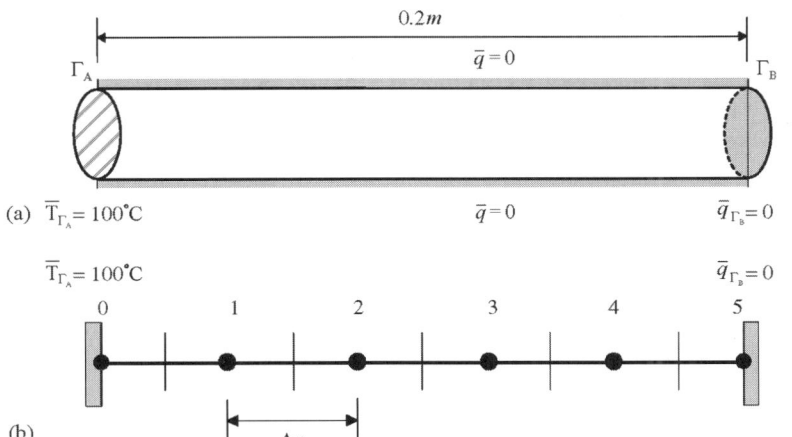

Figure 5.8. One-dimensional steady-state heat conduction problems: (a) the geometry (b) the grids used in the simulation.

Substitution of numerical values into equation (2) gives the following resulting set of algebraic equations for this problem.

$$\begin{bmatrix} -2500 & 1250 & 0 & 0 & 0 \\ 1250 & -2500 & 1250 & 0 & 0 \\ 0 & 1250 & -2500 & 1250 & 0 \\ 0 & 0 & 1250 & -2500 & 1250 \\ 0 & 0 & 0 & 1250 & -1250 \end{bmatrix} \begin{Bmatrix} T_1 \\ T_2 \\ T_3 \\ T_4 \\ T_5 \end{Bmatrix} = \begin{Bmatrix} -129000 \\ -4000 \\ -4000 \\ -4000 \\ -2000 \end{Bmatrix}. \quad (3)$$

The above set of equations yields the steady-state temperature distribution in the steel rod. Equation (3) can easily be solved by the standard Gauss elimination method and the solution is

$$\mathbf{T} = (T_1, T_2, T_3, T_4, T_5)^\mathrm{T} = (114.4, 125.6, 133.6, 138.4, 140.0)^\mathrm{T}. \quad (4)$$

5.5.2 Example 5.2

Let us now consider a one-dimensional steady-state heat conduction in a large steel plate ($\lambda_{\Omega_1} = 50\,\mathrm{W/m \cdot K}$) with a thickness of 0.05 m. One side of the steel plate is maintained at a constant temperature of $\bar{T}_{\Gamma_A} = 900\,^\circ\mathrm{C}$ (Γ_0-type boundary) and the other side is covered by a ceramic plate ($\lambda_{\Omega_2} = 0.5\,\mathrm{W/m \cdot K}$) with a thickness of 0.08 m. The ceramic plate is exposed to an ambient temperature of $T_{\mathrm{air}} = 25\,^\circ\mathrm{C}$ with a convection heat transfer coefficient of $h_{\mathrm{air}} = 50\,\mathrm{W/m^2 \cdot K}$ (Γ_{III}-type boundary). The interfacial heat transfer coefficient at the steel/ceramic plate interface is $h_{\mathrm{int}} = 500\,\mathrm{W/m^2 \cdot K}$ (Γ_V-type boundary).

Develop the finite difference equations for this problem and calculate the temperature distributions in the steel and ceramic plates.

Solution
Let us divide the regions of the steel and the ceramic plates into five and four equal control volumes having six and five nodes, respectively. Then, we have $\Delta x_{\Omega_1} = 0.01\,\mathrm{m}$ for the sub-domain Ω_1 and $\Delta x_{\Omega_2} = 0.02\,\mathrm{m}$ for the sub-domain Ω_2, as shown in figure 5.9.

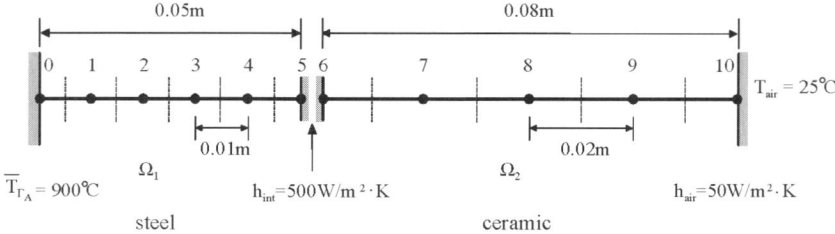

Figure 5.9. One-dimensional steady-state heat conduction problems with two sub-domains.

Firstly, consider the derivation of the finite difference equations for nodes 1 through 5 in the sub-domain Ω_1. By applying the boundary conditions given by equations (5.2.7b)–(5.2.7f) into equation (5.5.2) for the control volumes 1 through 5, we can derive the finite difference equations as follows.

For node i ($i = 1$–4), each control volume has two control-volume faces of Γ_1-type boundary (equation (5.2.7b)), thus the finite difference equation is

$$\lambda_{\Omega_1} \frac{T_{i-1} - T_i}{\Delta x_{\Omega_1}} + \lambda_{\Omega_1} \frac{T_{i+1} - T_i}{\Delta x_{\Omega_1}} = 0 \tag{1a}$$

where $T_0 = \bar{T}_{\Gamma_A}$.

For node 5, we have

$$\lambda_{\Omega_1} \frac{T_4 - T_5}{\Delta x_{\Omega_1}} + h_{\text{int}}(T_6 - T_5) = 0. \tag{1b}$$

Now, consider the derivation of the finite difference equations for nodes 6 through 10 in the sub-domain Ω_2. Similarly, by applying the boundary conditions given by equations (5.2.7b)–(5.2.7f) into equation (5.5.2) for the control volumes 6 through 10, we can derive the finite difference equations as follows.

For node 6, we have

$$h_{\text{int}}(T_5 - T_6) + \lambda_{\Omega_2} \frac{T_7 - T_6}{\Delta x_{\Omega_2}} = 0. \tag{1c}$$

For node i ($i = 7$–9), we have

$$\lambda_{\Omega_2} \frac{T_{i-1} - T_i}{\Delta x_{\Omega_2}} + \lambda_{\Omega_2} \frac{T_{i+1} - T_i}{\Delta x_{\Omega_2}} = 0. \tag{1d}$$

For node 10, we have

$$\lambda_{\Omega_2} \frac{T_9 - T_{10}}{\Delta x_{\Omega_2}} + h_{\text{air}}(T_{\text{air}} - T_{10}) = 0. \tag{1e}$$

Rearranging the above equations for nodes 1 through 10 gives the following resulting set of algebraic equations for this problem.

$$-2T_1 + T_2 = -\bar{T}_{\Gamma_A}$$
$$T_1 - 2T_2 + T_3 = 0$$
$$T_2 - 2T_3 + T_4 = 0$$
$$T_3 - 2T_4 + T_5 = 0$$
$$T_4 - \left(1 + \frac{h_{\text{int}} \cdot \Delta x_{\Omega_1}}{\lambda_{\Omega_1}}\right)T_5 + \left(\frac{h_{\text{int}} \cdot \Delta x_{\Omega_1}}{\lambda_{\Omega_1}}\right)T_6 = 0 \tag{2}$$

$$T_5 - \left(1 + \frac{\lambda_{\Omega_2}}{h_{\text{int}} \cdot \Delta x_{\Omega_2}}\right) T_6 + \left(\frac{\lambda_{\Omega_2}}{h_{\text{int}} \cdot \Delta x_{\Omega_2}}\right) T_7 = 0$$

$$T_6 - 2T_7 + T_8 = 0$$

$$T_7 - 2T_8 + T_9 = 0$$

$$T_8 - 2T_9 + T_{10} = 0$$

$$T_9 - \left(1 + \frac{h_{\text{air}} \cdot \Delta x_{\Omega_2}}{\lambda_{\Omega_2}}\right) T_{10} = -\left(\frac{h_{\text{air}} \cdot \Delta x_{\Omega_2}}{\lambda_{\Omega_2}}\right) T_{\text{air}}. \quad (2)$$

By substituting numerical values into equation (2), we obtain

$$\begin{aligned} -2T_1 + T_2 &= -900 \\ T_1 - 2T_2 + T_3 &= 0 \\ T_2 - 2T_3 + T_4 &= 0 \\ T_3 - 2T_4 + T_5 &= 0 \\ T_4 - 1.1T_5 + 0.1T_6 &= 0 \\ 20T_5 - 21T_6 + T_7 &= 0 \\ T_6 - 2T_7 + T_8 &= 0 \\ T_7 - 2T_8 + T_9 &= 0 \\ T_8 - 2T_9 + T_{10} &= 0 \\ T_9 - 3T_{10} &= -50. \end{aligned} \quad (3)$$

The above set of equations yields the steady-state temperature distributions in the steel and the ceramic plates. The results are

$$\mathbf{T} = (T_1, T_2, T_3, T_4, T_5, T_6, T_7, T_8, T_9, T_{10})^{\text{T}}$$
$$= (899.0, 898.1, 897.1, 896.2, 895.2, 885.7, 694.4, 503.1, 311.9, 120.6)^{\text{T}}. \quad (4)$$

5.5.3 Example 5.3

Consider one-dimensional steady-state heat conduction problems in a steel (a) plate, (b) hollow cylinder and (c) hollow sphere. The geometry, dimension and boundary conditions are described in figure 5.10. The thermal conductivity of the steel is $\lambda = 25 \, \text{W/m} \cdot \text{K}$. The inner and outer radii are $r_a = 0.05 \, \text{m}$ and $r_a = 0.06 \, \text{m}$ for both of the hollow cylinder and sphere, and the plate thickness is $0.01 \, \text{m}$. One face of the plate is kept at $\bar{T}_{\Gamma_A} = 100 \,°\text{C}$ (Γ_0-type boundary), and the other face is exposed to air with a temperature of $T_{\text{air}} = 25 \,°\text{C}$ and a convection heat transfer coefficient of $h_{\text{air}} = 100 \, \text{W/m}^2 \cdot \text{K}$ (Γ_{III}-type boundary). The inner surfaces of the hollow

Worked examples

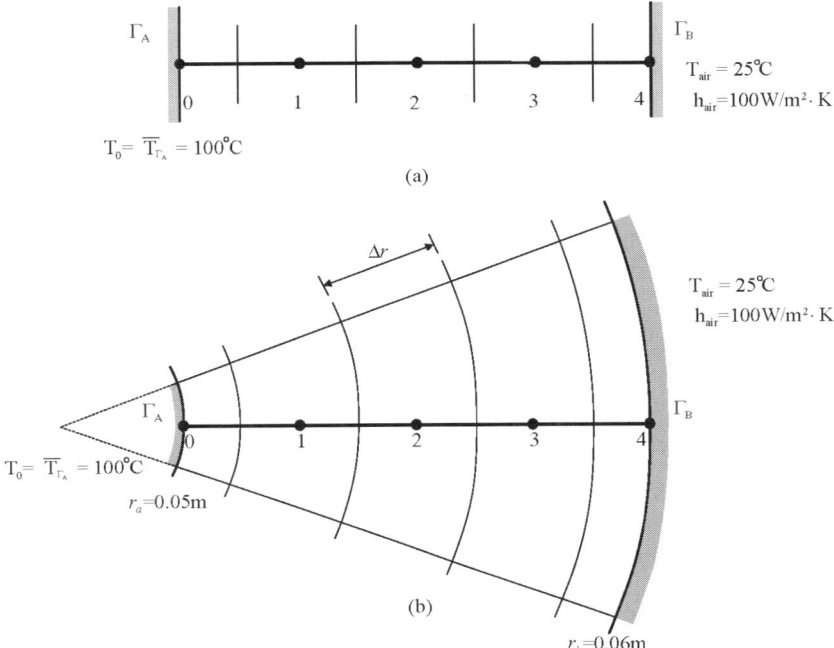

Figure 5.10. One-dimensional steady-state heat conduction problems: (a) a plane wall (b) a hollow cylinder or a hollow sphere.

cylinder and sphere are held at a constant temperature of $\bar{T}_{\Gamma_A} = 100\,°C$ (Γ_0-type boundary), and the outer surfaces are exposed to air with a temperature of $T_{\text{air}} = 25\,°C$ and a convection heat transfer coefficient of $h_{\text{air}} = 100\,\text{W/m}^2 \cdot \text{K}$ (Γ_{III}-type boundary).

Develop the finite difference equations and calculate the temperature distributions in the steel for three cases.

Solution
Let us divide the computational domains into four equal control volumes having five nodes with $\Delta x (\equiv \Delta r) = 0.0025\,\text{m}$, as shown in figure 5.10.

The temperature at node 0 is held at a constant temperature of $T_0 = \bar{T}_{\Gamma_A} = 100\,°C$ (Γ_0-type boundary). Therefore, the finite difference equations for nodes 1 through 4 are directly derived using equation (5.5.1). Since there is no heat generation in the domain, equation (5.5.1) is simplified to

$$\sum_{m=1}^{nf} (\hat{q}_{\Gamma_m} \times A_{\Gamma_m}) + V_i \dot{g} = (\hat{q}_{\Gamma_{i-1/2}} \times A_{\Gamma_{i-1/2}} + \hat{q}_{\Gamma_{i+1/2}} \times A_{\Gamma_{i+1/2}}) = 0. \quad (1)$$

By applying the boundary conditions given by equations (5.2.7b)–(5.2.7f) into equation (1), we obtain the finite difference equations for nodes 1 through 4 as follows.

For node i ($i = 1$–3), each control volume has two control-volume faces of Γ_1-type boundary (equation (5.2.7b)), thus the finite difference equation is

$$\lambda A_{\Gamma_{i-1/2}} \frac{T_{i-1} - T_i}{\Delta x} + \lambda A_{\Gamma_{i+1/2}} \frac{T_{i+1} - T_i}{\Delta x} = 0 \tag{2a}$$

where $T_0 = \bar{T}_{\Gamma_A} = 100\,°\text{C}$.

For node 4,

$$\lambda A_{\Gamma_{i-1/2}} \frac{T_3 - T_4}{\Delta x} + h_{\text{air}} A_{\Gamma_{i+1/2}} (T_{\text{air}} - T_4) = 0. \tag{2b}$$

Application of equations (2a) and (2b) to nodes 1 through 4 gives the following resulting set of algebraic equations.

$$-(A_{\Gamma_{i-1/2}} + A_{\Gamma_{i+1/2}})T_1 + A_{\Gamma_{i+1}} \cdot T_2 = -A_{\Gamma_{i-1/2}} \cdot \bar{T}_{\Gamma_A}$$

$$A_{\Gamma_{i-1/2}} \cdot T_1 - (A_{\Gamma_{i-1/2}} + A_{\Gamma_{i+1/2}})T_2 + A_{\Gamma_{i+1/2}} \cdot T_3 = 0$$

$$A_{\Gamma_{i-1/2}} \cdot T_2 - (A_{\Gamma_{i-1/2}} + A_{\Gamma_{i+1/2}})T_3 + A_{\Gamma_{i+1/2}} \cdot T_4 = 0 \tag{3}$$

$$A_{\Gamma_{i-1/2}} \cdot T_3 - \left(A_{\Gamma_{i-1/2}} + \frac{h_{\text{air}} A_{\Gamma_{i+1/2}} \Delta x}{\lambda} \right) T_4 = -\left(\frac{h_{\text{air}} A_{\Gamma_{i+1/2}} \Delta x \, T_{\text{air}}}{\lambda} \right).$$

The values of $A_{\Gamma_{i-1/2}}$ and $A_{\Gamma_{i+1/2}}$ for a hollow cylinder and a hollow sphere can be calculated by equations (5.3.1) and (5.3.2).

By substituting numerical values into equation (3) and solving them for three coordinate systems, we obtain

For a plane plate,

$$\begin{array}{rl} -2T_1 + T_2 = & -100 \\ T_1 - 2T_2 + T_3 = & 0 \\ T_2 - 2T_3 + T_4 = & 0 \\ 100T_3 - 101T_4 = & -25 \end{array} \begin{bmatrix} T_1 \\ T_2 \\ T_3 \\ T_4 \end{bmatrix} = \begin{bmatrix} 99.28 \\ 98.56 \\ 97.84 \\ 97.12 \end{bmatrix}. \tag{4}$$

For a hollow cylinder,

$$\begin{array}{rl} -10.5T_1 + 5.375T_2 = & -512.5 \\ 5.375T_1 - 11.0T_2 + 5.625T_3 = & 0 \\ 5.625T_2 - 11.5T_3 + 5.875T_4 = & 0 \\ 587.5T_3 - 593.5T_4 = & -150 \end{array} \begin{bmatrix} T_1 \\ T_2 \\ T_3 \\ T_4 \end{bmatrix} = \begin{bmatrix} 99.16 \\ 98.36 \\ 97.59 \\ 96.86 \end{bmatrix}. \tag{5}$$

For a hollow sphere,

$$\begin{matrix} -5.516T_1 + 2.889T_2 = & -262.6 \\ 2.889T_1 - 6.053T_2 + 3.164T_3 = & 0 \\ 3.164T_2 - 6.616T_3 + 3.452T_4 = & 0 \\ 345.2T_3 - 348.8T_4 = & -90 \end{matrix} \begin{bmatrix} T_1 \\ T_2 \\ T_3 \\ T_4 \end{bmatrix} = \begin{bmatrix} 99.02 \\ 98.13 \\ 97.31 \\ 96.57 \end{bmatrix}. \quad (6)$$

5.6 Case study: one-dimensional steady state heat conduction problems

5.6.1 Description of the problem

Consider one-dimensional steady-state heat conduction problems in a steel (a) plate, (b) hollow cylinder and (c) hollow sphere. The geometry, dimension and boundary conditions are described in figure 5.10. The thermal conductivity of the steel is $\lambda = 25\,\text{W/m} \cdot \text{K}$. The inner and outer radii are $r_a = 0.05\,\text{m}$ and $r_a = 0.06\,\text{m}$ for both of the hollow cylinder and sphere, and the plate thickness is $0.01\,\text{m}$. One face of the plate is kept at $\bar{T}_{\Gamma_A} = 100\,°\text{C}$ (Γ_0-type boundary), and the other face is exposed to air with a temperature of $T_{\text{air}} = 25\,°\text{C}$ and a convection heat transfer coefficient of $h_{\text{air}} = 100\,\text{W/m}^2 \cdot \text{K}$ (Γ_{III}-type boundary). The inner surfaces of the hollow cylinder and sphere are held at a constant temperature of $\bar{T}_{\Gamma_A} = 100\,°\text{C}$ (Γ_0-type boundary), and the outer surfaces are exposed to air with a temperature of $T_{\text{air}} = 25\,°\text{C}$ and a convection heat transfer coefficient of $h_{\text{air}} = 100\,\text{W/m}^2 \cdot \text{K}$ (Γ_{III}-type boundary).

Develop the finite difference equations and calculate the temperature distributions in the steel for three cases.

5.6.2 Glossary of FORTRAN notation

FORTRAN name	Meaning
A(I, NL)	The left boundary area of node i
A(I, NR)	The right boundary area of node i
DX	Distance between nodes
HAIR	Heat transfer coefficient at the material/air interface
ICOOR	Integer variable which selects a coordination system: 1 for rectangular, 2 for cylindrical and 3 for spherical coordinates, respectively
I	Index denoting node i
N	The number of nodes
RA	Inner radius
RB	Outer radius

FORTRAN name	Meaning
T(I)	Temperature of node i
T0	The prescribed temperature of node 0
TAIR	The temperature of air
V(I)	Volume of node i

5.6.3 Simulations

Let us simulate the temperature distributions for three coordinate systems using the execution file [sheat.exe].

(1) For a plane plate

Data input
```
> SELECT COORD. SYSTEM(1:RECT, 2:CYLINDER, 3:SPHERE) : 1
> ENTER RA AND RB : 0.05 0.06
```

Results
```
-- SOLUTION --
T( 1) = 99.28
T( 2) = 98.56
T( 3) = 97.84
T( 4) = 97.12
```

(2) For a hollow cylinder

Data input
```
> SELECT COORD. SYSTEM(1:RECT, 2:CYLINDER, 3:SPHERE) : 2
> ENTER RA AND RB : 0.05 0.06
```

Results
```
-- SOLUTION --
T( 1) = 99.16
T( 2) = 98.36
T( 3) = 97.59
T( 4) = 96.86
```

(3) For a hollow sphere

Data input
```
> SELECT COORD. SYSTEM(1:RECT, 2:CYLINDER, 3:SPHERE) : 3
> ENTER RA AND RB : 0.05 0.06
```

Results
```
-- SOLUTION --
T( 1) = 99.02
T( 2) = 98.13
T( 3) = 97.31
T( 4) = 96.57
```

5.6.4 Program list

```
C ****************************************************
C *                                                  *
C * A COMPUTER PROGRAM FOR THE SIMULATION OF 1-D     *
C * STEADY STATE HEAT CONDUCTION PROBLEMS            *
C *                                                  *
C ****************************************************
      PROGRAM SHEAT
      PARAMETER(N=5)
      PARAMETER(NL=1, NR=2)
      PARAMETER(IRECT=1, ICYLIN=2, ISPHER=3)
      COMMON
     &/VAR/T(N), V(N), A(N, 2)
     &/PROPERTY/TCON
     &/GEOMETRY/RA, RB, DX, ICOOR
     &/BOUNDARY/HAIR, TAIR, T0
     &/COEFFICIENT/C(N, N+1)
C
   10 WRITE(6, 20)
   20 FORMAT(' > SELECT COORD.SYSTEM(1:RECT,2:CYLINDER,
     &       3:SPHERE) : ',\)
      READ(5, *) ICOOR
      IF(ICOOR.LE.0.OR. ICOOR.GE.4) GOTO 10
C
C---- INPUT GEOMETRICAL VARIABLES
      WRITE(6, 30)
   30 FORMAT(' > ENTER RA AND RB : ',\)
      READ(5,*) RA, RB
C
C---- INITIALIZE VARIABLES
      CALL INIT
C
C---- THE SOLVER OF GAUSS ELIMINATION METHOD
      CALL GAUSS
C
C---- PRINT RESULTS
      WRITE(6, *)' -- SOLUTION --'
      DO 400 I=1, N-1
      WRITE(6, 300) I, T(I+1)
```

```
  300 FORMAT(5X,'T(',I2,') =', F8.2)
  400 CONTINUE
      PAUSE
      END
C
C     SUBROUTINE FOR THE INITIALIZATION OF VARIABLES
C
      SUBROUTINE INIT
      PARAMETER(N=5)
      PARAMETER(NL=1, NR=2)
      PARAMETER(IRECT=1, ICYLIN=2, ISPHER=3)
      COMMON
     &/VAR/T(N), V(N), A(N, 2)
     &/PROPERTY/TCON
     &/GEOMETRY/RA, RB, DX, ICOOR
     &/BOUNDARY/HAIR, TAIR, T0
     &/COEFFICIENT/C(N, N+1)
      REAL PI
C
      PI=ATAND(45.0)
C
C---- THERMAL AND PHYSICAL PROPERTIES
      TCON=25.0
      HAIR=100.0
      TAIR=25.0
      T0=100.0
C
      DX = (RB-RA)/FLOAT(N-1)
C
C---- CALCULATE VOLUMES AND SURFACE AREAS
C
      SELECT CASE (ICOOR)
      CASE (IRECT)
      DO 100 I=1, N
      V(I) = DX
      A(I, NL) = 1.0
      A(I, NR) = 1.0
  100 CONTINUE
C
      V(1) = DX/2.0
      V(N) = DX/2.0
C
      CASE (ICYLIN)
      DO 200 I=1, N
      A(I,NL)=2.0*PI*(RA+FLOAT(I-1)*DX-DX/2.0)
      A(I,NR)=2.0*PI*(RA+FLOAT(I-1)*DX+DX/2.0)
      V(I)=PI*((RA+FLOAT(I-1)*DX+DX/2.0)**2
     &      -(RA+FLOAT(I-1)*DX-DX/2.0)**2)
```

```
  200 CONTINUE
      A(1,NL)=2.0*PI*RA
      A(N,NR)=2.0*PI*RB
      V(1) = PI*((RA+DX/2.0)**2-RA**2)
      V(N) = PI*(RB**2-(RB-DX/2.0)**2)
C
      CASE (ISPHER)
      DO 300 I=1, N
      A(I,NL)= 4*PI*(RA+FLOAT(I-1)*DX-DX/2.0)**2
      A(I,NR)= 4*PI*(RA+FLOAT(I-1)*DX+DX/2.0)**2
      V(I) = 4.0*PI/3.0*((RA+FLOAT(I-1)*DX+DX/2.0)**3
     &       -(RA+FLOAT(I-1)*DX-DX/2.0)**3)
  300 CONTINUE
      A(1,NL)= 4.0*PI*RA**2
      A(N,NR)= 4.0*PI*RB**2
      V(1) = 4.0*PI/3.0*((RA+DX/2.0)**3-RA**3)
      V(N) = 4.0*PI/3.0*(RB**3-(RB-DX/2.0)**3)
      END SELECT
C
      DO 400 I=1, N
      DO 400 J=1, N+1
      C(I,J)=0.0
  400 CONTINUE
C
C---- ASSEMBLY OF COEFFICIENTS
      DO 500 I=2, N-1
      C(I,I-1) = A(I,NL)
      C(I,I)  =-A(I,NL)-A(I,NR)
      C(I,I+1) = A(I,NR)
  500 CONTINUE
C
      C(1, 1) = 1.0
      C(1,N+1) = T0
      C(N,N-1) = A(N,NL)
      C(N,N)  =-A(N,NL)-HAIR*A(N,NR)*DX/TCON
      C(N,N+1) =-HAIR*A(N,NR)*DX*TAIR/TCON
      END
C
C     SUBROUTINE OF GAUSS ELIMINATION METHOD
C
      SUBROUTINE GAUSS
      PARAMETER(N=5)
      PARAMETER(NL=1, NR=2)
      COMMON
     &/VAR/T(N), V(N), A(N, 2)
     &/PROPERTY/TCON
     &/GEOMETRY/RA, RB, DX, ICOOR
     &/BOUNDARY/HAIR, TAIR, T0
```

```
      &/COEFFICIENT/C(N, N+1)
C
      NM1=N-1
      L=N+1
      DO 10 K=1,NM1
         J=K
         KP1=K+1
         DO 20 I=KP1,N
            IF(ABS(C(I,J)).LT.ABS(C(I,J))) J=I
 20      CONTINUE
         IF(J.EQ.K) GOTO 100
         DO 30 I=1,L
            TEMP=C(K,I)
            C(K,I)=C(J,I)
            C(J,I)=TEMP
 30      CONTINUE
 100     DO 40 I=KP1,N
            D=C(I,K)/C(K,K)
            DO 50 J=1,L
               C(I,J)=C(I,J)-D*C(K,J)
 50         CONTINUE
 40      CONTINUE
 10   CONTINUE
C
      T(N)=C(N,L)/C(N,N)
      DO 60 J=NM1,1,-1
         SUM=0.0
         JJ = J + 1
         DO 70 K=JJ,N
            SUM=SUM+C(J,K)*T(K)
 70      CONTINUE
         T(J)=(C(J,L)-SUM)/C(J,J)
 60   CONTINUE
      END
```

Chapter 6

Transient heat conduction

The transient heat conduction and mass diffusion problems have numerous important applications in materials processing, such as melting and solidification, crystal growth, casting, heat treatment, and welding processes, etc. In chapter 5 we studied the finite volume method for steady-state potential flow problems. In this chapter, we will examine how to solve transient problems, in which the temperature or species concentration within a domain varies with both position and time. Because of the similarities between heat and species transfer, it might be enough only to consider heat conduction problems here.

6.1 Mathematical formulation

6.1.1 Governing equation

The integral form of transient heat conduction equation, which will be used in the finite volume approach, is given by

$$\frac{\partial}{\partial t} \iiint_\Omega (\rho C_v T) \, d\Omega = - \iint_\Gamma (\mathbf{q} \cdot \mathbf{n}) \, d\Gamma + \iiint_\Omega \dot{g} \, d\Omega. \quad (6.1.1)$$

6.1.2 Initial and boundary conditions

In case of transient problems we need an initial condition which describes the state of the physical domain at time $t = 0$ as follow.

$$T(\Omega) = T_0(\Omega) \quad \text{at } t = 0. \quad (6.1.2)$$

$T_0(\Omega)$ is usually assumed to be a constant temperature in the domain Ω.

The boundary conditions frequently encountered in heat transfer have already been mentioned in chapters 2 and 5, and will not be described here.

6.2 Finite volume approach for transient problems

6.2.1 Computational grids

The first step in the finite volume method is to divide the computational domain into discrete control volumes. Let us consider the one-dimensional control volumes defined at the internal and outer boundary points shown in figure 6.1.

6.2.2 Derivation of finite difference equations

As described in chapter 5, we can derive the finite difference equations by applying the integral form of transient heat conduction equation into the control volumes in figure 6.1. Integration of equation (6.1.1) with respect to time over a time interval from t to $t + \Delta t$ gives

$$\int_t^{t+\Delta t} \frac{\partial}{\partial t}\left\{\iiint_\Omega (\rho C_v T)\, d\Omega\right\} dt = -\int_t^{t+\Delta t} \iint_\Gamma (\mathbf{q}\cdot\mathbf{n})\, d\Gamma\, dt$$
$$+ \int_t^{t+\Delta t} \iiint_\Omega \dot{g}\, d\Omega\, dt. \qquad (6.2.1)$$

If we assume that the temperature of the node (i), T_i, prevails throughout the control volume (i), the integral term on the left-hand side of equation (6.2.1) can be expressed by

$$\int_t^{t+\Delta t} \frac{\partial}{\partial t}\left\{\iiint_\Omega (\rho C_v T)\, d\Omega\right\} dt = \rho C_v V_i (T_i^{t+\Delta t} - T_i^t). \qquad (6.2.2)$$

Here T_i^t and $T_i^{t+\Delta t}$ are the temperatures of the control volume (i) at the time levels t and $t + \Delta t$, respectively.

The first integral term on the right-hand side of equation (6.2.1), which is called *surface phenomena* as mentioned in chapter 2, representing the

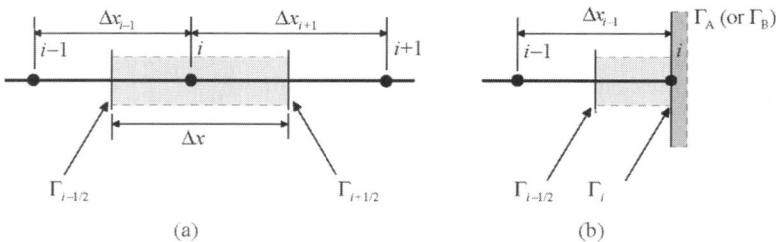

Figure 6.1. A schematic representation of control volumes: (a) a control volume in the internal region and (b) a control volume on the outer boundary.

total heat influx into the control volume (*i*) through its control-volume faces, can be integrated in a similar manner to equation (5.2.5) as mentioned in chapter 5. When we evaluate the time integral, we need to make an assumption about the variation of **q**, which is a function of temperature, with time. We could use **q** (or temperatures) at time t or $t + \Delta t$ to calculate the time integral or, alternatively, a combination of **q** (or temperatures) at time t and $t + \Delta t$. We can generalize the approach by employing a weighting parameter (W) between 0 and 1 at time t and $t + \Delta t$ simultaneously. Then, we have

$$-\int_t^{t+\Delta t} \iint_\Gamma (\mathbf{q} \cdot \mathbf{n}) \, d\Gamma \, dt = \left\{ W \sum_{m=1}^{nf} (\hat{q}_{\Gamma_m}^{t+\Delta t} A_{\Gamma_m}) + (1-W) \sum_{m=1}^{nf} (\hat{q}_{\Gamma_m}^{t} A_{\Gamma_m}) \right\} \Delta t \tag{6.2.3}$$

where *nf* is the number of control-volume faces of the control volume (*i*) and *m* indicates the *m*th control-volume face of the control volume (*i*). It must be noted that the control-volume face might be a part of the inner boundary between the control volume (*i*) and its neighboring control volume or a part of the outer boundary of the computational domain (Ω). The terms $\hat{q}_{\Gamma_m}^{t}$ and $\hat{q}_{\Gamma_m}^{t+\Delta t}$ indicate the net heat influxes through the *m*th control-volume face into the control volume (*i*) at time level t and $t + \Delta t$, respectively.

Finally, the second term on the right-hand side of equation (6.2.1) is given by

$$\int_t^{t+\Delta t} \iiint_\Omega \dot{g} \, d\Omega \, dt = V_i \dot{g}_i \Delta t. \tag{6.2.4}$$

By substituting equations (6.2.2) through (6.2.4) into equation (6.2.1) and rearranging it, we obtain the following general form of finite difference equation for transient heat conduction problems.

$$\rho C_v V_i \frac{T_i^{t+\Delta t} - T_i^t}{\Delta t} = W \sum_{m=1}^{nf} (\hat{q}_{\Gamma_m}^{t+\Delta t} A_{\Gamma_m}) + (1-W) \sum_{m=1}^{nf} (\hat{q}_{\Gamma_m}^{t} A_{\Gamma_m}) + V_i \dot{g}_i \tag{6.2.5}$$

where W is a weighting parameter, $0 \leq W \leq 1$. It is to be noted that the general form of finite difference equation, equation (6.2.5), is available for any type of coordinate system and for two- or three-dimensional problems.

6.3 Solving schemes

Let us now consider the solving scheme of equation (6.2.5). As described in section 5.2.2, the term $\hat{q}_{\Gamma_m}$ can be evaluated using equations (5.2.7b) through (5.2.7f) according to the boundary conditions.

Transient heat conduction

As an example, consider the finite difference equation of an internal node (i) in one-dimensional rectangular coordinate system. By substituting equation (5.2.7b) into equation (6.2.5) and rearranging it, we obtain

$$\frac{T_i^{t+\Delta t} - T_i^t}{\Delta t} = \alpha \left\{ W \frac{T_{i-1}^{t+\Delta t} - 2T_i^{t+\Delta t} + T_{i+1}^{t+\Delta t}}{\Delta x^2} + (1-W) \frac{T_{i-1}^t - 2T_i^t + T_{i+1}^t}{\Delta x^2} \right\}. \quad (6.3.1)$$

According to the weighting parameter we have the following three types of finite difference schemes. When $W = 0$, the temperature at an old time level t is used, and the scheme is called the *fully explicit method*. When $W = 1$, the temperature at a new time level $t + \Delta t$ is used, and the scheme is called the *fully implicit method*. And if $W = 1/2$, the temperatures at t and $t + \Delta t$ are equally weighed, and the resulting scheme is called the *Crank–Nicolson method*.

In this section, we will consider one-dimensional transient heat conduction problems in case of no heat generation.

6.3.1 Fully explicit method

For the sake of simplicity, consider the finite difference equation for an inner node (i), which has two Γ_I-type control-volume faces, as shown in figure 6.2.

If we take $W = 0$ in equation (6.2.5), the general discretization equation reduces to the fully explicit scheme as follows.

$$\rho C_v V_i \frac{T_i^{t+\Delta t} - T_i^t}{\Delta t} = \sum_{m=1}^{nf} (\hat{q}_{\Gamma_m}^t A_{\Gamma_m}). \quad (6.3.2)$$

Since the control volume shown in figure 6.2 has two Γ_I-type control-volume faces, $\Gamma_{i-1/2}$ and $\Gamma_{i+1/2}$, we have

$$\frac{T_i^{t+\Delta t} - T_i^t}{\Delta t} = \frac{1}{\rho C_v V_i} \{\hat{q}_{\Gamma_{i-1/2}}^t A_{\Gamma_{i-1/2}} + \hat{q}_{\Gamma_{i+1/2}}^t A_{\Gamma_{i+1/2}}\}. \quad (6.3.3)$$

By substituting equation (5.2.7b) into equation (6.3.3) and rearranging it, we obtain the following finite difference equation for an internal node (i).

$$T_i^{t+\Delta t} = \frac{\alpha \Delta t}{V_i \Delta x} A_{\Gamma_{i-1/2}} T_{i-1}^t + \left\{1 - \frac{\alpha \Delta t}{V_i \Delta x}(A_{\Gamma_{i-1/2}} + A_{\Gamma_{i+1/2}})\right\} T_i^t$$

$$+ \frac{\alpha \Delta t}{V_i \Delta x} A_{\Gamma_{i+1/2}} T_{i+1}^t. \quad (6.3.4)$$

Equation (6.3.4) is called the fully explicit form of finite difference equation for one-dimensional transient heat conduction problems since it has only one unknown temperature $T_i^{t+\Delta t}$ at the future time level $t + \Delta t$. Roache [1] has given the name FTCS method to the above scheme obtained by applying

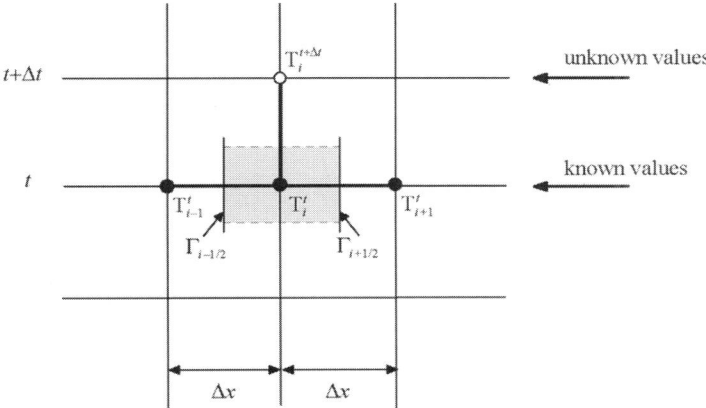

Figure 6.2. The simple explicit scheme.

the Forward-Time and Centered-Space differences to the heat conduction equation. The future temperature $T_i^{t+\Delta t}$ can be directly evaluated using the known temperatures T_{i-1}^t, T_i^t and T_{i+1}^t at the present time level t, as described in figure 6.2.

In case of the rectangular coordinate system, $V_i = \Delta x$ and $A_{\Gamma_{i-1/2}} = A_{\Gamma_{i+1/2}} = 1$. Then, equation (6.3.4) is simply expressed as

$$\frac{T_i^{t+\Delta t} - T_i^t}{\Delta t} = \alpha \left\{ \frac{T_{i-1}^t - 2T_i^t + T_{i+1}^t}{\Delta x^2} \right\} \quad (6.3.5a)$$

or in the form

$$T_i^{t+\Delta t} = \gamma T_{i-1}^t + (1 - 2\gamma) T_i^t + \gamma T_{i+1}^t \quad (6.3.5b)$$

where $\gamma \ (\equiv \alpha \Delta t / \Delta x^2)$ is the Fourier number and also known as the diffusion number.

Let us now consider how to decide the time step Δt for the calculation of the explicit finite difference equation, such as equation (6.3.5b). The new temperature $T_i^{t+\Delta t}$ of the node (i) is positively proportional to the present temperatures of the node (i) and of its two neighboring nodes $(i-1)$ and $(i+1)$, i.e., T_{i-1}^t, T_i^t and T_{i+1}^t, with the proportional coefficients γ, $(1-2\gamma)$ and γ in equation (6.3.5b). The proportional coefficient can be called the weighting factor in regard to their role. It is also seen that the sum of all the weighting factors are equal to 1.0, and that the weighting factors of the neighboring nodes $(i-1)$ and $(i+1)$ are positive. If the weighting factor of the node (i), $(1-2\gamma)$, is negative, the new temperature $T_i^{t+\Delta t}$ will oscillate and finally diverge. Thus, all the weighting factors should always be positive for obtaining the stable solution of equation

(6.3.5b). Based on this concept, the value of the Fourier number γ should be restricted to

$$0 < \gamma \leq \tfrac{1}{2} \qquad (6.3.6a)$$

or in the form of the time step as

$$\Delta t \leq \frac{\Delta x^2}{2\alpha}. \qquad (6.3.6b)$$

In the case of the general form of finite difference equation, equation (6.3.4), the weighting factor of the term T_i^t should be restricted as

$$\left\{ 1 - \frac{\alpha \Delta t}{V_i \Delta x} (A_{\Gamma_{i-1/2}} + A_{\Gamma_{i+1/2}}) \right\} \geq 0 \qquad (6.3.7a)$$

or

$$\Delta t \leq \frac{V_i \Delta x}{\alpha (A_{\Gamma_{i-1/2}} + A_{\Gamma_{i+1/2}})}. \qquad (6.3.7b)$$

The stability criterion implies that the magnitude of the time step, which is dependent upon the size of the control volume and thermo-physical properties, cannot exceed the limit given by equations (6.3.6) and (6.3.7) not to cause divergence in the solution.

In the above discussion, we have only considered the finite difference equation for an inner nodal point and its related stability conditions. The finite difference equations for the outer boundary nodal points and their stability conditions can also be considered based on the similar concept.

6.3.2 Fully implicit method

The maximum size of the stable time step for the repeated calculation in the fully explicit method is restricted by the stability criterion, resulting in an increase of the computational time in some cases. Implicit methods have been developed in order to alleviate this difficulty.

In the case of $W = 1$, equation (6.2.5) reduces to

$$\rho C_v V_i \frac{T_i^{t+\Delta t} - T_i^t}{\Delta t} = \sum_{m=1}^{nf} (\hat{q}_{\Gamma_m}^{t+\Delta t} A_{\Gamma_m}). \qquad (6.3.8)$$

In a similar procedure, the finite difference equation of an internal node (i) in the rectangular coordinate system can be derived by substituting equation (5.2.7b) into equation (6.3.8) as follows.

$$\frac{T_i^{t+\Delta t} - T_i^t}{\Delta t} = \alpha \left\{ \frac{T_{i-1}^{t+\Delta t} - 2 T_i^{t+\Delta t} + T_{i+1}^{t+\Delta t}}{\Delta x^2} \right\}. \qquad (6.3.9)$$

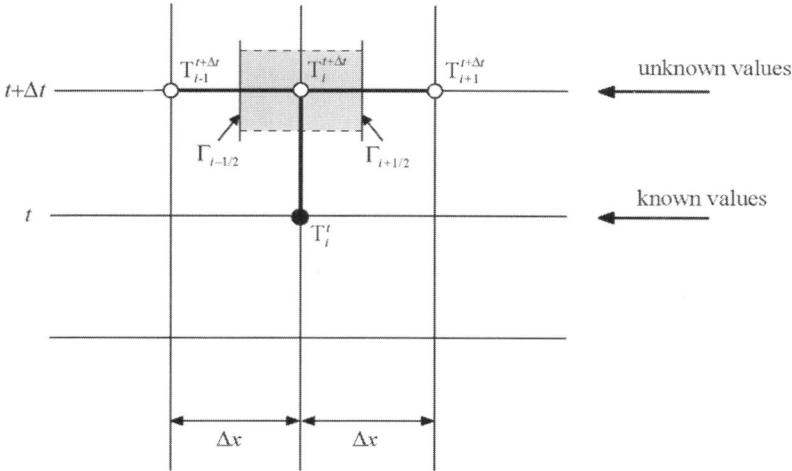

Figure 6.3. The simple implicit scheme.

This equation is an implicit form since all the unknown temperatures at each time level should be solved simultaneously.

Figure 6.3 indicates the finite difference representation for the fully implicit scheme. If the problem involves N unknown nodal temperatures, N simultaneous equations at each time level should be solved. In this method, there is no restriction on the time step for iterative calculation, which will be described in section 6.4.

6.3.3 Crank–Nicolson method

In the case of $W = 1/2$, equation (6.2.5) reduces to

$$\rho C_v V_i \frac{T_i^{t+\Delta t} - T_i^t}{\Delta t} = \frac{1}{2} \left\{ \sum_{m=1}^{nf} (\hat{q}_{\Gamma_m}^{t+\Delta t} A_{\Gamma_m}) + \sum_{m=1}^{nf} (\hat{q}_{\Gamma_m}^{t} A_{\Gamma_m}) \right\}. \quad (6.3.10)$$

The finite difference equation for an internal node (i) in the rectangular coordinate system is given by

$$\frac{T_i^{t+\Delta t} - T_i^t}{\Delta t} = \frac{\alpha}{2} \left\{ \frac{T_{i-1}^{t+\Delta t} - 2T_i^{t+\Delta t} + T_{i+1}^{t+\Delta t}}{\Delta x^2} + \frac{T_{i-1}^t - 2T_i^t + T_{i+1}^t}{\Delta x^2} \right\}. \quad (6.3.11)$$

This equation is also an implicit form since all the unknown temperatures at each time level should be solved simultaneously. If the problem involves N unknown nodal temperatures, N simultaneous equations at each time level should be solved. Similar to the fully implicit method, there is no restriction on the time step. A schematic of the finite difference representation for Crank–Nicolson scheme is shown in figure 6.4.

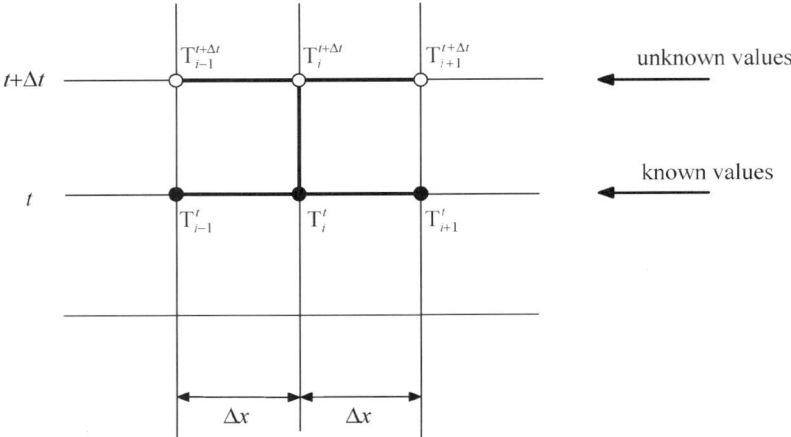

Figure 6.4. The finite difference molecules for the Crank–Nicolson scheme.

6.4 Stability analysis—von Neumann stability analysis

In general, three types of error are introduced into the solution of finite difference equations: truncation error, round-off and discretization errors. The truncation error mentioned in chapter 4 is the difference between the partial derivative and its finite-difference approximation. Suppose that the exact solution of a finite difference equation would be obtained using a computer with infinite accuracy, and that the numerical solution be computed using a real machine with finite accuracy. Then, the round-off error is defined as the difference between the exact solution of the finite difference equation and the numerical solution, and the discretization error as the difference between the exact solution of the finite difference equation and the analytical solution of the partial differential equation.

However, computers cannot perform calculations to infinite accuracy. Therefore, in the numerical solution of finite difference equations with a digital computer, round-off errors are introduced during computation. The difference between the analytical solution of the partial differential equation and the computer solution to the finite difference equation would be equal to the sum of the discretization error and the round-off error associated with the finite difference calculation. If the errors introduced into the finite difference equation are not controlled, the growth of errors will result in an unstable solution. The question of stability of a numerical method examines the error growth in the sequences of numerical procedures as the calculation proceeds. Stability analysis is essential for obtaining a successful solution of an finite difference equation.

There are several methods [2] available for stability analysis, such as the discrete perturbation stability analysis, the von Neumann stability analysis and the matrix method, etc. In the discrete perturbation stability analysis, a disturbance is introduced at a point, and its effect on neighboring points is evaluated. If the disturbance decays as the computation proceeds, the numerical method is stable. On the other hand, when the disturbance grows with the computation, the method is unstable. The von Neumann stability analysis was originally suggested by Neumann, and first described in a complete form by O'Brien et al [3]. The von Neumann method is one of the mostly widely used methods for determining the stability (or instability) of a finite difference approximation, in which a solution of the finite difference equation is expanded in a Fourier series. The decay or growth of the amplification factor indicates whether or not the numerical method is stable. In the matrix method, the finite difference representations of the partial differential equation and the boundary conditions are expressed in a matrix form and the problem of stability is transformed to the investigation of the eigenvalues of the coefficient matrix.

In this section, we will briefly describe the von Neumann stability analysis and its application to the stability of finite difference equations. In the von Neumann method, the errors are expressed in a finite Fourier series and the propagation of errors with time are investigated. It must be noted that this method applies only to linear, constant coefficient, finite difference approximations and does not accommodate the effects of boundary conditions on the stability of the solution.

As a starting point for stability analysis, let us consider one-dimensional heat conduction equation,

$$\frac{\partial T}{\partial t} = \alpha \frac{\partial^2 T}{\partial x^2}. \tag{6.4.1}$$

The fully explicit form of finite difference equation for equation (6.4.1) is given by the FTCS method as

$$\frac{T_i^{n+1} - T_i^n}{\Delta t} = \alpha \left\{ \frac{T_{i-1}^n - 2T_i^n + T_{i+1}^n}{\Delta x^2} \right\}. \tag{6.4.2}$$

In this case the numerical solution (N) actually computed can be written as the sum of the exact solution (E) and an error term (ε) as

$$N = E + \varepsilon. \tag{6.4.3}$$

This computed numerical solution must satisfy the finite difference equation, equation (6.4.2). Substituting equation (6.4.3) into the finite difference equation, equation (6.4.2), yields

$$\frac{E_i^{n+1} + \varepsilon_i^{n+1} - E_i^n - \varepsilon_i^n}{\Delta t}$$
$$= \alpha \left\{ \frac{E_{i-1}^n + \varepsilon_{i-1}^n - 2(E_i^n + \varepsilon_i^n) + E_{i+1}^n + \varepsilon_{i+1}^n}{\Delta x^2} \right\}. \tag{6.4.4}$$

Since the exact solution (E) must satisfy the finite difference equation, the same is true of the error (ε). Thus, we obtain the following difference equation for the error (ε).

$$\frac{\varepsilon_i^{n+1} - \varepsilon_i^n}{\Delta t} = \alpha \left\{ \frac{\varepsilon_{i-1}^n - 2\varepsilon_i^n + \varepsilon_{i+1}^n}{\Delta x^2} \right\} \qquad (6.4.5)$$

or

$$\varepsilon_i^{n+1} = (1 - 2\gamma)\varepsilon_i^n + \gamma \{\varepsilon_{i-1}^n + \varepsilon_{i+1}^n\}. \qquad (6.4.6)$$

To examine the propagation of the errors introduced during computation, we need to consider a single term in the Fourier series since the finite difference equation is linear. In this case, a Fourier component of the form is assumed as

$$\varepsilon_i^n = V^n e^{Ik_x(\Delta x)i} \qquad (6.4.7)$$

where V^n is the amplitude at time level n, k_x is the wave number in the x direction, i.e., $\Lambda = 2\pi/k_x$, where Λ is the wavelength, and $I = \sqrt{-1}$.

If a phase angle $\theta = k_x \Delta x$ is defined, equation (6.4.7) is rewritten as

$$\varepsilon_i^n = V^n e^{I\theta i}. \qquad (6.4.8)$$

Similarly, we have the following relations for ε_i^{n+1} and $\varepsilon_{i\pm 1}^n$.

$$\varepsilon_i^{n+1} = V^{n+1} e^{I\theta i} \qquad (6.4.9)$$

and

$$\varepsilon_{i\pm 1}^n = V^n e^{I\theta(i\pm 1)}. \qquad (6.4.10)$$

In order to determine the stability criterion, let us substitute equations (6.4.8), (6.4.9) and (6.4.10) into the FTCS form of finite difference equation, equation (6.4.6). Then we obtain

$$V_i^{n+1} e^{I\theta i} = (1 - 2\gamma) V_i^n e^{I\theta i} + \gamma \{V_i^n e^{I\theta(i-1)} + V_i^n e^{I\theta(i+1)}\}. \qquad (6.4.11)$$

Cancellation of the common term $e^{I\theta i}$ yields

$$V_i^{n+1} = \{(1 - 2\gamma) + \gamma \{e^{-I\theta} + e^{I\theta}\}\} V_i^n. \qquad (6.4.12)$$

Now, using the definition $\cos\theta = (e^{-I\theta} + e^{I\theta})/2$, the above equation becomes

$$\frac{V_i^{n+1}}{V_i^n} = 1 - 2\gamma(1 - \cos\theta). \qquad (6.4.13)$$

By introducing an amplification factor such that $V_i^{n+1}/V_i^n = G$, we have an expression for G as follows.

$$G = 1 - 2\gamma(1 - \cos\theta). \qquad (6.4.14)$$

For a stable solution, $|G|$ must be less than or equal to 1 for all values of θ. Thus,

$$-1 \leq 1 - 2\gamma(1 - \cos\theta) \leq 1. \qquad (6.4.15)$$

This is the requirement of stability for the FTCS form of finite difference approximation, equation (6.4.2).

In evaluating the inequality in equation (6.4.15), two possible cases must be considered. The inequality on the right-hand side of the above equation is always satisfied for all values of θ if $\gamma \geq 0$. With the maximum value of $(1 - \cos\theta) = 2$, the inequality of the left-hand side is satisfied only if

$$\gamma \leq 1/2 \quad \text{or} \quad \Delta t \leq \frac{\Delta x^2}{2\alpha}. \qquad (6.4.16)$$

This condition, which is identical to the result of equation (6.3.6), numerically places a constraint on the size of the time step Δt relative to the size of the mesh spacing Δx and the thermal diffusivity α. If this condition is violated, physically unrealistic results could emerge.

For a second application of the von Neumann stability analysis, let us consider the fully implicit form of finite difference approximation for equation (6.4.1) as

$$\frac{T_i^{n+1} - T_i^n}{\Delta t} = \alpha \left\{ \frac{T_{i-1}^{n+1} - 2T_i^{n+1} + T_{i+1}^{n+1}}{\Delta x^2} \right\}. \qquad (6.4.17)$$

Similarly, we have the following difference equation for the error (ε).

$$\frac{\varepsilon_i^{n+1} - \varepsilon_i^n}{\Delta t} = \alpha \left\{ \frac{\varepsilon_{i-1}^{n+1} - 2\varepsilon_i^{n+1} + \varepsilon_{i+1}^{n+1}}{\Delta x^2} \right\}. \qquad (6.4.18)$$

By substituting equations (6.4.8) through (6.4.10) into equation (6.4.18) and rearranging it, we obtain

$$\frac{V_i^{n+1}}{V_i^n} = 1/(1 + 2\gamma - 2\gamma \cos\theta). \qquad (6.4.19)$$

Using the trigonometric identity $\sin^2(\theta/2) = (1 - \cos\theta)/2$, we have the amplification factor

$$G = \frac{1}{1 + 4\gamma \sin^2(\theta/2)}. \qquad (6.4.20)$$

In this case the condition for stability $|G| \leq 1$ is satisfied for all γ. There is no limitation on the size of time step, even though a practical limit is necessary because of truncation error.

Following a similar procedure, the amplification factor for the Crank–Nicolson method is given as

$$G = \frac{1 - \gamma(1 - \cos\theta)}{1 - \gamma(1 + \cos\theta)}. \qquad (6.4.21)$$

Similar to the case of the fully implicit method, it is seen from equation (6.4.21) that the condition for stability $|G| \leq 1$ is satisfied for all γ.

The von Neumann method is probably the most widely used to determine the stability of numerical schemes. However, it must be noted that this method applies only to linear, constant coefficient, finite difference approximations, and does not accommodate the effects of boundary conditions on the stability of the solution.

6.5 Multi-dimensional problems

In cases of multi-dimensional problems, a control volume has more than two control-volume faces. The general form of finite difference equation (6.2.5) can also be used for two- or three-dimensional problems. The number of control-volume faces (nf) in two and three dimensions are generally 4 and 6, respectively.

$$\rho C_v V_i \frac{T_i^{t+\Delta t} - T_i^t}{\Delta t} = W \sum_{m=1}^{nf} (\hat{q}_{\Gamma_m}^{t+\Delta t} A_{\Gamma_m}) + (1 - W) \sum_{m=1}^{nf} (\hat{q}_{\Gamma_m}^t A_{\Gamma_m}) + V_i \dot{g}_i. \tag{6.2.5}$$

Figure 6.5 indicates the nomenclature for finite difference representations for two-dimensional transient heat conduction. The physical domain is divided into (M, N) finite control volumes. The explicit form of finite difference equation for a two-dimensional control volume (i, j) having four

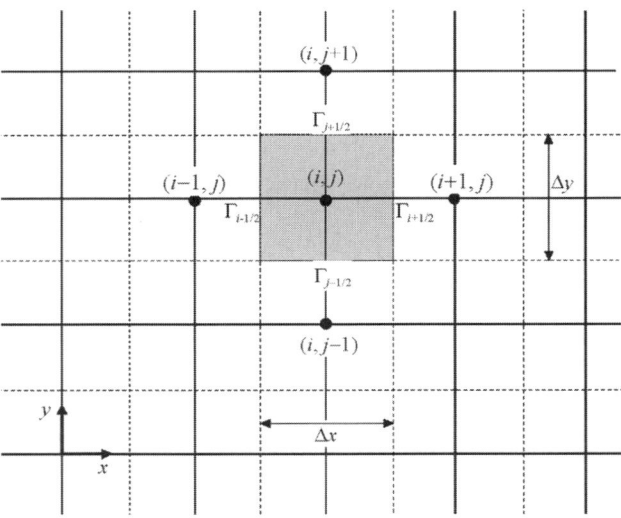

Figure 6.5. A part of two-dimensional structured grids.

control-volume faces is given by

$$\rho C_v V_{i,j} \frac{T_{i,j}^{t+\Delta t} - T_{i,j}^t}{\Delta t} = \sum_{m=1}^{nf} (\hat{q}_{\Gamma_m}^t A_{\Gamma_m}) + V_{i,j}\dot{g}_{i,j}$$

$$= \hat{q}_{\Gamma_{i-1/2}}^t A_{\Gamma_{i-1/2}} + \hat{q}_{\Gamma_{i+1/2}}^t A_{\Gamma_{i+1/2}} + \hat{q}_{\Gamma_{j-1/2}}^t A_{\Gamma_{j-1/2}}$$

$$+ \hat{q}_{\Gamma_{j+1/2}}^t A_{\Gamma_{j+1/2}} + V_{i,j}\dot{g}_{i,j} \quad (6.5.1)$$

where $\hat{q}_{\Gamma_m}^t$ indicates the net heat flux rates into a control volume (i,j) through its mth control-volume face at time level t, which can be evaluated using equations (5.2.7b) through (5.2.7f) according to the boundary conditions of the control-volume faces.

Similarly for three-dimensional problems, we have the explicit form of finite difference equation for control volume (i,j,k) having six control-volume faces as

$$\rho C_v V_{i,j,k} \frac{T_{i,j,k}^{t+\Delta t} - T_{i,j,k}^t}{\Delta t} = \sum_{m=1}^{nf} (\hat{q}_{\Gamma_m}^t A_{\Gamma_m}) + V_{i,j,k}\dot{g}_{i,j,k}$$

$$= \hat{q}_{\Gamma_{i-1/2}}^t A_{\Gamma_{i-1/2}} + \hat{q}_{\Gamma_{i+1/2}}^t A_{\Gamma_{i+1/2}}$$

$$+ \hat{q}_{\Gamma_{j-1/2}}^t A_{\Gamma_{j-1/2}} + \hat{q}_{\Gamma_{j+1/2}}^t A_{\Gamma_{j+1/2}}$$

$$+ \hat{q}_{\Gamma_{k-1/2}}^t A_{\Gamma_{k-1/2}} + \hat{q}_{\Gamma_{k+1/2}}^t A_{\Gamma_{k+1/2}} + V_{i,j,k}\dot{g}_{i,j,k}. \quad (6.5.2)$$

6.6 Worked examples

In this section we will consider the application of the finite volume method to one-dimensional transient heat conduction problems having various types of boundary conditions. For the sake of simplicity, the fully explicit method is adopted. In one-dimensional problems, each control volume has two control-volume faces, $\Gamma_{i-1/2}$ and $\Gamma_{i+1/2}$. Thus equation (6.2.5) can be expressed by

$$\rho C_v V_i \frac{T_i^{t+\Delta t} - T_i^t}{\Delta t} = \sum_{m=1}^{nf} (\hat{q}_{\Gamma_m}^t A_{\Gamma_m}) + V_i \dot{g}_i$$

$$= \hat{q}_{\Gamma_{i-1/2}}^t A_{\Gamma_{i-1/2}} + \hat{q}_{\Gamma_{i+1/2}}^t A_{\Gamma_{j+1/2}} + V_i \dot{g}_i. \quad (6.6.1)$$

In the case of the rectangular coordinate system, the cross-sectional areas $A_{\Gamma_{i-1/2}}$ and $A_{\Gamma_{i+1/2}}$ are equal to unity. Then, equation (6.6.1) is simplified to

$$\rho C_v V_i \frac{T_i^{t+\Delta t} - T_i^t}{\Delta t} = (\hat{q}_{\Gamma_{i-1/2}}^t + \hat{q}_{\Gamma_{i+1/2}}^t) + V_i \dot{g}_i. \quad (6.6.2)$$

86   Transient heat conduction

The values of the net heat fluxes $\hat{q}^t_{\Gamma_{i-1/2}}$ and $\hat{q}^t_{\Gamma_{i+1/2}}$ can be evaluated using equations (5.2.7b) through (5.2.7f) according to the boundary conditions of the control-volume faces $\Gamma_{i-1/2}$ and $\Gamma_{i+1/2}$, respectively.

6.6.1 Example 6.1

Consider one-dimensional transient heat conduction in an insulated steel rod as indicated in figure 6.6. The length of the rod is 0.2 m. The steel rod is initially at a uniform temperature of $\bar{T}_{\Gamma_A} = 100\,°C$ and heated electrically by the passage of electric current, which generates energy at a rate of $\dot{g} = 10^5\,W/m^3$. The temperature of one end of the rod is kept at 100 °C (Γ_0-type boundary), and the other end is insulated, $\bar{q}_{\Gamma_B} = 0$ (Γ_{II}-type boundary). The density, specific heat and thermal conductivity are $\rho = 7800\,kg/m^3$, $C_v = 500\,J/kg \cdot K$ and $\lambda = 50\,W/m \cdot K$, respectively. Develop the finite difference equations for this problem and calculate the temperature variations in the steel rod with time.

Solution
Let us divide the length of the rod into five equal control volumes having six nodes with $\Delta x = 0.04\,m$, as shown in figure 6.6.

Control volumes 1 through 4 have two control-volume faces, $\Gamma_{i-1/2}$ and $\Gamma_{i+1/2}$. Therefore the finite difference equations for nodes 1 through 5 can be derived using the following equation.

$$\rho C_v V_i \frac{T_i^{t+\Delta t} - T_i^t}{\Delta t} = (\hat{q}^t_{\Gamma_{i-1/2}} + \hat{q}^t_{\Gamma_{i+1/2}}) + V_i \dot{g}_i. \qquad (1)$$

In the case of a one-dimensional rectangular coordinate system, $V_i = \Delta x$.

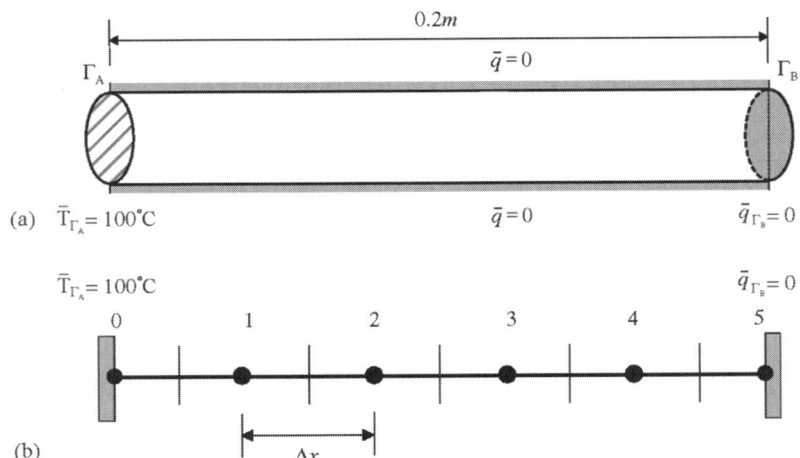

Figure 6.6. One-dimensional transient heat conduction problems: (a) the geometry and (b) the grids used in the simulation.

By applying the boundary conditions given by equations (5.2.7b)–(5.2.7f) into equation (1), we have the finite difference equations for nodes 1 through 5 as follows.

For node i ($i = 1$–4), the finite difference equation is

$$\rho C_v \Delta x \frac{T_i^{t+\Delta t} - T_i^t}{\Delta t} = \lambda \frac{T_{i-1}^t - T_i^t}{\Delta x} + \lambda \frac{T_{i+1}^t - T_i^t}{\Delta x} + \Delta x \dot{g} \quad (2a)$$

where $T_0^t = \bar{T}_{\Gamma_A} = 100\,°C$. Equation (2a) is rearranged as

$$T_i^{t+\Delta t} = \gamma(T_{i-1}^t + T_{i+1}^t) + (1 - 2\gamma)T_i^t + \frac{\Delta t}{\rho C_v}\dot{g} \quad (2b)$$

where $\gamma = \alpha \Delta t / \Delta x^2$ and $\alpha = \lambda / \rho C_v$.

For node 5,

$$\rho C_v \Delta x \frac{T_5^{t+\Delta t} - T_5^t}{\Delta t} = \lambda \frac{T_4^t - T_5^t}{\Delta x} + 0(\bar{q}_{\Gamma_B} = 0) + \Delta x \dot{g}. \quad (3a)$$

Similarly, equation (3a) is rearranged as

$$T_5^{t+\Delta t} = \gamma T_4^t + (1 - \gamma)T_5^t + \frac{\Delta t}{\rho C_v}\dot{g}. \quad (3b)$$

Equations (2b) and (3b) are the simple explicit forms of finite difference approximation, which involve only one unknown temperature, $T_i^{t+\Delta t}$ for the time level $t + \Delta t$, and can be evaluated from the known temperatures, T_i^t, T_{i-1}^t and T_{i+1}^t at the previous time level t.

According to the stability criterion mentioned in section 6.4, the stable time step for the iterative calculation of equations (2b) and (3b) is determined by the following condition.

$$0 \leq \gamma \leq \tfrac{1}{2}. \quad (4)$$

Thus, the maximum size of the stable time step is given by

$$\Delta t \leq \frac{\Delta x^2}{2\alpha} \leq 62.4\,\text{s}.$$

Let us select $\Delta t = 60\,\text{s}$. By substitution of numerical values and known temperatures to equations (2b) and (3b) for every 60 s, we obtain the temperature distributions of nodes 1 through 5 with time as shown in table 6.1.

6.6.2 Example 6.2

Let us now consider one-dimensional transient heat conduction in a large steel plate ($\rho_{\Omega_1} = 7800\,\text{kg/m}^3$, $(C_v)_{\Omega_1} = 500\,\text{J/kg} \cdot \text{K}$ and $\lambda_{\Omega_1} = 50\,\text{W/m} \cdot \text{K}$) with a thickness of 0.05 m. One side of the steel plate is maintained at a constant temperature of $\bar{T}_{\Gamma_A} = 900\,°C$ (Γ_0-type boundary) and the other side is covered by a ceramic plate with a thickness of 0.08 m ($\rho_{\Omega_2} = 2000\,\text{kg/m}^3$,

88 Transient heat conduction

Table 6.1. Results of Example 6.1.

	Node number				
Time (s)	1 $x = 0.04$	2 $x = 0.08$	3 $x = 0.12$	4 $x = 0.16$	5 $x = 0.20$
0	100.0	100.0	100.0	100.0	100.0
60	101.5	101.5	101.5	101.5	101.5
120	102.3	103.1	103.1	103.1	103.1
180	103.1	104.3	104.6	104.6	104.6
240	103.7	105.4	106.0	106.2	106.2
300	104.3	106.4	107.3	107.6	107.7
600	106.5	110.6	113.0	114.2	114.5
1200	109.5	116.4	120.9	123.4	124.3
1800	111.4	119.9	125.7	129.2	130.3
2400	112.6	122.1	128.7	132.7	134.0
3000	113.3	123.4	130.6	134.9	136.3
6000	114.3	125.4	133.3	138.1	139.7
12000	114.4	125.6	133.6	138.4	140.0

$(C_v)_{\Omega_2} = 1000\,\text{J/kg} \cdot \text{K}$ and $\lambda_{\Omega_2} = 0.5\,\text{W/m} \cdot \text{K})$. The ceramic plate is exposed to an ambient temperature of $T_{\text{air}} = 25\,°\text{C}$ with a convection heat transfer coefficient of $h_{\text{air}} = 50\,\text{W/m}^2 \cdot \text{K}$ (Γ_{III}-type boundary). The interfacial heat transfer coefficient at the steel/ceramic plate interface is $h_{\text{int}} = 500\,\text{W/m}^2 \cdot \text{K}$ (Γ_{V}-type boundary). The initial temperatures of the steel and ceramic plates are $900\,°\text{C}$ and $25\,°\text{C}$, respectively.

Develop the finite difference equations for this problem and calculate the temperature distributions in the steel and ceramic plates with time.

Solution

Let us divide the regions of the steel and the ceramic plates into five and four equal control volumes having six and five nodes, respectively. Then, we have $\Delta x_{\Omega_1} = 0.01\,\text{m}$ for the sub-domain Ω_1 and $\Delta x_{\Omega_2} = 0.02\,\text{m}$ for the sub-domain Ω_2, as shown in figure 6.7.

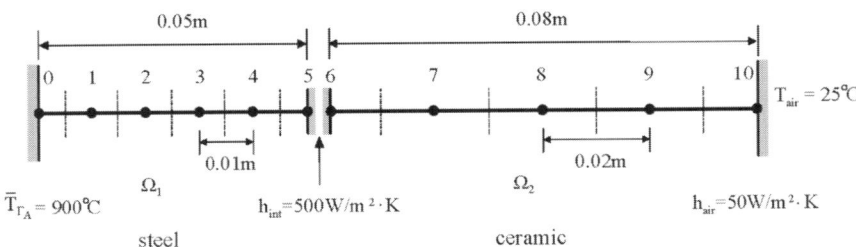

Figure 6.7. One-dimensional transient heat conduction problems with two sub-domains.

Firstly, consider the derivation of the finite difference equations for nodes 1 through 5 in the sub-domain Ω_1. By applying the boundary conditions given by equations (5.2.7b)–(5.2.7f) into equation (6.6.2) for the control volumes 1 through 5, we can derive the finite difference equations as follows.

For node i ($i = 1$–4), each control volume has two control-volume faces of Γ_1-type boundary (equation (5.2.7b)), thus the finite difference equation is

$$(\rho C_v)_{\Omega_1} \Delta x_{\Omega_1} \frac{T_i^{t+\Delta t} - T_i^t}{\Delta t} = \lambda_{\Omega_1} \frac{T_{i-1}^t - T_i^t}{\Delta x_{\Omega_1}} + \lambda_{\Omega_1} \frac{T_{i+1}^t - T_i^t}{\Delta x_{\Omega_1}} \quad (1)$$

where $T_0^t = \bar{T}_{\Gamma_A} = 900\,°\text{C}$.

For node 5, we have

$$(\rho C_v)_{\Omega_1} \frac{\Delta x_{\Omega_1}}{2} \frac{T_5^{t+\Delta t} - T_5^t}{\Delta t} = \lambda_{\Omega_1} \frac{T_4^t - T_5^t}{\Delta x_{\Omega_1}} + h_{\text{int}}(T_6^t - T_5^t). \quad (2)$$

Now, consider the derivation of the finite difference equations for nodes 6 through 10 in the sub-domain Ω_2. Similarly, by applying the boundary conditions given by equations (5.2.7b)–(5.2.7f) into equation (6.6.2) for the control volumes 6 through 10, we can derive the finite difference equations as follows.

For node 6, we have

$$(\rho C_v)_{\Omega_2} \frac{\Delta x_{\Omega_2}}{2} \frac{T_6^{t+\Delta t} - T_6^t}{\Delta t} = h_{\text{int}}(T_5^t - T_6^t) + \lambda_{\Omega_2} \frac{T_7^t - T_6^t}{\Delta x_{\Omega_2}}. \quad (3)$$

For node i ($i = 7$–9), we have

$$(\rho C_v)_{\Omega_2} \Delta x_{\Omega_2} \frac{T_i^{t+\Delta t} - T_i^t}{\Delta t} = \lambda_{\Omega_2} \frac{T_{i-1}^t - T_i^t}{\Delta x_{\Omega_2}} + \lambda_{\Omega_2} \frac{T_{i+1}^t - T_i^t}{\Delta x_{\Omega_2}}. \quad (4)$$

For node 10, we have

$$(\rho C_v)_{\Omega_2} \frac{\Delta x_{\Omega_2}}{2} \frac{T_{10}^{t+\Delta t} - T_{10}^t}{\Delta t} = \lambda_{\Omega_2} \frac{T_9^t - T_{10}^t}{\Delta x_{\Omega_2}} + h_{\text{air}}(T_{\text{air}} - T_{10}^t). \quad (5)$$

The stable time steps for iterative calculation of equations (1) through (5) can be evaluated by the stability criterion. The smallest stable time step which is determined by equation (1) should be used for the iterative calculation of temperature distributions for all nodes. The stable time step is given by

$$\Delta t \leq \frac{\Delta x_{\Omega_1}^2}{2\alpha_{\Omega_1}} \leq 3.5\,\text{s}. \quad (6)$$

Let us select $\Delta t = 3$ s. By substitution of numerical values and known temperatures to equations (1) through (5) for every 3 s, we obtain the temperature distributions of nodes 1 through 10 with time as shown in table 6.2.

Table 6.2 Results of Example 6.2.

Time(s)	Node number					
	1 $x = 0.01$	3 $x = 0.03$	5 $x = 0.05$	6 $x = 0.05$	8 $x = 0.09$	10 $x = 0.13$
0	900	900	900	25	25	25
30	892.0	859.5	786.3	276.8	25.0	25.0
150	876.2	831.0	793.2	650.2	26.2	25.0
300	887.2	863.4	844.3	774.7	31.4	25.0
1500	897.5	892.5	887.5	863.2	142.7	29.7
3000	898.2	894.6	891.0	873.3	270.2	49.6
6000	898.7	896.1	893.6	880.8	404.9	87.6
12000	899.0	897.0	894.9	884.8	485.3	114.5
18000	899.0	897.1	895.2	885.5	499.9	119.5
24000	899.0	897.1	895.2	885.6	502.6	120.4
30000	899.0	897.1	895.2	885.7	503.0	120.6
39000	899.0	897.1	895.2	885.7	503.1	120.6

6.6.3 Example 6.3

Consider one-dimensional transient heat conduction problems in a steel (a) plate, (b) hollow cylinder and (c) hollow sphere. The geometry, dimension and boundary conditions are described in figure 6.8. The density, the specific heat and the thermal conductivity of the steel are $\rho = 7800 \, \text{kg/m}^3$, $C_v = 500 \, \text{J/kg} \cdot \text{K}$ and $\lambda = 25 \, \text{W/m} \cdot \text{K}$, respectively. The inner and outer radii are $r_a = 0.05 \, \text{m}$ and $r_a = 0.06 \, \text{m}$ for both of the hollow cylinder and sphere, and the plate thickness is 0.01 m. One face of the plate is kept at $\bar{T}_{\Gamma_A} = 100 \, °\text{C}$ (Γ_0-type boundary), and the other face is exposed to air, with a temperature of $T_{\text{air}} = 25 \, °\text{C}$ and a convection heat transfer coefficient of $h_{\text{air}} = 100 \, \text{W/m}^2 \cdot \text{K}$ (Γ_{III}-type boundary). The inner surfaces of the hollow cylinder and sphere are held at a constant temperature $\bar{T}_{\Gamma_A} = 100 \, °\text{C}$ (Γ_0-type boundary), and the outer surfaces are exposed to air, with a temperature of $T_{\text{air}} = 25 \, °\text{C}$ and a convection heat transfer coefficient of $h_{\text{air}} = 100 \, \text{W/m}^2 \cdot \text{K}$ (Γ_{III}-type boundary). The initial temperature of the steel is 100 °C.

Develop the finite difference equations and calculate the temperature distributions in the steel for three cases.

Solution
Let us divide the computational domain into four equal control volumes having five nodes with $\Delta x (\equiv \Delta r) = 0.0025 \, \text{m}$, as shown in figure 6.8.

The finite difference equations for nodes 1 through 4 are directly derived using equation (6.6.1). Since there is no heat generation in the domain,

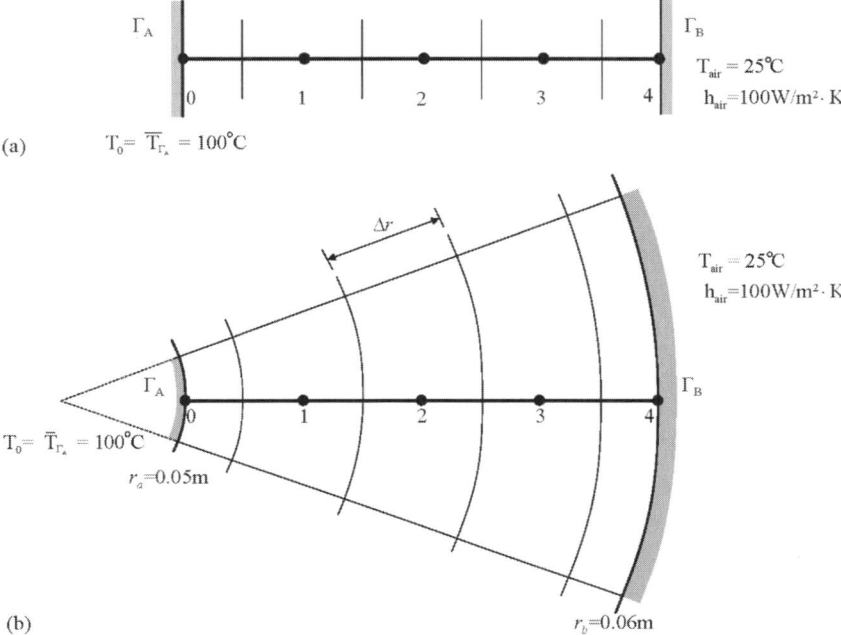

Figure 6.8. One-dimensional transient heat conduction problems: (a) a plane wall and (b) a hollow cylinder or a hollow sphere.

equation (6.6.1) is simplified to

$$\rho C_v V_i \frac{T_i^{t+\Delta t} - T_i^t}{\Delta t} = (\hat{q}_{\Gamma_{i-1/2}}^t A_{\Gamma_{i-1/2}} + \hat{q}_{\Gamma_{i+1/2}}^t A_{\Gamma_{i+1/2}}). \tag{1}$$

By applying the boundary conditions given by equations (5.2.7b)–(5.2.7f) into equation (1), we obtain the finite difference equations for nodes 1 through 4 as follows.

For node i ($i = 1$–3), each control volume has two control-volume faces of Γ_1-type boundary (equation (5.2.7b)), thus the finite difference equation is

$$\rho C_v V_i \frac{T_i^{t+\Delta t} - T_i^t}{\Delta t} = \lambda A_{\Gamma_{i-1/2}} \frac{T_{i-1}^t - T_i^t}{\Delta x} + \lambda A_{\Gamma_{i+1/2}} \frac{T_{i+1}^t - T_i^t}{\Delta x}. \tag{2a}$$

Equation (2a) is rearranged as

$$T_i^{t+\Delta t} = \left(\frac{\lambda \Delta t A_{\Gamma_{i-1/2}}}{\rho C_v V_i \Delta x} \right) T_{i-1}^t + \left\{ 1 - \Delta t \left(\frac{\lambda A_{\Gamma_{i-1/2}}}{\rho C_v V_i \Delta x} + \frac{\lambda A_{\Gamma_{i+1/2}}}{\rho C_v V_i \Delta x} \right) \right\} T_i^t$$

$$+ \left(\frac{\lambda \Delta t A_{\Gamma_{i+1/2}}}{\rho C_v V_i \Delta x} \right) T_{i+1}^t \tag{2b}$$

where $T_0 = \bar{T}_{\Gamma_A} = 100\,°C$.

Transient heat conduction

Table 6.3 Results of Example 6.3.

(a) In the case of a plane wall

Time (s)	Node number			
	1 $x = 0.0025$	2 $x = 0.005$	3 $x = 0.0075$	4 $x = 0.01$
0	100.00	100.00	100.00	100.00
4	99.75	99.42	98.96	98.32
8	99.52	99.00	98.41	97.73
12	99.40	98.79	98.13	97.43
16	99.34	98.67	97.99	97.28
20	99.31	98.62	97.91	97.20
24	99.30	98.59	97.88	97.16
28	99.29	98.57	97.86	97.14
32	99.28	98.57	97.85	97.13
36	99.28	98.56	97.84	97.12
40	99.28	98.56	97.84	97.12

(b) In the case of a hollow cylinder

Time (s)	Node number			
	1 $x = 0.0525$	2 $x = 0.055$	3 $x = 0.0575$	4 $x = 0.06$
0	100.00	100.00	100.00	100.00
4	99.72	99.38	98.90	98.26
8	99.46	98.90	98.29	97.61
12	99.32	98.65	97.97	97.26
16	99.25	98.52	97.79	97.07
20	99.21	98.44	97.70	96.97
24	99.18	98.40	97.65	96.92
28	99.17	98.38	97.62	96.89
32	99.17	98.37	97.61	96.87
36	99.16	98.36	97.60	96.87
40	99.16	98.36	97.60	96.86

(c) In the case of a hollow sphere

0	100.00	100.00	100.00	100.00
4	99.69	99.33	98.84	98.19
8	99.40	98.80	98.17	97.48
12	99.23	98.51	97.79	97.08
16	99.14	98.34	97.58	96.85
20	99.09	98.25	97.46	96.73
24	99.06	98.19	97.40	96.66
28	99.04	98.16	97.36	96.62
32	99.03	98.15	97.34	96.60
36	99.03	98.14	97.33	96.58
40	99.02	98.13	97.32	96.58

Case study: one-dimensional transient heat conduction problems 93

For node 4,

$$\rho C_v V_4 \frac{T_4^{t+\Delta t} - T_4^t}{\Delta t} = \lambda A_{\Gamma_{i-1/2}} \frac{T_3^t - T_4^t}{\Delta x} + h_{\text{air}} A_{\Gamma_{i+1/2}} (T_{\text{air}} - T_4^t). \quad (3a)$$

Equation (3a) is also rearranged as

$$T_4^{t+\Delta t} = \left(\frac{\lambda \Delta t A_{\Gamma_{i-1/2}}}{\rho C_v V_4 \Delta x}\right) T_3^t + \left\{1 - \Delta t \left(\frac{\lambda A_{\Gamma_{i-1/2}}}{\rho C_v V_4 \Delta x} + \frac{h_{\text{air}} A_{\Gamma_{i+1/2}}}{\rho C_v V_4}\right)\right\} T_4^t$$

$$+ \left(\frac{h_{\text{air}} \Delta t A_{\Gamma_{i+1/2}}}{\rho C_v V_4}\right) T_{\text{air}}. \quad (3b)$$

The values of $A_{\Gamma_{i-1/2}}$, $A_{\Gamma_{i+1/2}}$ and V_i for a plate, a hollow cylinder and a hollow sphere can be calculated by equations (5.3.1) and (5.3.2).

According to the stability criterion, the stable time steps for the iterative calculation of equations (2b) and (3b) are determined by the following conditions.

$$\Delta t \leq \frac{\rho C_v V_i \Delta x}{\lambda (A_{\Gamma_{i-1/2}} + A_{\Gamma_{i+1/2}})} \quad (4a)$$

and

$$\Delta t \leq \frac{1}{\frac{A_{\Gamma_{i-1/2}}}{V_i \Delta x} \frac{\lambda}{\rho C_v} + \frac{A_{\Gamma_{i+1/2}}}{V_i} \frac{h_{\text{air}}}{\rho C_v}}. \quad (4b)$$

The stable time steps for three coordinate systems evaluated by equation (4) are as follows.

For a plane plate, $\Delta t \leq 0.483 \cdots$
For a hollow cylinder, $\Delta t \leq 0.487 \cdots$
For a hollow sphere, $\Delta t \leq 0.487 \cdots$

Let us select $\Delta t = 0.4$ s for three coordinate systems. By substitution of numerical values and known temperatures to equations (2b) and (3b), we obtain the temperature distributions of nodes 1 through 4 with time for three coordinate systems as shown in table 6.3.

6.7 Case study: one-dimensional transient heat conduction problems

6.7.1 Description of the problem

Consider one-dimensional transient heat conduction problems in a steel (a) plate, (b) hollow cylinder and (c) hollow sphere. The geometry, dimension and boundary conditions are described in figure 6.8. The density, the specific heat and the thermal conductivity of the steel are $\rho = 7800 \text{ kg/m}^3$,

94 Transient heat conduction

$C_v = 500\,\text{J/kg} \cdot \text{K}$ and $\lambda = 25\,\text{W/m} \cdot \text{K}$, respectively. The inner and outer radii are $r_a = 0.05\,\text{m}$ and $r_a = 0.06\,\text{m}$ for both of the hollow cylinder and sphere, and the plate thickness is 0.01 m. One face of the plate is kept at $\bar{T}_{\Gamma_A} = 100\,°\text{C}$ (Γ_0-type boundary), and the other face is exposed to air, with a temperature of $T_{\text{air}} = 25\,°\text{C}$ and a convection heat transfer coefficient of $h_{\text{air}} = 100\,\text{W/m}^2 \cdot \text{K}$ (Γ_{III}-type boundary). The inner surfaces of the hollow cylinder and sphere are held at a constant temperature $\bar{T}_{\Gamma_A} = 100\,°\text{C}$ (Γ_0-type boundary), and the outer surfaces are exposed to air, with a temperature of $T_{\text{air}} = 25\,°\text{C}$ and a convection heat transfer coefficient of $h_{\text{air}} = 100\,\text{W/m}^2 \cdot \text{K}$ (Γ_{III}-type boundary). The initial temperature of the steel is 100 °C.

Develop the finite difference equations and calculate the temperature distributions in the steel for three cases.

6.7.2 Glossary of FORTRAN notation

FORTRAN name	Meaning
A(I, NL)	The left boundary area of node i
A(I, NR)	The right boundary area of node i
DEN	Density
DX	Distance between nodes
DT	Time increment
HAIR	Heat transfer coefficient at the material/air interface
ICOOR	Integer variable which selects a coordination system: 1 for rectangular, 2 for cylindrical and 3 for spherical coordinates, respectively
I	Index denoting node i
ITER	A running counter of the number of iterations performed
ITMAX	Total number of iterations performed
N	The number of nodes
PI	$\pi(3.1415\cdots)$
ITPRN	The interval of iterations for printing out the results
RA	Inner radius
RB	Outer radius
SPH	Specific heat capacity
T0	The prescribed temperature of node 0
TAIR	The temperature of air
TCON	The reference value of the thermal conductivity
TINI	Initial temperature of the material
TIME	Elapsed time
TNEW(I)	New temperature of node i at time $t + \Delta t$
TOLD(I)	Old temperature of node i at time t
V(I)	Volume of node i

Case study: one-dimensional transient heat conduction problems 95

6.7.3 Simulations

Let us simulate the variations of temperature distributions with time for three coordinate systems using the execution file [theat.exe].

(1) For a plane plate

Data input
```
> INPUT ITPRN AND ITMAX : 10 100
> SELECT COORD. SYSTEM(1:RECT, 2:CYLINDER, 3:SPHERE) : 1
> ENTER RA AND RB : 0.05 0.06
> INPUT TIME STEP (RECOMMENDED: <= .483) : 0.4
```

Results

TIME	NODE NUMBER			
	1	2	3	4
.00	100.00	100.00	100.00	100.00
4.00	99.75	99.42	98.96	98.32
8.00	99.52	99.00	98.41	97.73
12.00	99.40	98.79	98.13	97.43
16.00	99.34	98.67	97.99	97.28
20.00	99.31	98.62	97.91	97.20
24.00	99.30	98.59	97.88	97.16
28.00	99.29	98.57	97.86	97.14
32.00	99.28	98.57	97.85	97.13
36.00	99.28	98.56	97.84	97.12
40.00	99.28	98.56	97.84	97.12

(2) For a hollow cylinder

Data input
```
> INPUT ITPRN AND ITMAX : 10 100
> SELECT COORD. SYSTEM(1:RECT, 2:CYLINDER, 3:SPHERE) : 2
> ENTER RA AND RB : 0.05 0.06
> INPUT TIME STEP (RECOMMENDED: <= .483) : 0.4
```

Results

TIME	NODE NUMBER			
	1	2	3	4
.00	100.00	100.00	100.00	100.00
4.00	99.72	99.38	98.90	98.26
8.00	99.46	98.90	98.29	97.61
12.00	99.32	98.65	97.97	97.26
16.00	99.25	98.52	97.79	97.07
20.00	99.21	98.44	97.70	96.97
24.00	99.18	98.40	97.65	96.92
28.00	99.17	98.38	97.62	96.89
32.00	99.17	98.37	97.61	96.87
36.00	99.16	98.36	97.60	96.87
40.00	99.16	98.36	97.60	96.86

96 Transient heat conduction

(3) For a hollow sphere

Data input
```
> INPUT ITPRN AND ITMAX : 10 100
> SELECT COORD. SYSTEM(1:RECT, 2:CYLINDER, 3:SPHERE) : 3
> ENTER RA AND RB : 0.05 0.06
> INPUT TIME STEP (RECOMMENDED: <= .483) : 0.4
```

Results

TIME	NODE NUMBER			
	1	2	3	4
.00	100.00	100.00	100.00	100.00
4.00	99.69	99.33	98.84	98.19
8.00	99.40	98.80	98.17	97.48
12.00	99.23	98.51	97.79	97.08
16.00	99.14	98.34	97.58	96.85
20.00	99.09	98.25	97.46	96.73
24.00	99.06	98.19	97.40	96.66
28.00	99.04	98.16	97.36	96.62
32.00	99.03	98.15	97.34	96.60
36.00	99.03	98.14	97.33	96.58
40.00	99.02	98.13	97.32	96.58

6.7.4 Program list

```
C     **************************************************
C     *                                                *
C     *  A COMPUTER PROGRAM FOR THE SIMULATION OF 1-D  *
C     *  TRANSIENT HEAT CONDUCTION PROBLEMS            *
C     *                                                *
C     **************************************************
C
      PROGRAM THEAT
      PARAMETER(N=5)
      PARAMETER(NL=1, NR=2)
      PARAMETER(IRECT=1, ICYLIN=2, ISPHER=3)
      COMMON
     &/VAR/TOLD(N), TNEW(N), V(N), A(N, 2), DT, TIME
     &/PROPERTY/TCON, DEN, SPH
     &/GEOMETRY/RA, RB, DX, ICOOR
     &/BOUNDARY/HAIR, TAIR, T0
     &/INITIAL/TINI
      INTEGER ITER, ITPRN, ITMAX
C
      WRITE(6, 100)
  100 FORMAT(1X,'> INPUT ITPRN AND ITMAX : ',\)
      READ(5, *) ITPRN, ITMAX
C
```

```
C---- SELECT A COORDINATE SYSTEM
  110 WRITE(6, 200)
  200 FORMAT(' > SELECT COORD. SYSTEM(1:RECT,2:CYLINDER,
     3:SPHERE) : ',\)
      READ(5, *) ICOOR
      IF(ICOOR.LE.0.OR. ICOOR.GE.4) GOTO 110
C
C---- INPUT GEOMETRICAL VARIABLES
      WRITE(6, 210)
  210 FORMAT(1X, '> ENTER RA AND RB : ',\)
      READ(5,*) RA, RB
C
C---- INITIALIZE VARIABLES
      CALL INIT
C---- CALCULATE TIME STEP
      CALL SETDT
C
      TIME=0.0
      WRITE(6, 220) (I, I=1, N-1)
  220 FORMAT(/,1X,' TIME NODE NUMBER', /,7X,4I10)
      WRITE(6, 300) TIME, (TNEW(I), I=2, N)
  300 FORMAT(1X, F6.2, 4F10.2)
C
C---- START OF ITERATIVE CALCULATION ----
      DO 1000 ITER=1, ITMAX
      TIME=TIME+DT
C
      DO 10 I=2, N-1
      R=TCON*DT/DEN/SPH/V(I)/DX
      TNEW(I)=R*A(I,NL)*TOLD(I-1)+(1-R*(A(I,NL)+A(I,NR)))*
     TOLD(I)
     &         +R*A(I,NR)*TOLD(I+1)
   10 CONTINUE
C
C---- BOUNDARY CONDITIONS
      R=TCON*DT/DEN/SPH/V(N)/DX
      TNEW(N)=R*A(I,NL)*TOLD(N-1)
     & +(1-(R*A(N,NL)+DT*HAIR*A(N,NR)/DEN/SPH/V(N)))*TOLD(N)
     & +HAIR*DT*A(I,NR)/DEN/SPH/V(N)*TAIR
C
C---- PRINT OUT TEMPERATURES BY ITPRN
      IF(MOD(ITER, ITPRN).EQ. 0) THEN
      WRITE(6, 300) TIME, (TNEW(I), I=2, N)
      ENDIF
C
C---- UPDATE TEMPERATURES
      DO 20 I=2, N
      TOLD(I)=TNEW(I)
```

```
   20 CONTINUE
 1000 CONTINUE
C---- END OF ITERATIVE CALCULATION ----
      PAUSE
      STOP
      END
C
C     SUBROUTINE FOR THE INITIALIZATION OF VARIABLES
C
      SUBROUTINE INIT
      PARAMETER(N=5)
      PARAMETER(NL=1, NR=2)
      PARAMETER(IRECT=1, ICYLIN=2, ISPHER=3)
      COMMON
     &/VAR/TOLD(N), TNEW(N), V(N), A(N, 2), DT, TIME
     &/PROPERTY/TCON, DEN, SPH
     &/GEOMETRY/RA, RB, DX, ICOOR
     &/BOUNDARY/HAIR, TAIR, T0
     &/INITIAL/TINI
      REAL PI
C
      PI=ATAND(45.0)
C
C---- THERMAL AND PHYSICAL PROPERTIES
      TCON=25.0
      DEN=7800.0
      SPH=500.0
      HAIR=100.0
      TAIR=25.0
      T0=100.0
      TINI=100.0
C
      DX=(RB-RA)/FLOAT(N-1)
C
C---- SETTING VOLUMES AND SURFACE AREAS
C
      SELECT CASE (ICOOR)
      CASE (IRECT)
      DO 100 I=1, N
      V(I)=DX
      A(I, NL)=1.0
      A(I, NR)=1.0
  100 CONTINUE
      V(1)=DX/2.0
      V(N)=DX/2.0
C
      CASE (ICYLIN)
      DO 200 I=1, N
```

```
            A(I,NL)=2.0*PI*(RA+FLOAT(I-1)*DX-DX/2.0)
            A(I,NR)=2.0*PI*(RA+FLOAT(I-1)*DX+DX/2.0)
            V(I)=PI*((RA+FLOAT(I-1)*DX+DX/2.0)**2
     &          -(RA+FLOAT(I-1)*DX-DX/2.0)**2)
  200     CONTINUE
            A(1,NL)=2.0*PI*RA
            A(N,NR)=2.0*PI*RB
            V(1)=PI*((RA+DX/2.0)**2-RA**2)
            V(N)=PI*(RB**2-(RB-DX/2.0)**2)
C
          CASE (ISPHER)
          DO 300 I=1, N
            A(I,NL)=4*PI*(RA+FLOAT(I-1)*DX-DX/2.0)**2
            A(I,NR)=4*PI*(RA+FLOAT(I-1)*DX+DX/2.0)**2
            V(I)=4.0*PI/3.0*((RA+FLOAT(I-1)*DX+DX/2.0)**3
     &          -(RA+FLOAT(I-1)*DX-DX/2.0)**3)
  300     CONTINUE
            A(1,NL)=4.0*PI*RA**2
            A(N,NR)=4.0*PI*RB**2
            V(1)=4.0*PI/3.0*((RA+DX/2.0)**3-RA**3)
            V(N)=4.0*PI/3.0*(RB**3-(RB-DX/2.0)**3)
          END SELECT
C
C---- INITIAL CONDITIONS
      DO 400 I=1, N
        TOLD(I)=TINI
        TNEW(I)=TINI
  400 CONTINUE
C---- BOUNDARY CONDITIONS
      TOLD(1)=T0
      TNEW(1)=T0
      END
C
C     TIME STEP FOR EXPLICIT SCHEME
      SUBROUTINE SETDT
      PARAMETER(N=5)
      PARAMETER(NL=1, NR=2)
      COMMON
     &/VAR/TOLD(N), TNEW(N), V(N), A(N, 2), DT, TIME
     &/PROPERTY/TCON, DEN, SPH
     &/GEOMETRY/RA, RB, DX, ICOOR
     &/BOUNDARY/HAIR, TAIR, T0
     &/INITIAL/TINI
      REAL T1, T2, TEMP
C
      T1=1E10
      DO 10 I=2, N-1
        TEMP=DEN*SPH*V(I)*DX/TCON/(A(I,NL)+A(I,NR))
```

```
      IF(TEMP.LT.T1) T1=TEMP
   10 CONTINUE
C
      T2=V(N)*DEN*SPH/(A(N, NL)*TCON/DX+A(N,NR)*HAIR)
      DT=MIN(T1, T2)
C
      WRITE(6, 100) DT
  100 FORMAT(1X, '> INPUT TIME STEP (RECOMMENDED: <=', F6.3,
     ') : ',\)
      READ(5, *) DT
      END
```

References

[1] Roache P J 1972 *Computational Fluid Dynamics* (Albuquerque, New Mexico: Hermosa).
[2] Richtmyer R D and Morton K W 1967 *Difference Methods for Initial-Value Problems* 2nd edition (New York: Interscience Publishers, Wiley)
[3] O'Brien G G, Hyman M A and Kaplan S 1950 *J. Math. Phys.* **29** 223

Chapter 7

Phase change problems

Materials processing problems often involve phase transformations from liquid to solid, solid to solid, or gas to solid, etc. Transport phenomena significantly affect the morphology and the velocity in phase transformations, and therefore the quality of the resultant materials. In this chapter, numerical methods for solving heat transfer problems involving liquid-to-solid phase change, which is mostly important in materials processing, will be described briefly.

7.1 Introduction

Heat transfer problems involving melting or solidification are generally referred to as phase change or moving boundary problems, and also as Stefan problems. Most metallic materials are produced through melting and solidification processes. Representative materials processing involving melting or solidification include such technologies as crystal growth, rapid solidification, continuous casting, shape casting, welding, powder processing, surface modification, etc.

Three important phenomena should be considered in solidification of alloys. First, most metals and alloys shrink on solidification giving rise to the formation of shrinkage defects. In addition, it causes fluid flow near the solid/liquid interface and affects the formation of segregation. Second, the latent heat of fusion is released at the solid/liquid interface, which determines the solidification velocity and the resultant microstructures. And third, in the case of alloys the solute is rejected or absorbed in liquid at the solid/liquid interface during solidification, which causes the formation of micro- and macro-segregation in the final products. Figure 7.1 indicates the heat and solute balance near the solid/liquid interface during solidification. The heat balance equation at the solid/liquid interface can be expressed as follows.

$$-(q_l) + \rho_s H_f R_x = -(q_s) \qquad (7.1.1\text{a})$$
$$\quad (1) \qquad (3) \qquad\quad (2)$$

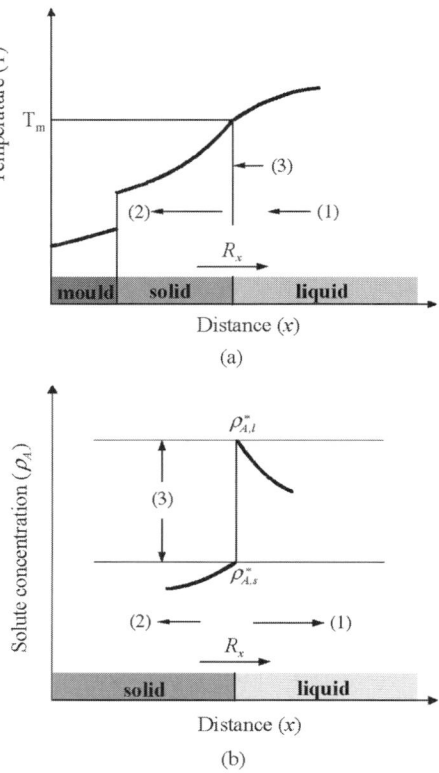

Figure 7.1. (a) Temperature and (b) solute concentration profiles near the solid/liquid interface during the solidification of an alloy.

$$-\left(-\lambda \frac{\partial T}{\partial x}\right)_l + \rho_s H_f R_x = -\left(-\lambda \frac{\partial T}{\partial x}\right)_s. \qquad (7.1.1b)$$

where ρ_s is the density of solid, R_x is the solid/liquid interface velocity in the x direction, H_f is the latent heat of freezing, and (q_l) and (q_s) are the heat fluxes in liquid and solid in the x direction. In the above equation, the sum of the heat flux in liquid toward the solid/liquid interface in the $(-x)$ direction and the latent heat of freezing released at the solid/liquid interface must be balanced to the heat flux in solid toward the mould in the $(-x)$ direction, i.e., term (1) + term (3) = term (2).

Similarly, the solute balance equation at the solid/liquid interface can be expressed as follows.

$$\rho_{A,l}^*(1-k_0)R_x\{\equiv(\rho_{A,l}^* - \rho_{A,s}^*)R_x\} = (J_A)_l - (J_A)_s \qquad (7.1.2a)$$
$$\phantom{\rho_{A,l}^*(1-k_0)R_x\{\equiv(\rho_{A,l}^* - \rho_{A,s}^*)R_x\} =}(3)\phantom{\rho_{A,l}^* - \rho_{A,s}^*)R_x\}}(1)(2)$$

$$\rho_{A,l}^*(1-k_0)R_x = \left(-D_A \frac{\partial \rho_A}{\partial x}\right)_l - \left(-D_A \frac{\partial \rho_A}{\partial x}\right)_s \quad (7.1.2b)$$

where k_0 is the partition coefficient, D_A is the diffusion coefficient of solute A, and $\rho_{A,l}^*$ and $\rho_{A,s}^*$ are the concentrations of solute A in liquid and solid at the solid/liquid interface. $(J_A)_l$ and $(J_A)_s$ are the solute fluxes in liquid and solid in the x direction. In the above equation, the sum of the solute flux in liquid in the $(+x)$ direction and the solute flux in solid in the $(-x)$ direction must be balanced to the amount of solute released at the solid/liquid interface during solidification, i.e., term (3)=term (1)+term (2).

7.2 Methods of solution for phase change

In this section, we will describe numerical methods for treating latent heat of fusion in solidification heat transfer problems. The mode of latent heat evolution in solidification of a material depends upon whether the material is a pure metal which solidifies at its freezing temperature or an alloy which solidifies over a range of temperature. Solidification morphology may also affect the mode of latent heat evolution. The latent heat associated with the phase change is evolved at the solid/liquid interface when liquid is transformed to solid. Thermal and physical properties such as thermal conductivity, density and specific heat are varied as a function of temperature. As a result, melting or solidification problems are nonlinear and their analytical solutions are very difficult to obtain. A limited number of analytical solutions have been reported for one-dimensional solidification problems of pure metals, such as Neumann's solution [1] and Schwarz's solution [2]. However, melting or solidification problems in most materials processing can be considered to be multi-dimensional and multi-components problems, which cannot be solved analytically. Several numerical methods have also been developed for solving moving boundary problems [3], but those methods still have limited applications except for one-dimensional simple problems.

7.2.1 Numerical methods

Typical numerical methods for solving phase change problems can be categorized into three typical methods: (1) fixed grid methods, (2) variable grid methods and (3) transformed grid methods.

7.2.1.1 Fixed grid methods

In the fixed grid methods, the computational domain is divided into a finite number of uniform, orthogonal grids, which remain invariant throughout the calculation for all times. The latent heat evolution in the phase-change region is incorporated into the governing equation as a suitable volume source term. The evolution of latent heat of fusion in solidification of pure metals or alloys is generally evaluated by the equivalent specific heat method, the temperature recovery method and the enthalpy method, which will be described later. In this method, the location of the moving solid/liquid interface cannot be solved directly, but can be interpolated using the calculated temperature distributions. The fixed grid methods are generally used in most casting solidification problems including fluid flow since this method can be simply applied to multi-dimensional and multi-components problems.

7.2.1.2 Variable grid methods

In the case of variable grid methods, one of the grids Δx or Δt in the space–time domain is fixed and the other is variable to be determined so that the moving boundary (the solid/liquid interface) always remains at a grid point. Murray and Landis [4] suggested a method in which the size of time interval Δt is fixed for all times, and the number of space grids is fixed but the size of space interval Δx is changed (decreased or increased) as the solid/liquid interface moves. Alternatively, the space domain can be divided into fixed space interval and the time interval Δt is allowed to vary in such a manner that the moving interface always remains at a grid point [5]. In this method an appropriate energy balance equation at the moving solid/liquid interface is needed to account for the liberation of latent heat. The exact position of the solid/liquid interface can be evaluated by this method. However, the application of this method is limited to one-dimensional pure substance phase change problems.

7.2.1.3 Transformed grid methods

The conventional finite difference methods have computational simplicity when they are applied for solving problems having regular geometry with a uniform, orthogonal grid system. However, they may encounter some difficulties in dealing with a complex geometric domain. Recently, transformed or adaptive grid methods [6, 7] based on the coordinate transformation technique [8] have been proposed for solving multi-dimensional moving boundary problems. In these methods, the numerical grid generation technique is applied to map the irregular region into a regular shaped region in the computational domain where the problem is solved by the conventional finite difference methods and the results are transformed back into the physical domain. Since the grids are generated at each successive time step, the method needs considerable amounts of computational time. This method may not be available [9] in such problems that involve multiple solid/liquid fronts and coupling with additional transport fields, appearing in most practical casting solidification problems. The coordinate transformation technique can be extended to multi-component phase change problems in which the material does not have a distinct solid/liquid interface; instead melting or solidification occurs over a temperature range.

7.2.2 Alloy solidification

In the case of alloys, melting or solidification takes place over an extended range of temperatures. The solid and liquid phases are separate by a two-phase moving region, the 'mushy zone'. For such situations, the methods such as the variable grid or adaptive grid methods mentioned in the previous section used for solving the problems involving a single discrete phase change temperature are not applicable. The fixed grid method has been widely used to solve phase change problems for such situations, which appear in most practical casting and solidification problems.

In this section, we will describe the methods for solving alloy solidification problems in multi-dimensional coordinate systems. For the sake of simplicity, we will consider solidification problems without fluid flow.

7.2.2.1 Governing equation involving latent heat

As described in chapter 2, the differential energy balance equation involving the liberation of latent heat of freezing is given by

$$\rho C_v \frac{\partial T}{\partial t} = \nabla \cdot (\lambda \nabla T) + \rho H_f \frac{\partial f_s}{\partial t}. \quad (7.2.1)$$

Here, $\rho H_f (\partial f_s / \partial t)$ is the energy generation term, H_f is the latent heat of freezing and f_s is the fraction solid. By reforming the energy generation term, we obtain

$$\rho H_f \frac{\partial f_s}{\partial t} = \rho H_f \frac{\partial f_s}{\partial T} \frac{\partial T}{\partial t}. \quad (7.2.2)$$

By substituting equation (7.2.2) into equation (7.2.1) and rearranging it, we obtain

$$\rho \left(C_v - H_f \frac{\partial f_s}{\partial T} \right) \frac{\partial T}{\partial t} = \nabla \cdot (\lambda \nabla T). \quad (7.2.3)$$

In order to solve the above equation numerically, we need to evaluate the variation of fraction solid as a function of temperature $(\partial f_s / \partial T)$.

7.2.2.2 Fraction solid and temperature

The relation between the fraction solid and the temperature in solidification of alloys may be evaluated from the phase diagrams, except for congruent melting or solidification, which appears in pure metals, eutectic or peritectic reaction. The fraction solid f_s varies from 0 to 1 between the liquidus T_l to the

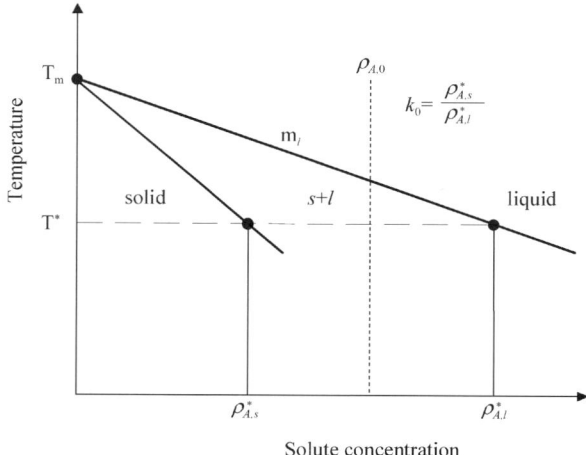

Figure 7.2. Phase diagram and equilibrium partition ratio.

solidus T_s in the mushy zone. As shown in figure 7.2 for a binary alloy system, the liquidus temperature is a function of solute concentration in liquid, which depends upon the solute redistribution models [10]. The liquidus temperature is given by the following equation.

$$T_l = T_m + m_l \rho_{A,l}. \qquad (7.2.4)$$

Here, T_m is the melting temperature of a pure component and m_l is the slope of the liquidus line at $\rho_{A,l}$.

Several models, which are generally used to evaluate the relation between the fraction solid and the temperature in practical solidification analyses, are summarized as follows.

(i) Lever rule (equilibrium solidification model)

Complete mixing of solute both in liquid and in solid is assumed, i.e., equilibrium solidification. The volume fraction of solid is given as a function of solute concentration or temperature (see figure 7.2).

$$f_s = \frac{\rho_{A,l} - \rho_{A,0}}{\rho_{A,l}(1-k_0)} = \frac{T_l - T}{(1-k_0)(T_m - T)}. \qquad (7.2.5)$$

Differentiating f_s by T gives

$$-H_f \frac{\partial f_s}{\partial T} = \frac{H_f}{(1-k_0)} \frac{T_m - T_l}{(T_m - T)^2} \qquad (7.2.6)$$

where $\rho_{A,0}$ is the initial solute concentration.

(ii) Scheil model

Complete mixing of solute in liquid and no mixing in solid are assumed in the Scheil model [11]. In this case, the volume fraction of solid as a function of temperature is given by

$$f_s = 1 - \left(\frac{\rho_{A,s}}{k_0 \rho_{A,0}}\right)^{1/(k_0-1)} = 1 - \left(\frac{T_m - T}{T_m - T_l}\right)^{1/(k_0-1)}. \qquad (7.2.7)$$

Similarly we have

$$-H_f \frac{\partial f_s}{\partial T} = \frac{H_f}{1-k_0} \frac{(T_m - T)^{(2-k_0)/(k_0-1)}}{(T_m - T_l)^{1/(k_0-1)}}. \qquad (7.2.8)$$

(iii) Brody–Flemings model

Complete mixing in liquid and some mixing in solid are assumed in this model. The volume fraction of solid as a function of temperature is given by

$$f_s = (1 + \alpha_t k_0)\left[1 - \left(\frac{T_m - T_l}{T_m - T}\right)^{1/(1-k_0)}\right]. \qquad (7.2.9)$$

Here, $\alpha_t \ (\equiv 4D_s t_f/\lambda_a^2)$ is the Brody–Flemings [12] constant relating to diffusion in the solid state, D_s is the diffusion coefficient of solute atoms in solid, t_f is the local solidification time and λ_a is the dendrite arm spacing. Similarly, we have

$$-H_f \frac{\partial f_s}{\partial T} = \frac{H_f(1+\alpha_t k_0)}{1-k_0} \frac{(T_m - T_l)^{1/(1-k_0)}}{(T_m - T)^{(2-k_0)/(1-k_0)}}. \qquad (7.2.10)$$

(iv) Linear distribution of latent heat of freezing between T_l and T_s

When f_s cannot easily be evaluated as a function of temperature since phase diagrams are not known for certain multi-component alloys, it is sometimes assumed that the latent heat of freezing is distributed linearly over the solidification range between T_l and T_s. Then, we have

$$f_s = \frac{T_l - T}{T_l - T_s}. \qquad (7.2.11)$$

Similarly we have

$$-H_f \frac{\partial f_s}{\partial T} = \frac{H_f}{(T_l - T_s)}. \qquad (7.2.12)$$

7.2.2.3 Treatment of latent heat of freezing

(i) Temperature recovery method

This method [13] consists of transforming the latent heat into an equivalent number of degrees by dividing the latent heat by the specific heat. The

108   *Phase change problems*

temperature at a node is calculated in the absence of latent heat, and after each time step the temperature that falls below the freezing point is reset to the freezing point until the solidification is over.

In this method the liberated latent heat, ΔH_f, is evaluated as a function of the corresponding solid fraction increment, Δf_s, over a time step, Δt, as follows.

$$\Delta H_f = \rho C_v V_i \Delta T = \rho \Delta H_f V_i \Delta f_s \tag{7.2.13}$$

where V_i is the volume of the element considered and ΔT is the temperature drop from the freezing point over a time step Δt. The fraction solid is

$$f_s = \sum \Delta f_s. \tag{7.2.14}$$

This procedure may be repeated until the solidification is over at $f_s = 1.0$. The temperature variations at nodes over a time step are calculated without considering the latent heat using equation (7.2.15), and then the true temperatures at nodes can be evaluated by the temperature recovery method using the results obtained without considering the latent heat.

$$\rho C_v \frac{\partial T}{\partial t} = \nabla \cdot (\lambda \nabla T). \tag{7.2.15}$$

The above mentioned conventional temperature recovery method has been used for the evaluation of latent heat of freezing of congruently melting materials.

A more generalized temperature recovery method was developed [13]. One can consider a unit volume element for which the temperature is in the solidification range. Then, the temperature variations with time are considered in two different steps, as illustrated in figure 7.3. Step I represents the imaginary case in which the temperature variation is analyzed involving the latent heat of freezing. Step II shows the case in which the temperature

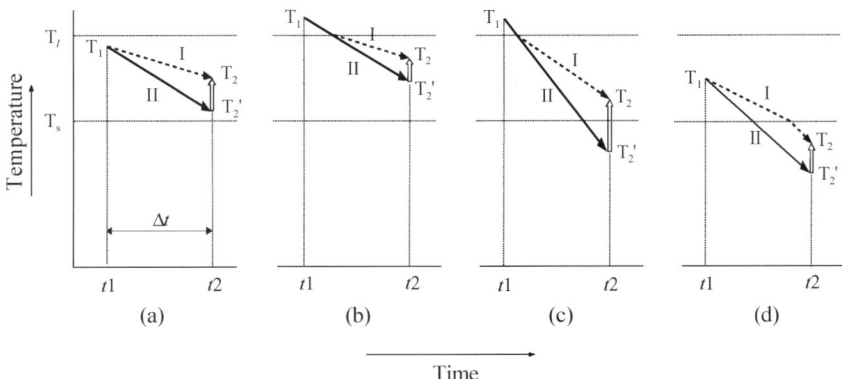

Figure 7.3. Temperature recovery method [13].

variation is analyzed without considering the latent heat using equation (7.2.15), therefore the calculated temperature T_2' may be lower than T_2 which is obtained in Step I. In the case of (a) in figure 7.3, one may express the liberated heat for Step I as follows.

$$Q_{\text{StepI}} = \int_{t_1}^{t_2} \rho C_v \frac{\partial T}{\partial t} dt + \Delta H_f = \int_{T_1}^{T_2} \rho C_v dT - \int_{T_1}^{T_2} \rho H_f \frac{\partial f_s}{\partial T} dT. \quad (7.2.16)$$

For Step II, equation (7.2.17) is available.

$$Q_{\text{StepII}} = \int_{t_1}^{t_2} \rho C_v \frac{\partial T}{\partial t} dt = \int_{T_1}^{T_2'} \rho C_v dT. \quad (7.2.17)$$

Provided the liberated (or absorbed) heat over a time interval Δt for Step I is equal to that for Step II, T_2 is easily calculated by the following relationships. Combining equations (7.2.16) and (7.2.17) gives

$$\int_{T_1}^{T_2'} \rho C_v dT = \int_{T_1}^{T_2} \rho C_v dT - \int_{T_1}^{T_2} \rho H_f \frac{\partial f_s}{\partial T} dT. \quad (7.2.18)$$

Similar procedures for (b), (c) and (d) in figure 7.3 give the following relationships. For cases (b) and (c)

$$\int_{T_l}^{T_2'} \rho C_v dT = \int_{T_l}^{T_2} \rho C_v dT - \int_{T_l}^{T_2} \rho H_f \frac{\partial f_s}{\partial T} dT \quad (7.2.19)$$

and for case (d)

$$\int_{T_1}^{T_2'} \rho C_v dT = \int_{T_1}^{T_2} \rho C_v dT - \int_{T_1}^{T_s} \rho H_f \frac{\partial f_s}{\partial T} dT. \quad (7.2.20)$$

It can be seen from the definition of this scheme that this method can be applied to solidification simulations, not only for congruently melting materials, but also for alloys, which solidify over a wide temperature range.

(ii) Equivalent specific heat method

This method [14, 15] is one of the most widely used techniques to handle the latent heat of freezing. The latent heat of freezing is converted into appropriate units for specific heat and added to the specific heat term over a temperature range at which solidification occurs. The term in the parentheses on the left-hand side of equation (7.2.3) may now be expressed as

$$C_E = C_v - H_f \frac{\partial f_s}{\partial T} \quad (7.2.21)$$

where C_E represents the equivalent specific heat which includes the latent heat term and varies over the solidification range. The second term on the

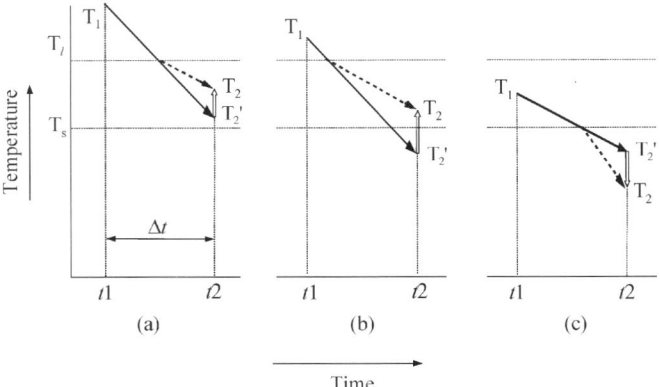

Figure 7.4. Compensation of temperature in the equivalent specific heat method.

right-hand side of equation (7.2.21) can be evaluated as a function of fraction solid (or temperature) using equations (7.2.5) through (7.2.12). Then, equation (7.2.3) becomes

$$\rho C_E \frac{\partial T}{\partial t} = \nabla \cdot (\lambda \nabla T). \quad (7.2.22)$$

Thus, the solution of equation (7.2.22) by the finite volume method is simplified to a simple transient heat conduction problem. However, in cases of pure metals or congruently melting materials, this method has limitations such as the term $(-H_f(\partial f_s/\partial T))$ becomes infinite since $T_l - T_s = 0$. In this case $(T_l - T_s)$ is assumed to be 0.1–1.0 °C.

In cases of alloys having a narrow solidification range, this method may cause some inaccuracy in the results if the equivalent specific heat C_E at time level t is used during the time interval $\Delta t (t \approx t + \Delta t)$ without considering whether the temperature at time level t is in the solidification range. The inaccuracy becomes large in the case of alloys having a narrow solidification range, in particular when Δt is large. The basic concept of the modified temperature recovery method can also be used for the compensation of this problem. There are three cases shown in figure 7.4.

Case (a) indicates that the equivalent specific heat C_E at time level t, which is equal to that of liquid, is used in the solidification range $(T_1 \sim T_2')$ without compensating the latent heat evolution, leading to shorten the solidification time. In this case, the true temperature T_2 at time level $t + \Delta t$ can be evaluated by the following energy balance equation.

$$\int_{T_l}^{T_2'} \rho C_v \, dT = \int_{T_l}^{T_2} \rho C_E \, dT. \quad (7.2.23)$$

Equation (7.2.23) can also be available for case (b). In case (c) the equivalent specific heat C_E at time level t, which includes the latent heat term, is used even after the solidification is over, leading to an increase of the solidification time. Similarly, we have the following energy balance equation.

$$\int_{T_1}^{T_2'} \rho C_E \, dT = \int_{T_1}^{T_s} \rho C_E \, dT + \int_{T_s}^{T_2} \rho C_v \, dT. \qquad (7.2.24)$$

(iii) Enthalpy method

This technique [16, 17] is based on the observation that enthalpy, the integral of heat capacity with respect to temperature, is a smooth function of temperature in the solidification range, even though specific heat shows a rapid variation over the same range, which causes some difficulties in numerical procedures. In general, the enthalpy related to phase change could be a function of variables such as temperature, concentration, cooling condition, etc. In many solidification models, however, the enthalpy in the mushy region can be assumed to be a function of temperature alone [17]. Figure 7.5 shows the enthalpy/temperature relationships for three cases: (a) pure metals and eutectics, (b) glassy substances and alloys such as carbon steels and (c) alloys involving eutectic reaction such as Al–7.0 mass% Si. The enthalpy involving phase change is given as a function of temperature.

$$H = H_0 + \int_{T_0}^{T} \rho C_v \, dT + (1 - f_s)\rho H_f \qquad (7.2.25)$$

where H_0 is the enthalpy at an arbitrary reference temperature T_0. By differentiating equation (7.2.25) by T, we have

$$\frac{\partial H}{\partial T} = C_v - H_f \frac{\partial f_s}{\partial T}. \qquad (7.2.26)$$

Thus, equation (7.2.3) can be expressed by

$$\rho \frac{\partial H}{\partial t} = \nabla \cdot (\lambda \nabla T). \qquad (7.2.27)$$

Equation (7.2.27) is valid over the entire computational domain, including both the solid and liquid phases as well as the mushy region. Numerical methods based on the finite difference and finite element methods related to the solution of equation (7.2.27) have been reported in the literature [16–19].

For the simple case of constant thermo-physical properties, let us consider the following governing equation for one-dimensional solidification or melting problems.

$$\rho \frac{\partial H}{\partial t} = \lambda \frac{\partial^2 T}{\partial x^2} \quad \text{in } 0 < x < L, \quad t > 0. \qquad (7.2.28)$$

Both explicit and implicit time integration schemes can be adopted to discretize equation (7.2.28). Similar to equation (6.3.1) in chapter 6, the general form of finite difference equation for equation (7.2.28) is given by

$$\frac{H_i^{t+\Delta t} - H_i^t}{\Delta t} = \frac{\lambda}{\rho} \left\{ W \frac{T_{i-1}^{t+\Delta t} - 2T_i^{t+\Delta t} + T_{i+1}^{t+\Delta t}}{\Delta x^2} \right.$$
$$\left. + (1 - W) \frac{T_{i-1}^t - 2T_i^t + T_{i+1}^t}{\Delta x^2} \right\} \quad (7.2.29)$$

where W is the weighting parameter. According to the value of the weighting parameter W, the above equation can be simplified as follows.

In the case of $W = 0$, equation (7.2.29) simplifies to the following fully explicit scheme.

$$\frac{H_i^{t+\Delta t} - H_i^t}{\Delta t} = \frac{\lambda}{\rho} \left\{ \frac{T_{i-1}^t - 2T_i^t + T_{i+1}^t}{\Delta x^2} \right\}. \quad (7.2.30)$$

In the case of $W = 1$, equation (7.2.29) simplifies to the following fully implicit scheme.

$$\frac{H_i^{t+\Delta t} - H_i^t}{\Delta t} = \frac{\lambda}{\rho} \left\{ \frac{T_{i-1}^{t+\Delta t} - 2T_i^{t+\Delta t} + T_{i+1}^{t+\Delta t}}{\Delta x^2} \right\}. \quad (7.2.31)$$

Let us now summarize briefly the solution procedure using the fully explicit scheme. Then, equation (7.2.30) is rearranged as

$$H_i^{t+\Delta t} = H_i^t + \frac{\lambda}{\rho} \eta (T_{i-1}^t - 2T_i^t + T_{i+1}^t) \quad (7.2.32)$$

where $\eta = \Delta t / \Delta x^2$. We need the following condition for stability in solving equation (7.2.32).

$$\eta = \frac{\Delta t}{\Delta x^2} \leq \frac{\rho C_v}{2\lambda} \quad (7.2.33)$$

which is equivalent to the stability condition, $\gamma = \alpha \Delta t / \Delta x^2 \leq 1/2$, given by equation (6.3.6). The explicit finite difference equation, equation (7.2.32), can be solved according to the following procedure. It is assumed that the values of H_i^t and T_i^t for nodes ($0 \leq i \leq M$) at an arbitrary time level t are known. Then, the new $H_i^{t+\Delta t}$ for nodes ($0 \leq i \leq M$) at the next time level $t + \Delta t$ can be calculated using equation (7.2.32). The new $T_i^{t+\Delta t}$ for nodes ($0 \leq i \leq M$) at the next time level $t + \Delta t$ is then calculated using the enthalpy/temperature relationships given by figure 7.5. The calculations are started with $t = 0$, which corresponds to the initial condition, and the above procedure is repeated for each successive time step.

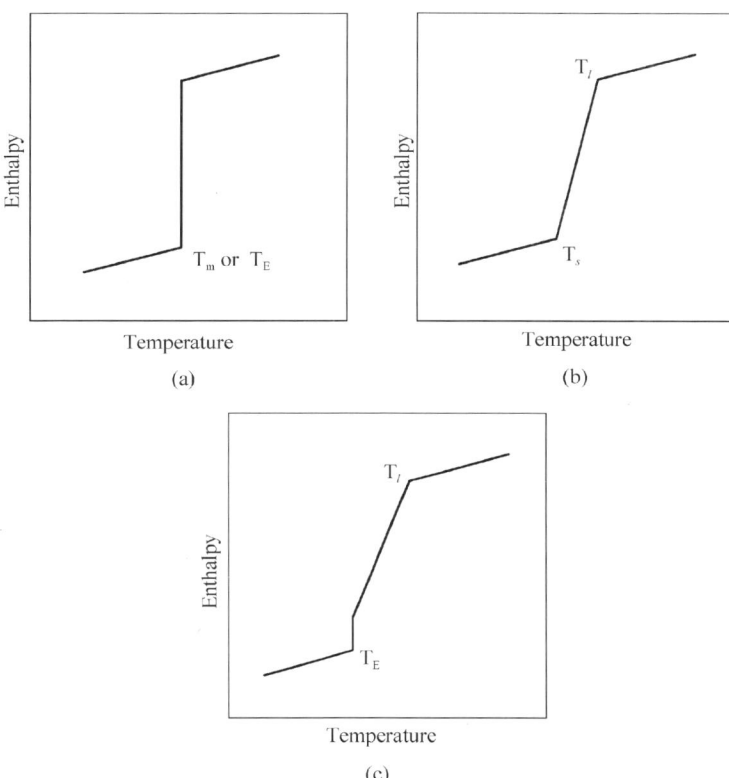

Figure 7.5. Enthalpy/temperature relationships for: (a) pure metals and eutectics, (b) carbon steels, and (c) Al–7.0 mass% Si alloy.

7.3 Case study: one-dimensional phase change problems

7.3.1 Description of the problem

Let us consider one-dimensional heat conduction problems including phase change, i.e., solidification of metals. Figure 7.6 indicates the geometry,

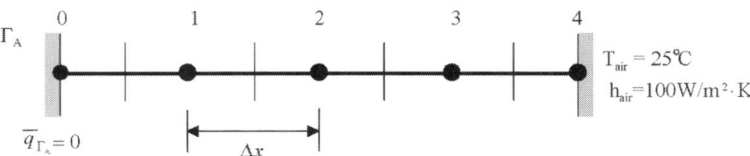

Figure 7.6. Control volumes for one-dimensional phase change problems.

114   *Phase change problems*

dimension and boundary conditions. The density, the specific heat and the thermal conductivity of an Al–4.5 mass% Cu alloy used in the simulation are $\rho = 2780\,\text{kg/m}^3$, $C_v = 1086\,\text{J/kg}\cdot\text{K}$ and $\lambda = 192.5\,\text{W/m}\cdot\text{K}$, respectively. One face of the plate is insulated, i.e., zero flux, $\bar{q}_{\Gamma_A} = 0$ (Γ_{II}-type boundary), and the other face is exposed to air, with a temperature of $T_{\text{air}} = 25\,°\text{C}$ and a convective heat transfer coefficient of $h_{\text{air}} = 100\,\text{W/m}^2\cdot\text{K}$ (Γ_{III}-type boundary). The liquidus and the solidus temperatures are 649 °C and 580 °C, respectively. The latent heat of freezing is $3.9 \times 10^5\,\text{J/kg}$. It is assumed that the latent of freezing is distributed linearly over the solidification range. The initial temperature of the liquid Al–4.5 mass% Cu is 680 °C.

Examine the three types of methods for solving the latent heat of freezing, mentioned in section 7.2.

7.3.2. Glossary of FORTRAN notation

FORTRAN name	Meaning
CLEN	The length of the computational domain
DEN	Density
DX	Distance between nodes
DT	Time increment
HNEW(I)	New enthalpy of node i at time $t + \Delta t$
H0	Enthalpy at the standard state
HAIR	Heat transfer coefficient at the material/air interface
HF	Latent heat of freezing
HLIQ	Enthalpy at $f_s = 0.0$
HOLD(I)	Old enthalpy of node i at time t
HSOL	Enthalpy at $f_s = 1.0$
I	Index denoting node i
ITER	A running counter of the number of iterations performed
ITMAX	Total number of iterations performed
ITPRN	The interval of iterations for printing out results
N	The number of nodes
SPH	The specific heat capacity
TAIR	The temperature of air
TCON	The reference value of the thermal conductivity
TH0	Temperature of the standard state
TINI	Initial temperature of the material
TIME	Elapsed time
TLIQ	Liquidus temperature
TSOL	Solidus temperature
TNEW(I)	New temperature of node i at time $t + \Delta t$
TOLD(I)	Old temperature of node i at time t

7.3.3 Simulation

Let us simulate the temperature distributions in one-dimensional phase change problems with time based on the three methods for treating the latent heat of freezing using the execution file [phase.exe].

Data input
```
> INPUT ITPRN AND ITMAX : 500 10000
> INPUT THE NODE NUMBER TO MONITOR : 3
> INPUT THE LENGTH OF MATERIAL : 0.02
> INPUT TIME STEP (RECOMMENDED: <= .196) : 0.15
```

Results

TIME	TRM	SHM	EM
.00	680.00	680.00	680.00
75.00	640.77	640.77	640.77
150.00	628.61	628.61	628.61
225.00	616.69	616.69	616.68
300.00	605.01	605.01	605.00
375.00	593.56	593.56	593.54
450.00	582.33	582.33	582.31
525.00	531.29	531.28	531.16
600.00	472.33	472.33	472.22
675.00	420.24	420.23	420.14
750.00	374.22	374.21	374.13
825.00	333.55	333.54	333.47
900.00	297.62	297.61	297.55

7.3.4 Program List

```
C    ****************************************************
C    *                                                  *
C    *   A COMPUTER PROGRAM FOR THE SIMULATION OF       *
C    *   1-D PHASE CHANGE PROBLEMS                      *
C    *                                                  *
C    ****************************************************
C
      PROGRAM PHASE
      PARAMETER(N=5)
      PARAMETER(MTRM=1, MSHM=2, MEM=3)
      COMMON
     &/VAR/TOLD(N,3), TNEW(N,3), DT, TIME
     &/PROPERTY/TCON, DEN, SPH, TLIQ, TSOL, HF
     &/GEOMETRY/CLEN, DX
     &/BOUNDARY/HAIR, TAIR
     &/INITIAL/TINI
     &/ENTHALPY/HOLD(N), HNEW(N), H0, TH0, HLIQ, HSOL
      INTEGER ITER, ITPRN, ITMAX, MONI
```

```
C
      WRITE(6, 100)
  100 FORMAT(1X,'> INPUT ITPRN AND ITMAX : ',\)
      READ(5, *) ITPRN, ITMAX
      WRITE(6, 110)
  110 FORMAT(' > INPUT THE NODE NUMBER TO MONITOR : ',\)
      READ(5, *) MONI
      MONI=MONI+1
C
C---- INPUT GEOMETRICAL VARIABLES
      WRITE(6, 120)
  120 FORMAT(' > INPUT THE LENGTH OF MATERIAL : ',\)
      READ(5,*) CLEN
C---- INITIALIZE VARIABLES
      CALL INIT
C---- CALCULATE TIME STEP
      CALL SETDT
C
      TIME=0.0
      WRITE(6, *) '  TIME   TRM   SHM   EM'
      WRITE(6, 200) TIME, TNEW(MONI,1), TNEW(MONI,2), TNEW
     (MONI,3)
  200 FORMAT(1X, F10.2, 3F10.2)
C
C---- START OF ITERATIVE CALCULATION ----
      DO 1000 ITER=1, ITMAX
      TIME=TIME+DT
C---- TEMPERATURE RECOVERY METHOD
      CALL SUBTRM
C---- EQUIVALENT SPECIFIC HEAT METHOD
      CALL SUBSHM
C---- ENTHALPY METHOD
      CALL SUBEM
C
C---- PRINT OUT TEMPERATURES BY ITPRN
      IF(MOD(ITER, ITPRN).EQ. 0) THEN
      WRITE(6, 200) TIME, (TNEW(MONI, I), I=1, 3)
      ENDIF
C
C---- UPDATE TEMPERATURES AND ENTHALPY
      DO 20 I=1, N
      DO 20 J=1, 3
      TOLD(I, J)=TNEW(I, J)
   20 CONTINUE
C
      DO 30 I=1, N
      HOLD(I)=HNEW(I)
   30 CONTINUE
```

```
 1000 CONTINUE
C---- END OF ITERATIVE CALCULATION ----
      PAUSE
      STOP
      END
C
C     SUBROUTINE FOR THE INITIALIZATION OF VARIABLES
      SUBROUTINE INIT
      PARAMETER(N=5)
      PARAMETER(MTRM=1, MSHM=2, MEM=3)
      COMMON
     &/VAR/TOLD(N,3), TNEW(N,3), DT, TIME
     &/PROPERTY/TCON, DEN, SPH, TLIQ, TSOL, HF
     &/GEOMETRY/CLEN, DX
     &/BOUNDARY/HAIR, TAIR
     &/INITIAL/TINI
     &/ENTHALPY/HOLD(N), HNEW(N), H0, TH0, HLIQ, HSOL
C
      TCON=192.5
      DEN=2780.0
      SPH=1086.0
      HAIR=100.0
      TAIR=25.0
      HF=390000.0
      TINI=680.0
      TLIQ=649.0
      TSOL=580.0
      TH0=298.0
      H0=0.0
      HLIQ=H0+SPH*(TLIQ-TH0)+HF
      HSOL=H0+SPH*(TSOL-TH0)
      DX=CLEN/FLOAT(N-1)
C---- INITIAL CONDITIONS
      DO 400 I=1, N
      DO 400 J=1, 3
      TOLD(I, J)=TINI
      TNEW(I, J)=TINI
  400 CONTINUE
C
      DO 500 I=1, N
      HOLD(I)=H0+SPH*(TOLD(I, MEM)-TH0)+HF
      HNEW(I)=HOLD(I)
  500 CONTINUE
      END
C
C     TIME STEP FOR EXPLICIT SCHEME
      SUBROUTINE SETDT
      PARAMETER(N=5)
```

118   *Phase change problems*

```
      PARAMETER(MTRM=1, MSHM=2, MEM=3)
      COMMON
     &/VAR/TOLD(N,3), TNEW(N,3), DT, TIME
     &/PROPERTY/TCON, DEN, SPH, TLIQ, TSOL, HF
     &/GEOMETRY/CLEN, DX
     &/BOUNDARY/HAIR, TAIR
     &/INITIAL/TINI
     &/ENTHALPY/HOLD(N), HNEW(N), H0, TH0, HLIQ, HSOL
      REAL T1, T2
C
      T1=DEN*SPH*DX**2/TCON/2.0
      T2=(DX/2.0)*DEN*SPH/(TCON/DX+HAIR)
      DT=MIN(T1, T2)
C
      WRITE(6, 100) DT
  100 FORMAT(1X, '> INPUT TIME STEP (RECOMENDED: <=', F6.3,
     ') : ',\)
      READ(5, *) DT
      END
C
C     SUBROUTINE OF TEMPERATURE RECOVERY METHOD
      SUBROUTINE SUBTRM
      PARAMETER(N=5)
      PARAMETER(MTRM=1, MSHM=2, MEM=3)
      COMMON
     &/VAR/TOLD(N,3), TNEW(N,3), DT, TIME
     &/PROPERTY/TCON, DEN, SPH, TLIQ, TSOL, HF
     &/GEOMETRY/CLEN, DX
     &/BOUNDARY/HAIR, TAIR
     &/INITIAL/TINI
     &/ENTHALPY/HOLD(N), HNEW(N), H0, TH0, HLIQ, HSOL
      REAL DFS, T2P
C
C---- SOLIDIFICATION FRACTION
      DFS=-1.0/(TLIQ-TSOL)
C
      R=TCON*DT/DEN/SPH/DX**2
      DO 10 I=2, N-1
      TNEW(I, MTRM)=R*TOLD(I-1, MTRM)+(1-2*R)*TOLD(I, MTRM)
     &       +R*TOLD(I+1, MTRM)
   10 CONTINUE
C---- BOUNDARY CONDITIONS
      R=TCON*DT/DEN/SPH/(DX**2/2.0)
      TNEW(1, MTRM)=R*TOLD(2, MTRM)+(1.0-R)*TOLD(1, MTRM)
      TNEW(N, MTRM)=R*TOLD(N-1, MTRM)
     &       +(1-(R+DT*HAIR/DEN/SPH/(DX/2.0)))*TOLD(N, MTRM)
     &       +HAIR*DT/DEN/SPH/(DX/2.0)*TAIR
C
```

Case study: one-dimensional phase change problems

```
C---- TEMPERATURE CORRECTION ----
      DO 20 I=1, N
C---- CASE (A)
      IF(TOLD(I, MTRM).LE.TLIQ.AND. TNEW(I, MTRM).GT.TSOL)
     &TNEW(I,MTRM)=(SPH*TNEW(I,MTRM)-HF*DFS*TOLD(I,MTRM))/
      (SPH-HF*DFS)
C---- CASE (D)
      IF(TOLD(I, MTRM).LE.TLIQ.AND. TOLD(I, MTRM).GT.TSOL.
      AND.
     &TNEW(I, MTRM).LE.TSOL) THEN
      T2P=TNEW(I, MTRM)
      TNEW(I,MTRM)=TNEW(I,MTRM)+HF*DFS*(TSOL-TOLD(I,MTRM))/
      SPH
C---- CASE (C)
      IF(TNEW(I, MTRM).GT.TSOL) TNEW(I,MTRM)
     & =(SPH*T2P-HF*DFS*TOLD(I,MTRM))/(SPH-HF*DFS)
      ENDIF
C---- CASE (B)
      IF(TOLD(I, MTRM).GT.TLIQ.AND. TNEW(I, MTRM).LT.TLIQ)
     &TNEW(I,MTRM)=(SPH*TNEW(I,MTRM)-HF*DFS*TLIQ)/
      (SPH-HF*DFS)
   20 CONTINUE
      END
C
C     SUBROUTINE OF EQUIVALENT SPECIFIC HEAT METHOD
      SUBROUTINE SUBSHM
      PARAMETER(N=5)
      PARAMETER(MTRM=1, MSHM=2, MEM=3)
      COMMON
     &/VAR/TOLD(N,3), TNEW(N,3), DT, TIME
     &/PROPERTY/TCON, DEN, SPH, TLIQ, TSOL, HF
     &/GEOMETRY/CLEN, DX
     &/BOUNDARY/HAIR, TAIR
     &/INITIAL/TINI
     &/ENTHALPY/HOLD(N), HNEW(N), H0, TH0, HLIQ, HSOL
      REAL ESPH, CSPH
C
C---- EQUIVALENT SPECIFIC HEAT
      ESPH=SPH+HF/(TLIQ-TSOL)
C
      DO 10 I=2, N-1
      IF(TOLD(I, MSHM).LE.TLIQ.AND. TOLD(I, MSHM).GT.TSOL)
THEN
      CSPH=ESPH
      ELSE
      CSPH=SPH
      ENDIF
C
```

```
            R=TCON*DT/DEN/CSPH/DX**2
            TNEW(I, MSHM)=R*TOLD(I-1, MSHM)+(1-2*R)*TOLD(I, MSHM)
            +R*TOLD(I+1, MSHM)
   10    CONTINUE
C
C----   BOUNDARY CONDITIONS
            IF(TOLD(1, MSHM).LE.TLIQ.AND. TOLD(1, MSHM).GT.TSOL)
            THEN
            CSPH=ESPH
            ELSE
            CSPH=SPH
            ENDIF
            R=TCON*DT/DEN/CSPH/(DX**2/2.0)
            TNEW(1, MSHM)=R*TOLD(2, MSHM)+(1.0-R)*TOLD(1, MSHM)
C
            IF(TOLD(I, MSHM).LE.TLIQ.AND. TOLD(I, MSHM).GT.TSOL)
            THEN
            CSPH=ESPH
            ELSE
            CSPH=SPH
            ENDIF
            R=TCON*DT/DEN/CSPH/(DX**2/2.0)
            TNEW(N, MSHM)=R*TOLD(N-1, MSHM)
         &      +(1-(R+DT*HAIR/DEN/CSPH/(DX/2.0)))*TOLD(N, MSHM)
         &      +HAIR*DT/DEN/CSPH/(DX/2.0)*TAIR
C
C----   TEMPERATURE CORRECTION
            DO 20 I=1, N
            IF(TNEW(I, MSHM).LT.TLIQ.AND. TOLD(I, MSHM).GT.TLIQ)
            THEN
            TEMP=SPH/ESPH*(TLIQ-TNEW(I, MSHM))
            TNEW(I, MSHM)=TLIQ-SPH/ESPH*(TLIQ-TNEW(I, MSHM))
            ENDIF
            IF(TNEW(I, MSHM).LT.TSOL.AND. TOLD(I, MSHM).LE.TLIQ.
            AND.
         &TOLD(I, MSHM).GT.TSOL) THEN
            TNEW(I, MSHM)=(TSOL-ESPH/SPH*(TSOL-TNEW(I, MSHM)))
            ENDIF
   20    CONTINUE
            END
C
C       SUBROUTINE OF ENTHALPY METHOD
            SUBROUTINE SUBEM
            PARAMETER(N=5)
            PARAMETER(MTRM=1, MSHM=2, MEM=3)
            COMMON
         &/VAR/TOLD(N,3), TNEW(N,3), DT, TIME
         &/PROPERTY/TCON, DEN, SPH, TLIQ, TSOL, HF
```

```
      &/GEOMETRY/CLEN, DX
      &/BOUNDARY/HAIR, TAIR
      &/INITIAL/TINI
      &/ENTHALPY/HOLD(N), HNEW(N), H0, TH0, HLIQ, HSOL
       REAL R
C
       R=TCON*DT/DEN/DX**2
       DO 10 I=2, N-1
       HNEW(I)=HOLD(I)+R*(TOLD(I-1, MEM)-2*TOLD(I, MEM)+TOLD
      (I+1, MEM))
    10 CONTINUE
C
C---- BOUNDARY CONDITIONS
       R=TCON*DT/DEN/(DX**2/2.0)
       HNEW(1)=HOLD(1)+R*(TOLD(2, MEM)-TOLD(1, MEM))
       HNEW(N)=HOLD(N)+R*TOLD(N-1, MEM)
      &        -(R+DT*HAIR/DEN/(DX/2.0))*TOLD(N, MEM)
      &        +HAIR*DT/DEN/(DX/2.0)*TAIR
C---- CONVERT ENTHALPY TO TEMPERATURE
       DO 20 I=1, N
       IF(HNEW(I).LT.HSOL) THEN
       TNEW(I, MEM)=(HNEW(I)-H0+SPH*TH0)/SPH
       ELSEIF(HNEW(I).GT.HLIQ) THEN
       TNEW(I, MEM)=(HNEW(I)-H0+SPH*TH0-HF)/SPH
       ELSE
       R=HF/(TLIQ-TSOL)
       TNEW(I, MEM)=(HNEW(I)-H0+SPH*TH0+R*TSOL)/(SPH+R)
       ENDIF
    20 CONTINUE
       END
```

References

[1] Carslaw H S and Jaeger J C 1959 *Conduction of Heat in Solids* 2nd edition (London: Oxford University Press)
[2] Schwarz C 1933 *Z. Angew. Math. Mech.* **13** 202
[3] Crank J 1984 *Free and Moving Boundary Problems* (London: Oxford University Press)
[4] Murray W D and Landis F 1959 *J. Heat Transfer* **81** 106
[5] Gupta R S and Kumar D 1981 *Int. J. Heat Mass Transfer* **24** 251
[6] Lacroix M 1989 *Numerical Heat Transfer* Part B **15**(2) 191
[7] Brackbill J U and Saltzman J S 1982 *J. Computational Physics* **46** 342
[8] Thompson J F 1980 *Computational Fluid Dynamics* ed W Kollman (Washington, DC: Hemisphere) p 1
[9] Lacroix M and Voller V R 1990 *Numerical Heat Transfer* Part B **17** 25
[10] Flemings M C 1974 *Solidification Processing* (New York: McGraw-Hill)
[11] Scheil E 1942 *Z. Metallkd.* **34**, 70
[12] Brody H D and Flemings M C 1966 *Trans. AIME* **236** 615

[13] Hong C P, Umeda T and Kimura Y 1984 *Metall. Trans.* **15B** 91, 101
[14] Pashkis V 1945 *Trans. AFS* **53** 90
[15] Mizikar E A 1967 *Trans. AIME* **239** 1747
[16] Sarjant R J and Slack M R 1954 *J. Iron Steel Inst.* **177** 428
[17] Swaminathan C R and Voller V R 1992 *Metall. Trans.* **23B** 651
[18] Comini G, Del Giudice S, Lewis R W and Ziewkiewicz D C 1974 *Int. J. Numer. Methods Eng.* **8** 613
[19] Thomas B G, Samarasekera I V and Brimacombe J K 1984 *Metall Trans B* **15B** 307

Chapter 8

Discretization schemes for convection and diffusion terms

In chapters 4 through 7, we introduced the basics of the finite volume method for the discretization of the integral or differential equations containing the unsteady term, the diffusion term and the source term. The finite volume method was applied to the one-dimensional steady-state and transient heat conduction problems. In materials processing problems where fluid flow plays an important role we need to consider the effects of convection since the convection is created by fluid flow. Diffusion always occurs alongside convection in nature, so we need to handle the convection and diffusion terms together. In this chapter we describe the methods for handling combined convection/diffusion problems.

8.1 Introduction

The general integral form of transport equation for property Φ including the diffusion and convection terms is given by equation (3.3.6), and rewritten here as

$$\frac{\partial}{\partial t}\iiint_\Omega (\rho\Phi)\,d\Omega + \iint_\Gamma (\rho\mathbf{u}\Phi)\cdot\mathbf{n}\,d\Gamma = -\iint_\Gamma (\mathbf{f}_\Phi\cdot\mathbf{n})\,d\Gamma + \iiint_\Omega S\,d\Omega \quad (8.1.1)$$

where the transport property Φ indicates $\mathbf{u}$, $C_v T$ or ω_A and $\mathbf{f}_\Phi$ is τ, $\mathbf{q}$ or $\mathbf{j}_A$ for momentum, heat or species transfer, respectively. The above equation includes the rate of accumulation term and the convection term on the left-hand side, and the diffusion term and the source term on the right-hand side, respectively. In the above equation the diffusion flux term due to the gradient of the general transport property Φ would represent viscous stress, heat flux or species flux, and is expressed by $-K\nabla\Phi$. Then, the above equation can be expressed by

$$\frac{\partial}{\partial t}\iiint_\Omega (\rho\Phi)\,d\Omega + \iint_\Gamma (\rho\mathbf{u}\Phi)\cdot\mathbf{n}\,d\Gamma = \iint_\Gamma (K\nabla\Phi)\cdot\mathbf{n}\,d\Gamma + \iiint_\Omega S\,d\Omega \quad (8.1.2)$$

where K is the diffusion coefficient. The continuity equation is simply obtained by replacing $\Phi = 1$, $\mathbf{f}_\Phi = 0$ and $S = 0$ in equation (8.1.1).

In this chapter, we will describe the typical schemes for discretizing the convection and diffusion terms in combined convection/diffusion problems. As mentioned in chapter 4, equation (8.1.2) is the starting point for computational procedures in the finite volume method.

8.2 Steady one-dimensional convection and diffusion

8.2.1 Governing equations

For the sake of simplicity, let us consider one-dimensional steady convection/diffusion problems to explain the discretization methods for the convection term.

In the absence of source terms, the integral form of governing equation for the one-dimensional steady convection and diffusion of a transport property Φ in a given flow field is given by

$$\iint_\Gamma (\rho \mathbf{u}\Phi) \cdot \mathbf{n}\, d\Gamma = -\iint_\Gamma (\mathbf{f}_\Phi \cdot \mathbf{n})\, d\Gamma$$
$$= \iint_\Gamma (K\nabla\Phi) \cdot \mathbf{n}\, d\Gamma. \qquad (8.2.1a)$$

The differential form of equation (8.2.1a) is

$$\frac{d}{dx}(\rho u \Phi) = \frac{d}{dx}\left(K \frac{d\Phi}{dx}\right) \qquad (8.2.1b)$$

and the continuity equation becomes

$$\iint_\Gamma (\rho \mathbf{u}) \cdot \mathbf{n}\, d\Gamma = 0 \qquad \text{or} \qquad \rho u = \text{constant} \qquad (8.2.2)$$

where u represents the velocity in the x direction.

8.2.2 The analytical solution

Equation (8.2.1) can be solved numerically or analytically to calculate the distributions of a transport property Φ in the computational domain. In order to solve this problem numerically we need to discretize the convection and diffusion terms in equation (8.2.1). In case of the convection/diffusion problems, the discretization schemes are closely related to the accuracy of the results. In this section, we will consider the analytical solution first, which will be used for the comparison with numerical solutions.

If the diffusion coefficient K is assumed to be constant, the governing equation, equation (8.2.1b), can be solved analytically since ρu is already

Steady one-dimensional convection and diffusion   125

assumed to be constant, by equation (8.2.2). If a domain is given as $0 \leq x \leq L$, with the following boundary conditions, equations (8.2.3a) and (8.2.3b), the solution of equation (8.2.1b) is given by equation (8.2.4).

$$\Phi = \Phi_0 \quad \text{at} \quad x = 0 \tag{8.2.3a}$$

$$\Phi = \Phi_L \quad \text{at} \quad x = L \tag{8.2.3b}$$

$$\frac{\Phi - \Phi_0}{\Phi_L - \Phi_0} = \frac{\exp(P_e x/L) - 1}{\exp(P_e) - 1} \tag{8.2.4}$$

where P_e is the Peclet number defined by

$$P_e \equiv \frac{\rho u}{K/L}. \tag{8.2.5}$$

The Peclet number represents the relative strength of convection and diffusion in convection/diffusion problems. As can be seen from its definition, the influence of the convection and diffusion terms on the distribution of a transport property Φ in convection/diffusion problems is closely related to the Peclet number. By analyzing the analytical solution, equation (8.2.4), the influence of the Peclet number on the distribution of Φ can be estimated. Figure 8.1 indicates the variation of Φ with x for different values of the Peclet number. There are two extreme cases: (i) no convection and pure diffusion ($P_e = 0$) and (ii) no diffusion and pure convection ($P_e \to \infty$). When the

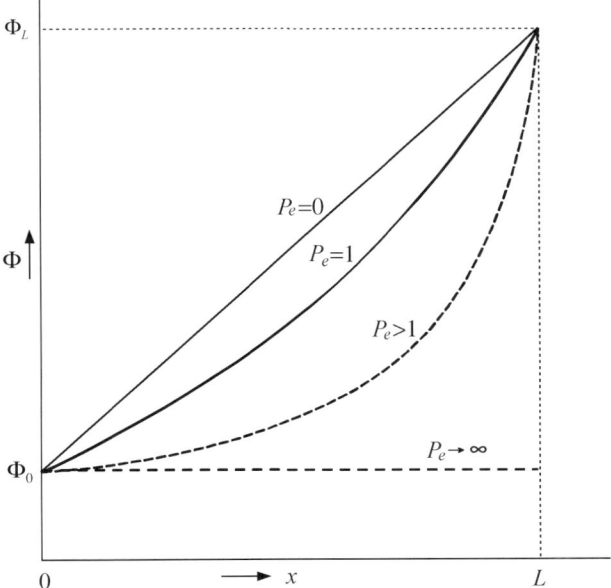

Figure 8.1. Variation of Φ with P_e.

flow is in the positive x direction ($P_e > 0$), the values of Φ in the domain seem to be more affected by the upstream value of Φ_0. With an increase of P_e, the value of Φ becomes closer to the upstream value of Φ_0 over much of the domain. The picture is reversed for negative values of P_e, and then Φ_L becomes the upstream value.

8.2.3 A control volume approach

In order to derive the discretization equation for equation (8.2.1), we consider a three-grid-point control volume for one-dimensional problems, as shown in figure 8.2. For the sake of simplicity, it is assumed that the control-volume faces $\Gamma_{i+1/2}$ and $\Gamma_{i-1/2}$ are located at the centers between the two nodes.

Let us now consider the expression of the integrals of equation (8.2.1). The direction of transfer of a transport property Φ by convection is determined only by the direction of flow. It is therefore reasonable to consider that the net influx into a control volume by convection is determined by the difference between the amount of inflow from the control-volume face $\Gamma_{i+1/2}$ and the amount of outflow to the control-volume face $\Gamma_{i-1/2}$. Thus, the integral of the left-hand side of equation (8.2.1a) can be expressed by

$$\iint_{\Gamma} (\rho \mathbf{u} \Phi) \cdot \mathbf{n} \, d\Gamma = \{(\rho u \Phi)_{\Gamma_{i+1/2}} \times A_{\Gamma_{i+1/2}} - (\rho u \Phi)_{\Gamma_{i-1/2}} \times A_{\Gamma_{i-1/2}}\}. \quad (8.2.6)$$

However, the direction of transfer of a transport property Φ by diffusion or other mechanisms given by equations (5.2.7b)–(5.2.7f) through control-volume faces is determined by the gradient $-K\nabla\Phi$ or the difference $(\Phi_{i\pm1} - \Phi_i)$ between the two neighboring control volumes. Similar to equation (5.2.4), the integral of the right-hand side of equation (8.2.1a) which indicates the net diffusion flux into a control volume (i) is given by

$$-\iint_{\Gamma} (\mathbf{f}_\Phi \cdot \mathbf{n}) \, d\Gamma = (\hat{f}_{\Phi\Gamma_{i+1/2}} \times A_{\Gamma_{i+1/2}}) + (\hat{f}_{\Phi\Gamma_{i-1/2}} \times A_{\Gamma_{i-1/2}}) \quad (8.2.7)$$

where $\hat{f}_{\Phi\Gamma_{i+1/2}}$ and $\hat{f}_{\Phi\Gamma_{i-1/2}}$ indicate the net diffusion influx into a control volume (i) across the control-volume faces $\Gamma_{i+1/2}$ and $\Gamma_{i-1/2}$, respectively.

In one-dimensional rectangular coordinate problems, $A_{\Gamma_{i+1/2}} = A_{\Gamma_{i-1/2}} = 1.0$. By substituting equations (8.2.6) and (8.2.7) into equation

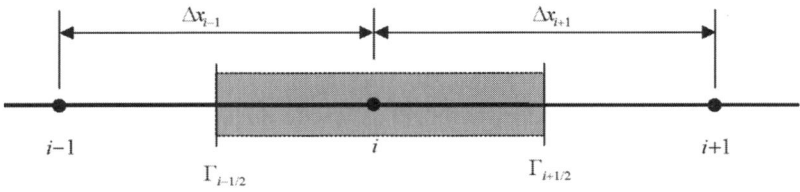

Figure 8.2. Control volumes for one-dimensional problems.

(8.2.1) we have

$$(\rho u \Phi)_{\Gamma_{i+1/2}} - (\rho u \Phi)_{\Gamma_{i-1/2}} = (\hat{f}_{\Phi \Gamma_{i+1/2}}) + (\hat{f}_{\Phi \Gamma_{i-1/2}}). \quad (8.2.8)$$

The integrated continuity equation becomes

$$(\rho u)_{\Gamma_{i+1/2}} - (\rho u)_{\Gamma_{i-1/2}} = 0. \quad (8.2.9)$$

If we apply equation (8.2.8) to figure 8.2 for the inner nodal point (i), we have

$$(\rho u)_{\Gamma_{i+1/2}} \Phi_{\Gamma_{i+1/2}} - (\rho u)_{\Gamma_{i-1/2}} \Phi_{\Gamma_{i-1/2}} = \left\{ \frac{K_{\Gamma_{i+1/2}}}{\Delta x_{i+1}} (\Phi_{i+1} - \Phi_i) \right\}$$
$$+ \left\{ \frac{K_{\Gamma_{i-1/2}}}{\Delta x_{i-1}} (\Phi_{i-1} - \Phi_i) \right\} \quad (8.2.10)$$

where the velocity field (ρu) was assumed to be given.

In order to solve equation (8.2.10), we need to evaluate the transport property Φ and the diffusion flux terms at the control-volume faces $\Gamma_{i+1/2}$ and $\Gamma_{i-1/2}$. A number of discretization methods for the convection terms will be presented in the following sections.

8.2.4 The central difference scheme

In chapters 5 and 6, the central difference scheme has been adopted to represent the diffusion terms, such as appeared on the right-hand side of equation (8.2.10).

Similar to the case of the diffusion terms if we adopt the central difference scheme in treating convection terms, the terms $\Phi_{\Gamma_{i+1/2}}$ and $\Phi_{\Gamma_{i-1/2}}$ on the left-hand side of this equation can be evaluated by the linear interpolation as follows.

$$\Phi_{\Gamma_{i+1/2}} = \tfrac{1}{2}(\Phi_i + \Phi_{i+1}) \quad (8.2.11a)$$

$$\Phi_{\Gamma_{i-1/2}} = \tfrac{1}{2}(\Phi_{i-1} + \Phi_i). \quad (8.2.11b)$$

Then, equation (8.2.10) can be written as

$$\tfrac{1}{2}(\rho u)_{\Gamma_{i+1/2}}(\Phi_i + \Phi_{i+1}) - \tfrac{1}{2}(\rho u)_{\Gamma_{i-1/2}}(\Phi_{i-1} + \Phi_i)$$
$$= \left\{ \frac{K_{\Gamma_{i+1/2}}}{\Delta x_{i+1}} (\Phi_{i+1} - \Phi_i) \right\} + \left\{ \frac{K_{\Gamma_{i-1/2}}}{\Delta x_{i-1}} (\Phi_{i-1} - \Phi_i) \right\}. \quad (8.2.12)$$

The diffusion coefficient K can be considered as a function of temperature [1]. In order to arrange the discretized equation more compactly, we define two new symbols F and D, which represent the convective mass flux and the diffusion conductance. It should be noted that F can take positive or negative values depending on the direction of the fluid flow, but D always remains positive.

$$F \equiv \rho u, \qquad D \equiv \frac{K}{\Delta x}. \tag{8.2.13}$$

The values of F and D at the control-volume faces can be written as

$$F_{\Gamma_{i-1/2}} \equiv (\rho u)_{\Gamma_{i-1/2}}, \qquad D_{\Gamma_{i-1/2}} \equiv \frac{K_{\Gamma_{i-1/2}}}{\Delta x_{i-1}} \tag{8.2.14a}$$

$$F_{\Gamma_{i+1/2}} \equiv (\rho u)_{\Gamma_{i+1/2}}, \qquad D_{\Gamma_{i+1/2}} \equiv \frac{K_{\Gamma_{i+1/2}}}{\Delta x_{i+1}}. \tag{8.2.14b}$$

Equation (8.2.12) can be rearranged as

$$C_i \Phi_i = C_{i-1} \Phi_{i-1} + C_{i+1} \Phi_{i+1} \tag{8.2.15}$$

where

$$C_{i-1} = \left(D_{\Gamma_{i-1/2}} + \frac{F_{\Gamma_{i-1/2}}}{2} \right)$$

$$C_{i+1} = \left(D_{\Gamma_{i+1/2}} - \frac{F_{\Gamma_{i+1/2}}}{2} \right) \tag{8.2.16}$$

$$C_i = C_{i-1} + C_{i+1} + (F_{\Gamma_{i+1/2}} - F_{\Gamma_{i-1/2}}).$$

In order to solve one-dimensional steady convection/diffusion problems we need to write discretized equations of the form of equation (8.2.15) for all grid nodes. The resulting algebraic equation is then solved to obtain the distribution of the transport property Φ by the similar methods as described in chapter 5 (section 5.2.4).

In solving the above discretized equation, we need to consider a simple basic rule on the coefficients, C_{i-1}, C_i and C_{i+1}. The value of Φ at a node is influenced by the values at its neighboring nodes only through the processes of convection and diffusion. It is reasonable to consider that an increase in the value of Φ at one node leads to an increase in the value at its neighboring nodes, i.e., the value of Φ_i increases or decreases, depending on whether Φ_{i-1} and Φ_{i+1} increase or decrease. It can be said that all the coefficients appeared in equation (8.2.16) must always be positive to obtain a realistic solution.

When the flow is in the positive direction ($F_{\Gamma_{i-1/2}} > 0$ and $F_{\Gamma_{i+1/2}} > 0$), C_{i-1} is always positive. According to equation (8.2.9), ($F_{\Gamma_{i+1/2}} - F_{\Gamma_{i-1/2}}$) is zero when the flow satisfies continuity, resulting in $C_i = C_{i-1} + C_{i+1}$. However, the coefficient C_{i+1} will be negative if P_e ($\equiv F_{\Gamma_{i+1/2}} / D_{\Gamma_{i+1/2}}$) is greater than 2. This will cause physically unrealistic results. This concept is also available when the flow is in the negative direction. The central difference scheme can only be available in diffusion-dominated low Reynolds number flows ($|P_e| < 0$). This is the reason why the central difference scheme is considered not to be a suitable discretization method for practical flow calculations.

8.2.5 The upwind difference scheme

In the central difference scheme, the value of transport property $\Phi_{\Gamma_{i-1/2}}$ at the control-volume face $\Gamma_{i-1/2}$ is assumed to be the average of Φ_i and Φ_{i-1}, indicating that it is equally influenced by both Φ_i and Φ_{i-1}. The above assumption is reasonable in the case of the diffusion process since the diffusive transfer of a transport property occurs along its gradients in all directions. However, in a strongly convective flow the above assumption is unreasonable because the transport property at the control-volume face $\Gamma_{i-1/2}$ must receive much stronger influence from the nodal value Φ_{i-1} than the nodal value Φ_i when the convective flow occurs in the positive direction.

The upwind difference or the donor-cell scheme takes into account the influence of the flow direction when evaluating the value of transport property Φ at a control-volume face. The value of transport property Φ at a control-volume face is equal to the value at the grid point on the upstream node. Thus, it can be summarized as follows.

When the flow is in the positive direction, $u_{\Gamma_{i-1/2}} > 0$, $u_{\Gamma_{i+1/2}} > 0$, the values of Φ at the control-volume faces $\Gamma_{i-1/2}$ and $\Gamma_{i+1/2}$ are $\Phi_{\Gamma_{i-1/2}} = \Phi_{i-1}$ and $\Phi_{\Gamma_{i+1/2}} = \Phi_i$. Then, the discretized equation (8.2.10) becomes

$$(\rho u)_{\Gamma_{i+1/2}} \Phi_i - (\rho u)_{\Gamma_{i-1/2}} \Phi_{i-1} = \left\{ \frac{K_{\Gamma_{i+1/2}}}{\Delta x_{i+1}} (\Phi_{i+1} - \Phi_i) \right\}$$
$$+ \left\{ \frac{K_{\Gamma_{i-1/2}}}{\Delta x_{i-1}} (\Phi_{i-1} - \Phi_i) \right\}. \quad (8.2.17)$$

When the flow is in the negative direction, $u_{\Gamma_{i-1/2}} < 0$, $u_{\Gamma_{i+1/2}} < 0$, the values of Φ are $\Phi_{\Gamma_{i-1/2}} = \Phi_i$ and $\Phi_{\Gamma_{i+1/2}} = \Phi_{i+1}$. And the discretized equation, equation (8.2.10), becomes

$$(\rho u)_{\Gamma_{i+1/2}} \Phi_{i+1} - (\rho u)_{\Gamma_{i-1/2}} \Phi_i = \left\{ \frac{K_{\Gamma_{i+1/2}}}{\Delta x_{i+1}} (\Phi_{i+1} - \Phi_i) \right\}$$
$$+ \left\{ \frac{K_{\Gamma_{i-1/2}}}{\Delta x_{i-1}} (\Phi_{i-1} - \Phi_i) \right\}. \quad (8.2.18)$$

Equations (8.2.17) and (8.2.18) can be simply expressed by

$$C_i \Phi_i = C_{i-1} \Phi_{i-1} + C_{i+1} \Phi_{i+1}. \quad (8.2.19)$$

The coefficients of equation (8.2.19) are given as follows.

For $u_{\Gamma_{i-1/2}} > 0$, $u_{\Gamma_{i+1/2}} > 0$,

$$C_{i-1} = (D_{\Gamma_{i-1/2}} + F_{\Gamma_{i-1/2}})$$
$$C_{i+1} = D_{\Gamma_{i+1/2}} \qquad (8.2.20a)$$
$$C_i = C_{i-1} + C_{i+1} + (F_{\Gamma_{i+1/2}} - F_{\Gamma_{i-1/2}}).$$

For $u_{\Gamma_{i-1/2}} < 0$, $u_{\Gamma_{i+1/2}} < 0$,

$$C_{i-1} = D_{\Gamma_{i-1/2}}$$
$$C_{i+1} = (D_{\Gamma_{i+1/2}} - F_{\Gamma_{i+1/2}}) \qquad (8.2.20b)$$
$$C_i = C_{i-1} + C_{i+1} + (F_{\Gamma_{i+1/2}} - F_{\Gamma_{i-1/2}}).$$

The coefficients of equation (8.2.20) can be summarized using a new operator as follows.

$$C_{i-1} = D_{\Gamma_{i-1/2}} + \max \|F_{\Gamma_{i-1/2}}, 0\|$$
$$C_{i+1} = D_{\Gamma_{i+1/2}} + \max \|-F_{\Gamma_{i+1/2}}, 0\| \qquad (8.2.21)$$
$$C_i = C_{i-1} + C_{i+1} + (F_{\Gamma_{i+1/2}} - F_{\Gamma_{i-1/2}}).$$

The new operator max $\| \ \|$ stands for the largest value of the quantities contained within it.

In this scheme the coefficients C_{i-1}, C_i and C_{i+1} are always positive, and the coefficient matrix is diagonally dominant since the term $(F_{\Gamma_{i+1/2}} - F_{\Gamma_{i-1/2}})$ in C_i is zero when the flow field satisfies continuity. Hence the solutions must be always physically reasonable. However, the upwind scheme is not suitable for accurate flow calculations because of *false diffusion* [1] when the influence of the diffusion flux term is much larger than that of convection, i.e., diffusion-dominated low Reynolds number flows.

8.2.6 The hybrid difference scheme

The hybrid difference scheme [2] is based on a combination of the central and upwind difference schemes. For small Peclet numbers ($|P_e| < 2$) the central difference scheme is employed for the convection and diffusion terms, and for large Peclet numbers ($|P_e| \geq 2$) the upwind difference scheme is employed for the convection term and the diffusion term is ignored. In order to adopt this concept, we need to define the Peclet numbers at the control-volume faces $\Gamma_{i-1/2}$ and $\Gamma_{i+1/2}$ of the control volume (i) as follows.

$$P_e|_{\Gamma_{i-1/2}} = \frac{F_{\Gamma_{i-1/2}}}{D_{\Gamma_{i-1/2}}} = \frac{(\rho u)_{\Gamma_{i-1/2}}}{K_{\Gamma_{i-1/2}}/\Delta x_{i-1}} \qquad (8.2.22a)$$

$$P_e|_{\Gamma_{i+1/2}} = \frac{F_{\Gamma_{i+1/2}}}{D_{\Gamma_{i+1/2}}} = \frac{(\rho u)_{\Gamma_{i+1/2}}}{K_{\Gamma_{i+1/2}}/\Delta x_{i+1}}. \qquad (8.2.22b)$$

Steady one-dimensional convection and diffusion

For the sake of simplicity we assume that there is no particular variation of the Peclet number in a control volume, i.e., P_e (at $\Gamma_{i-1/2}$) $\cong P_e$ (at $\Gamma_{i+1/2}$) $\equiv P_e$.

Based on this concept we have the discretized equations as follows.
For $-2 < P_e < 2$, the discretized equation becomes

$$\tfrac{1}{2}(\rho u)_{\Gamma_{i+1/2}}(\Phi_i + \Phi_{i+1}) - \tfrac{1}{2}(\rho u)_{\Gamma_{i-1/2}}(\Phi_{i-1} + \Phi_i)$$
$$= \left\{\frac{K_{\Gamma_{i+1/2}}}{\Delta x_{i+1}}(\Phi_{i+1} - \Phi_i)\right\} + \left\{\frac{K_{\Gamma_{i-1/2}}}{\Delta x_{i-1}}(\Phi_{i-1} - \Phi_i)\right\}. \tag{8.2.23a}$$

For $P_e \geq 2$, the discretized equation becomes

$$(\rho u)_{\Gamma_{i+1/2}}\Phi_i - (\rho u)_{\Gamma_{i-1/2}}\Phi_{i-1} = 0. \tag{8.2.23b}$$

For $P_e \leq -2$, the discretized equation becomes

$$(\rho u)_{\Gamma_{i+1/2}}\Phi_{i+1} - (\rho u)_{\Gamma_{i-1/2}}\Phi_i = 0. \tag{8.2.23c}$$

By substituting equation (8.2.22) into equation (8.2.23) and rearranging them we obtain the general form of the discretized equation, which is similar to equation (8.2.15).

$$C_i\Phi_i = C_{i-1}\Phi_{i-1} + C_{i+1}\Phi_{i+1}. \tag{8.2.24}$$

Similarly, the coefficients for the hybrid difference scheme for steady one-dimensional convection/diffusion problems can be summarized as follows.

$$C_{i-1} = \max\left\|\left(D_{\Gamma_{i-1/2}} + \frac{F_{\Gamma_{i-1/2}}}{2}\right), F_{\Gamma_{i-1/2}}, 0\right\|$$

$$C_{i+1} = \max\left\|\left(D_{\Gamma_{i+1/2}} - \frac{F_{\Gamma_{i+1/2}}}{2}\right), -F_{\Gamma_{i+1/2}}, 0\right\| \tag{8.2.25}$$

$$C_i = C_{i-1} + C_{i+1} + (F_{\Gamma_{i+1/2}} - F_{\Gamma_{i-1/2}}).$$

It is to be noted that the hybrid difference scheme adopts the favorable properties of the upwind and central difference schemes, and that it is unconditionally bounded since the coefficients are always positive. At high P_e numbers when the central scheme causes inaccurate results, the upwind scheme is applied to treat the convection terms. The hybrid scheme can also be extended to two- and three-dimensional problems [3]. The hybrid difference scheme has been widely used in various computational fluid dynamics procedures.

8.2.7 The power-law scheme

In the hybrid difference scheme the diffusion effects are assumed to be neglected when $|P_e|$ exceeds 2, which seems unreasonable. A more accurate approximation to one-dimensional convection/diffusion problems can be given by the power-law scheme [1]. In the power-law scheme the diffusion terms are set to zero when $|P_e|$ exceeds 10. If $|P_e| < 10$, the flux is evaluated using a polynomial expression.

Based on this concept we have the discretized equations as follows. For $P_e < -10$,

$$(\rho u)_{\Gamma_{i+1/2}} \Phi_i - (\rho u)_{\Gamma_{i-1/2}} \Phi_{i-1} = 0. \qquad (8.2.26a)$$

For $-10 \leq P_e \leq 0$,

$$(\rho u)_{\Gamma_{i+1/2}} \Phi_i - (\rho u)_{\Gamma_{i-1/2}} \Phi_{i-1} = \left\{ \beta \frac{K_{\Gamma_{i+1/2}}}{\Delta x_{i+1}} (\Phi_{i+1} - \Phi_i) \right\}$$

$$+ \left\{ \beta \frac{K_{\Gamma_{i-1/2}}}{\Delta x_{i-1}} (\Phi_{i-1} - \Phi_i) \right\}. \qquad (8.2.26b)$$

For $0 \leq P_e \leq 10$,

$$(\rho u)_{\Gamma_{i+1/2}} \Phi_{i+1} - (\rho u)_{\Gamma_{i-1/2}} \Phi_i = \left\{ \beta \frac{K_{\Gamma_{i+1/2}}}{\Delta x_{i+1}} (\Phi_{i+1} - \Phi_i) \right\}$$

$$+ \left\{ \beta \frac{K_{\Gamma_{i-1/2}}}{\Delta x_{i-1}} (\Phi_{i-1} - \Phi_i) \right\}. \qquad (8.2.26c)$$

For $P_e > 10$,

$$(\rho u)_{\Gamma_{i+1/2}} \Phi_{i+1} - (\rho u)_{\Gamma_{i-1/2}} \Phi_i = 0 \qquad (8.2.26d)$$

where $\beta = (1 - 0.1 P_e)^5$.

Similarly, we can summarize the coefficients for the power-law scheme for one-dimensional steady convection/diffusion problems as follows.

$$C_{i-1} = \max \|(\beta D_{\Gamma_{i-1/2}}), (\beta D_{\Gamma_{i-1/2}} + F_{\Gamma_{i-1/2}}), F_{\Gamma_{i-1/2}}, 0\|$$

$$C_{i+1} = \max \|(\beta D_{\Gamma_{i+1/2}}), (\beta D_{\Gamma_{i+1/2}} - F_{\Gamma_{i+1/2}}), -F_{\Gamma_{i+1/2}}, 0\| \qquad (8.2.27)$$

$$C_i = C_{i-1} + C_{i+1} + (F_{\Gamma_{i+1/2}} - F_{\Gamma_{i-1/2}}).$$

As mentioned in this section, the properties of the power-law scheme are similar to those of the hybrid difference scheme. The power-law scheme is known to be more accurate for one-dimensional problems than the hybrid difference scheme, and it can be used as an alternative to the hybrid difference scheme in many practical flow calculations.

Several other higher order schemes have been developed for the discretization of convective terms more accurately [4, 5].

8.3 Comparison among difference schemes

Now, let us consider a three-point grid system to compare the difference schemes in treating convection/diffusion problems, as shown in figure 8.3. The diffusion coefficient and the density of a material are assumed to be 10 and 1000, respectively.

The variation of the Peclet number corresponding to the flow velocity is given in table 8.1. The values of the transport property Φ_i are evaluated by various difference schemes mentioned in the previous section, as follows.

$$\Phi_i = \frac{1}{C_i}(C_{i+1}\Phi_{i+1} + C_{i-1}\Phi_{i-1}). \qquad (8.3.1)$$

Here the coefficients C_i, C_{i+1} and C_{i-1} are evaluated using equations (8.2.16), (8.2.21), (8.2.25) and (8.2.27). In this evaluation the flow velocity is the only variable that affects the Peclet number under constant values of ρ, u, Γ and Δx. In figure 8.4, the Φ_i values calculated from each scheme are shown

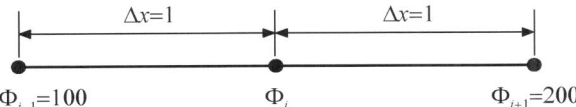

Figure 8.3. A three-point grid system.

Table 8.1. The values of Φ_i as a function of P_e for various difference schemes.

u	P_e [a]	Φ_i				
		Exact solution	Central difference	Upwind difference	Hybrid difference	Power-law
0	0	150.0	150.0	150.0	150.0	150.0
10^{-4}	0.01	150.3	150.3	150.2	150.0	150.2
5×10^{-4}	0.05	151.3	151.3	151.2	150.0	151.2
10^{-3}	0.10	152.5	152.5	152.4	152.5	152.5
5×10^{-3}	0.50	162.2	162.5	160.0	162.5	162.2
10^{-2}	1.0	173.1	175.0	166.7	175.0	172.9
5×10^{-2}	5.0	199.3	275.0	185.7	200.0	199.4
10^{-1}	10	200.0	400.0	191.7	200.0	200.0
5×10^{-1}	50	200.0	1400.0	198.1	200.0	200.0
1	100	200.0	2650.0	199.0	200.0	200.0
5	500	200.0	12650.0	199.8	200.0	200.0
10	1000	200.0	25150.0	199.9	200.0	200.0
50	5000	200.0	125150.0	200.0	200.0	200.0
100	10000	200.0	250150.0	200.0	200.0	200.0

[a] The values of P_e are evaluated using constant $\Gamma = 10$ and $\rho = 1000$.

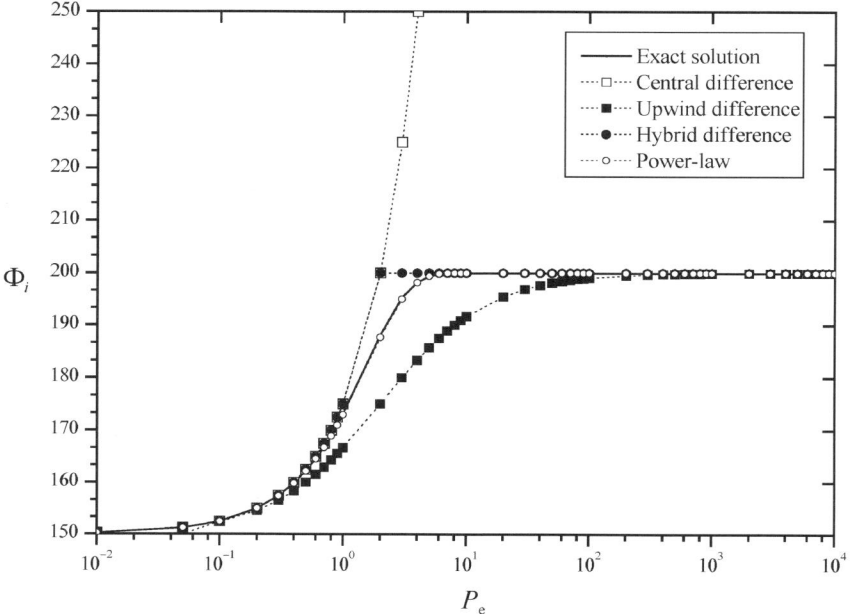

Figure 8.4. Comparison among difference schemes.

together with the exact solution with the x-axis representing the Peclet number in log scale. When the Peclet number is small, any of four schemes gives a satisfactory Φ_i value. It explains why the finer mesh guarantees the better solution. No other schemes except for the power-law scheme can provide correct solutions when the Peclet number has values between 0.1 and 10. The error of the upwind difference scheme becomes less when the Peclet number is extremely large. This is identical to the situation with a very high flow rate and a very small diffusion coefficient. The central difference scheme is not available when the Peclet number is larger than 2 or 3.

References

[1] Patankar S V 1980 *Numerical Heat Transfer and Fluid Flow* (New York: Hemisphere Publishing Co)
[2] Spalding D B 1972 *Int. J. Num. Methods Eng.* **4** 551
[3] Versteeg H K and Malalasekera W 1995 *An Introduction to Computational Fluid Dynamics—The Finite Volume Method* (London: Longman Scientific & Technical)
[4] Leonard B P 1979 *Comput. Methods Appl. Mech. Eng.* **19** 59
[5] Hirsch C 1990 *Numerical Computation of Internal and External Flows* vol. 2 (Chichester: Wiley)

Chapter 9

Solution algorithms for fluid flow analysis

In chapter 8, we introduced the basics of the discretization schemes to handle combined convection/diffusion problems. The convection of a scalar variable, transport property Φ, depends on the local velocity field. In the previous chapter we assumed that the velocity field was somehow known. In practical materials processing problems, the velocity field is not known, but given as a part of the overall solution process along with all other flow variables. The main purpose of this chapter is to describe fundamentals of solution schemes for computing the entire fluid flow field. Typical schemes for solving fluid flow problems will be briefly described in this chapter.

9.1 Governing equations

It is possible to assume that the flow encountered in most materials processing problems is incompressible. It is therefore important to introduce numerical schemes for solving incompressible flow in practical applications. The transport equations for a constant property flow without any source terms are given by

(continuity)
$$\nabla \cdot \mathbf{u} = 0 \qquad (9.1.1)$$

(momentum)
$$\frac{\partial \mathbf{u}}{\partial t} + \mathbf{u} \cdot \nabla \mathbf{u} = \nu \nabla^2 \mathbf{u} - \frac{1}{\rho} \nabla p \qquad (9.1.2)$$

(energy)
$$\frac{\partial T}{\partial t} + \mathbf{u} \cdot \nabla T = \alpha \nabla^2 T \qquad (9.1.3)$$

(species)
$$\frac{\partial \rho_A}{\partial t} + \mathbf{u} \cdot \nabla \rho_A = D_{AB} \nabla^2 \rho_A. \qquad (9.1.4)$$

These equations are a mixed set of elliptic/parabolic equations that contain the unknowns (one vector **u**, three scalars p, T, ρ_A). It is to be noted that temperature and solute concentration appear directly only in the energy and species equations, so that we do not need to couple these equations with the continuity and momentum equations if the fluid properties are assumed to be independent of temperature and species concentration. If we have solved **u**, the temperature and species distributions can be easily solved by the parabolic partial differential equations. With this in mind, let us focus our consideration on numerical schemes for solving the continuity and momentum equations.

9.2 Solving schemes

For the sake of simplicity, let us consider a two-dimensional fluid flow problem. The differential forms of the continuity and incompressible Navier–Stokes equations are given by

(Equation of Continuity)

$$\frac{\partial u_x}{\partial x} + \frac{\partial u_y}{\partial y} = 0 \qquad (9.2.1)$$

(Equation of Motion)

x-component:

$$\left\{\frac{\partial u_x}{\partial t} + u_x \frac{\partial u_x}{\partial x} + u_y \frac{\partial u_x}{\partial y}\right\} = \nu\left\{\frac{\partial^2 u_x}{\partial x^2} + \frac{\partial^2 u_x}{\partial y^2}\right\} - \frac{1}{\rho}\frac{\partial p}{\partial x} \qquad (9.2.2)$$

y-component:

$$\left\{\frac{\partial u_y}{\partial t} + u_x \frac{\partial u_y}{\partial x} + u_y \frac{\partial u_y}{\partial y}\right\} = \nu\left\{\frac{\partial^2 u_y}{\partial x^2} + \frac{\partial^2 u_y}{\partial y^2}\right\} - \frac{1}{\rho}\frac{\partial p}{\partial y}. \qquad (9.2.3)$$

It looks simple to solve the above equations because we have three equations, equations (9.2.1)–(9.2.3), with three unknown primitive variables (u_x, u_y and p). However, it is very complicated to solve the above equations because of the following reasons. (i) The convective terms in equations (9.2.2) and (9.2.3) contain non-linear quantities. (ii) Time derivatives of velocity but no time derivative of pressure appear in equations (9.2.2) and (9.2.3). Velocity components appear in the momentum and continuity equations, and pressure in both momentum equations, but there is no equation for pressure.

Let us now consider the solving schemes of incompressible flow problems, which are commonly encountered in materials processing. Methods for solving equations (9.2.1)–(9.2.3) can be classified into two groups: (i) vorticity-stream function approach and (ii) primitive-variable approach.

9.2.1 Vorticity-stream function approach

The vorticity-stream function approach has been one of the most popular methods for solving two-dimensional incompressible Navier–Stokes equations. In this approach, the primitive variables (the velocity components) are replaced by new variables such as the vorticity ω and the stream function ψ in order to eliminate the pressure p from the momentum equations.

Let us differentiate equation (9.2.3) and equation (9.2.2) with respect to x and y, respectively, and subtract one from the other. Then, we obtain the following equation.

$$\frac{\partial}{\partial t}\left(\frac{\partial u_y}{\partial x} - \frac{\partial u_x}{\partial y}\right) + u_x \frac{\partial}{\partial x}\left(\frac{\partial u_y}{\partial x} - \frac{\partial u_x}{\partial y}\right) + u_y \frac{\partial}{\partial y}\left(\frac{\partial u_y}{\partial x} - \frac{\partial u_x}{\partial y}\right)$$
$$= \nu\left\{\frac{\partial^2}{\partial x^2}\left(\frac{\partial u_y}{\partial x} - \frac{\partial u_x}{\partial y}\right) + \frac{\partial^2}{\partial y^2}\left(\frac{\partial u_y}{\partial x} - \frac{\partial u_x}{\partial y}\right)\right\}. \quad (9.2.4)$$

The vorticity ω for a two-dimensional rectangular coordinate system is defined as

$$\omega = \frac{\partial u_y}{\partial x} - \frac{\partial u_x}{\partial y}. \quad (9.2.5)$$

Substituting equation (9.2.5) into equation (9.2.4) gives

$$\left\{\frac{\partial \omega}{\partial t} + u_x \frac{\partial \omega}{\partial x} + u_y \frac{\partial \omega}{\partial y}\right\} = \nu\left\{\frac{\partial^2 \omega}{\partial x^2} + \frac{\partial^2 \omega}{\partial y^2}\right\}. \quad (9.2.6)$$

This parabolic partial differential equation is called the vorticity transport equation.

Also the stream function ψ is defined by the following equations, which satisfy the equation of continuity, equation (9.2.1).

$$\frac{\partial \psi}{\partial y} = u_x \quad (9.2.7a)$$

$$\frac{\partial \psi}{\partial x} = -u_y. \quad (9.2.7b)$$

By substituting equations (9.2.7a) and (9.2.7b) into equation (9.2.5), we obtain

$$\frac{\partial^2 \psi}{\partial x^2} + \frac{\partial^2 \psi}{\partial y^2} = -\omega. \quad (9.2.8)$$

This elliptic partial differential equation is the Poisson equation. As a result of the change of variables, the mixed elliptic/parabolic two-dimensional incompressible Navier–Stokes equations are separated into one parabolic equation (the vorticity transport equation) and one elliptic equation (the Poisson equation). The solving method based on equations (9.2.6) and (9.2.8) is called the vorticity-stream function approach.

The solving procedure of these equations is briefly summarized as follows:

1. Specify the initial values for ω and ψ at time level t ($t = 0$).
2. Solve the vorticity transport equation, equation (9.2.6), for new ω at time level $t + \Delta t$.
3. Solve the Poisson equation, equation (9.2.8), using new ω.
4. Evaluate the velocity components at time level $t + \Delta t$ using equation (9.2.7).
5. Repeat procedures 1–4 for the next calculation cycle.

However, the extension of the vorticity-stream function approach to a three-dimensional problem is complicated by the fact that a stream function does not exist for a truly three-dimensional flow [1].

9.2.2 Primitive variable approaches

Because of the limitation of extension to three-dimensional problems, the vorticity-stream function approach is not widely used in solving the incompressible flow. Consequently, two- and three-dimensional incompressible flows in most materials processing are most often solved using the primitive variables (u_x, u_y and p).

Methods for solving the incompressible Navier–Stokes equations in primitive variables can be categorized into two groups. The first approach is the artificial compressibility (pseudo-compressibility) method. This method is one of the early techniques proposed for solving the incompressible Navier–Stokes equations in primitive variable form [2]. In this method, the continuity equation is modified to include an artificial-time derivative of pressure (artificial compressibility term) that vanishes when the steady-state solution is reached. With the addition of this term to the continuity equation, the resulting Navier–Stokes equations are a mixed set of hyperbolic/parabolic equations, which can be solved using a standard time-dependent approach.

The second method is the pressure correction approach. This method is characterized by a formulation in which the momentum equations are solved sequentially for the velocity components using the best available estimate for the pressure distribution. Such a procedure will not yield a velocity field that satisfies the continuity equation unless the correct pressure distribution is employed. A Poisson equation for pressure derived using the continuity equation is used to determine the correct pressure, which will alter the velocity field in a direction such as to satisfy the continuity equation [1]. Pressure correction methods have been widely used for solving the incompressible Navier–Stokes equations. Some of the most widely used pressure-correction methods include the marker-and-cell (MAC) method [3], the simplified MAC (SMAC) method [4], the fractional-step method [5], the solution algorithm

(SOLA) method [6], the SIMPLE and SIMPLER methods [7], and the primitive-variable implicit split operator (PISO) method [8]. Some of the methods mentioned above will be briefly discussed here.

The continuity and Navier–Stokes equations are written here again to discuss the numerical methods.

(continuity)
$$\nabla \cdot \mathbf{u} = 0 \quad (9.2.9)$$

(momentum)
$$\frac{\partial \mathbf{u}}{\partial t} + \mathbf{u} \cdot \nabla \mathbf{u} = \nu \nabla^2 \mathbf{u} - \frac{1}{\rho} \nabla p. \quad (9.2.10)$$

We can solve equation (9.2.10) to obtain the velocity field by integrating it with respect to time. Three types of time integration can be considered as follows.

(fully explicit scheme)
$$\frac{\mathbf{u}^{n+1} - \mathbf{u}^n}{\Delta t} = -\mathbf{u}^n \cdot \nabla \mathbf{u}^n + \nu \nabla^2 \mathbf{u}^n - \frac{1}{\rho} \nabla p^n \quad (9.2.11)$$

(semi-implicit scheme)
$$\frac{\mathbf{u}^{n+1} - \mathbf{u}^n}{\Delta t} = -\mathbf{u}^n \cdot \nabla \mathbf{u}^n + \nu \nabla^2 \mathbf{u}^n - \frac{1}{\rho} \nabla p^{n+1} \quad (9.2.12)$$

(fully implicit scheme)
$$\frac{\mathbf{u}^{n+1} - \mathbf{u}^n}{\Delta t} = -\mathbf{u}^{n+1} \cdot \nabla \mathbf{u}^{n+1} + \nu \nabla^2 \mathbf{u}^{n+1} - \frac{1}{\rho} \nabla p^{n+1}. \quad (9.2.13)$$

Here the superscripts n and $n+1$ indicate the time levels at t and $t + \Delta t$, respectively. The velocity $\mathbf{u}^{n+1}$ at time level $t + \Delta t$ may be evaluated using one of the above equations. However, it is to be noted that p^{n+1} on the right-hand side of equations (9.2.12) and (9.2.13) are also unknown values.

Let us now consider the numerical methods to solve the above equations. First, when we adopt the fully explicit scheme given by equation (9.2.11), the velocity $\mathbf{u}^{n+1}$ at time level $t + \Delta t$ seems to be simply evaluated using $\mathbf{u}^n$ and p^n at time level t. In this scheme the calculated velocity field may not satisfy the continuity equation since the velocity after a time interval Δt is determined by the momentum equation only.

The second method often employed for solving the incompressible Navier–Stokes equations adopts the semi-implicit scheme given by equation (9.2.12). This scheme includes MAC, SMAC and SOLA methods, which have been developed in Los Alamos Scientific Laboratory.

The fully implicit scheme based on equation (9.2.13) includes the SIMPLE algorithm which stands for the Semi-Implicit Method for Pressure-Linked Equations.

9.2.2.1 Marker-and-cell (MAC) family of methods

(i) MAC method

The MAC (marker-and-cell) method is the earliest pressure-correction scheme for solving the incompressible Navier–Stokes equations introduced by Harlow and Welch in 1965 [9]. As mentioned above, in this scheme the velocity $\mathbf{u}^{n+1}$ at time level $t + \Delta t$ is calculated explicitly using the momentum equation, equation (9.2.12).

$$\frac{\mathbf{u}^{n+1} - \mathbf{u}^n}{\Delta t} = -\mathbf{u}^n \cdot \nabla \mathbf{u}^n + \nu \nabla^2 \mathbf{u}^n - \frac{1}{\rho} \nabla p^{n+1}. \tag{9.2.12}$$

But, p^{n+1} on the right-hand side of this equation at time level $t + \Delta t$ is also an unknown value to be determined. Without evaluating p^{n+1} we cannot carry out this procedure.

If we take divergence of equation (9.2.12), we obtain

$$\frac{\nabla \cdot (\mathbf{u}^{n+1} - \mathbf{u}^n)}{\Delta t} = -\nabla \cdot \{\mathbf{u}^n \cdot \nabla \mathbf{u}^n\} + \nabla \cdot (\nu \nabla^2 \mathbf{u}^n) - \frac{1}{\rho} \nabla^2 p^{n+1}. \tag{9.2.14}$$

Rearranging the above equation gives

$$\frac{\nabla \cdot \mathbf{u}^{n+1} - \nabla \cdot \mathbf{u}^n}{\Delta t} = -\nabla \cdot \{\mathbf{u}^n \cdot \nabla \mathbf{u}^n\} + \nabla^2(\nu \nabla \cdot \mathbf{u}^n) - \frac{1}{\rho} \nabla^2 p^{n+1}. \tag{9.2.15}$$

By substituting $D \equiv \nabla \cdot \mathbf{u}$, we obtain the following form of equation.

$$\frac{D^{n+1} - D^n}{\Delta t} = -\nabla \cdot \{\mathbf{u}^n \cdot \nabla \mathbf{u}^n\} + \nu \nabla^2 D^n - \frac{1}{\rho} \nabla^2 p^{n+1}. \tag{9.2.16}$$

According to the continuity equation, equation (9.2.9), D^{n+1} or D^n is assumed to be 0. However, it may not be easy to satisfy the above condition in numerical calculation. In order to satisfy the continuity equation at time level $t + \Delta t$, we assume that $D^{n+1} = 0$ and $D^n \neq 0$ because of some errors induced by numerical calculation in equation (9.2.16). Then, we obtain a Poisson equation for pressure as follow.

$$\frac{1}{\rho} \nabla^2 p^{n+1} = -\nabla \cdot \{\mathbf{u}^n \cdot \nabla \mathbf{u}^n\} + \nu \nabla^2 D^n + \frac{D^n}{\Delta t}. \tag{9.2.17}$$

Since all the variables on the right-hand side of equation (9.2.17) are known values at time level t, p^{n+1} at time level $t + \Delta t$ can be evaluated using equation (9.2.17). By substituting p^{n+1} into equation (9.2.12), the velocity $\mathbf{u}^{n+1}$ at time level $t + \Delta t$ can be calculated explicitly.

The solving procedure is briefly summarized as follows:

1. Specify the initial values for $\mathbf{u}^n$ at time level t ($t = 0$).
2. Evaluate p^{n+1} at time level $t + \Delta t$ by solving the Poisson equation, equation (9.2.17).

3. Evaluate $\mathbf{u}^{n+1}$ at time level $t + \Delta t$ by solving equation (9.2.12) explicitly.
4. Repeat procedures 1–3 for the next calculation cycle.

Since the MAC method takes the simple explicit scheme with regard to time, the time step Δt for iterative calculation should be restricted by the stability criterion, leading to an increase of computation time. In addition, a large computational time is also necessary for solving the Poisson equation for pressure.

(ii) SMAC method

The SMAC (simplified MAC) method is a simplified marker-and-cell method, which was developed in order to enhance the accuracy of the results at Los Alamos Scientific Laboratory [10].

In this approach the calculation procedure for solving equation (9.2.12) is divided into two steps with regard to time, in order to enhance the accuracy of the results and to simplify the Poisson equation for pressure.

$$\frac{\mathbf{u}^{n+1} - \mathbf{u}^n}{\Delta t} = -\mathbf{u}^n \cdot \nabla \mathbf{u}^n + \nu \nabla^2 \mathbf{u}^n - \frac{1}{\rho} \nabla p^{n+1}. \tag{9.2.12}$$

First, a tentative (provisional) velocity $\tilde{\mathbf{u}}$, which is defined in the middle stage between $\mathbf{u}^{n+1}$ and $\mathbf{u}^n$, is evaluated explicitly based on the present values of velocity and pressure as follows.

$$\frac{\tilde{\mathbf{u}} - \mathbf{u}^n}{\Delta t} = -\mathbf{u}^n \cdot \nabla \mathbf{u}^n + \nu \nabla^2 \mathbf{u}^n - \frac{1}{\rho} \nabla p^n. \tag{9.2.18}$$

By subtracting equation (9.2.18) from equation (9.2.12), we obtain

$$\frac{\mathbf{u}^{n+1} - \tilde{\mathbf{u}}}{\Delta t} = -\frac{1}{\rho} \nabla \{p^{n+1} - p^n\} = -\frac{1}{\rho} \nabla \delta p. \tag{9.2.19}$$

Here,

$$p^{n+1} = p^n + \delta p \tag{9.2.20}$$

where δp indicates the pressure change. However, the tentative velocity $\tilde{\mathbf{u}}$ obtained from equation (9.2.18) may not satisfy the continuity equation, equation (9.2.9). If we take divergence of equation (9.2.19) and assume that $D^{n+1} = 0$ in order to satisfy the continuity equation, we obtain a simple Poisson equation for pressure change δp as follows.

$$\nabla^2 \delta p = \frac{\rho \nabla \cdot \tilde{\mathbf{u}}}{\Delta t}. \tag{9.2.21}$$

The solving procedure is briefly summarized as follows:

1. Specify the initial values for $\mathbf{u}^n$ at time level t $(t = 0)$.
2. Evaluate the provisional velocity $\tilde{\mathbf{u}}$ by solving equation (9.2.18) explicitly.
3. Evaluate the pressure change δp by solving the Poisson equation, equation (9.2.21).

4. Evaluate the velocity $\mathbf{u}^{n+1}$ and the pressure p^{n+1} at time level $t + \Delta t$ by solving equations (9.2.19) and (9.2.20) explicitly.
5. Repeat procedures 1–4 for the next calculation cycle.

It is to be noted that in both the MAC and SMAC methods, the Poisson equation is to be solved for evaluating the pressure terms. However, the solving procedure in the SMAC method is much simplified compared to the MAC method, as can be seen from equations (9.2.17) and (9.2.21).

(iii) SOLA (solution algorithm) method

The SOLA (solution algorithm) method is a highly simplified MAC method developed at Los Alamos Scientific Laboratory [6].

Similar to the SMAC scheme, equation (9.2.12) is divided into two steps with regard to time as follows.

$$\frac{\tilde{\mathbf{u}} - \mathbf{u}^n}{\Delta t} = -\mathbf{u}^n \cdot \nabla \mathbf{u}^n + \nu \nabla^2 \mathbf{u}^n - \frac{1}{\rho} \nabla p^n \qquad (9.2.22)$$

$$\frac{\mathbf{u}^{n+1} - \tilde{\mathbf{u}}}{\Delta t} = -\frac{1}{\rho} \nabla \{p^{n+1} - p^n\} = -\frac{1}{\rho} \nabla \delta p. \qquad (9.2.23)$$

If we take divergence of equation (9.2.23) and assume that $D^{n+1} = 0$ in order to satisfy the continuity equation, we have

$$\nabla^2 \delta p = \frac{\rho \nabla \cdot \tilde{\mathbf{u}}}{\Delta t}. \qquad (9.2.24)$$

In the SMAC scheme, the pressure change δp required to make $D^{n+1} = 0$ is calculated according to the Poisson equation, equation (9.2.24). However, in the SOLA scheme, the pressure change δp is obtained by solving the above equation approximately. If we take the second-order accurate central-difference scheme for the second derivative of this equation, we have

$$\nabla^2 \delta p_{i,j} = \frac{\delta p_{i+1,j} - 2\delta p_{i,j} + \delta p_{i-1,j}}{\Delta x^2} + \frac{\delta p_{i,j+1} - 2\delta p_{i,j} + \delta p_{i,j-1}}{\Delta y^2}. \qquad (9.2.25)$$

If we remove all the terms except the diagonal term, $\delta p_{i,j}$, we obtain the following approximation.

$$\nabla^2 \delta p \approx -2 \left(\frac{1}{\Delta x^2} + \frac{1}{\Delta y^2} \right) \delta p. \qquad (9.2.26)$$

By substituting equation (9.2.26) into equation (9.2.24), we obtain a simple expression for the pressure change, δp.

$$\delta p \approx -\frac{\rho \nabla \cdot \tilde{\mathbf{u}}}{2\Delta t \left(\frac{1}{\Delta x^2} + \frac{1}{\Delta y^2} \right)}. \qquad (9.2.27)$$

Equation (9.2.27) indicates that how the pressure change, δp, varies according to the divergence ($\tilde{D} = \nabla \cdot \tilde{\mathbf{u}}$). If $\tilde{D} > 0$, δp should be negative, which means that fluid flows into the cell. Similarly, if $\tilde{D} < 0$, δp should be positive, which means that fluid flows out from the cell.

The velocity change corresponding to this pressure change can be evaluated using equation (9.2.23) as follow.

$$\delta \mathbf{u} = \mathbf{u}^{n+1} - \tilde{\mathbf{u}} = -\frac{\Delta t}{\rho} \nabla \delta p. \quad (9.2.28)$$

Now, we need to adjust the correct pressure and velocity values iteratively to satisfy the continuity equation since the pressure and velocity changes calculated according to equations (9.2.27) and (9.2.28) are approximate values.

$$\delta p^{(k)} \approx -\frac{\omega \rho \tilde{D}^{(k)}}{2 \Delta t \left(\frac{1}{\Delta x^2} + \frac{1}{\Delta y^2} \right)} \quad (9.2.29)$$

$$\left. \begin{array}{l} p^{(k+1)} \approx p^{(k)} + \delta p^{(k)} \\ \tilde{\mathbf{u}}^{(k+1)} \approx \tilde{\mathbf{u}}^{(k)} + \delta \mathbf{u}^{(k)} \end{array} \right\} \quad (k = 1, 2, 3, \ldots). \quad (9.2.30)$$

The superscript (k) indicates the values adjusted at the kth iterative calculation, and ω is the relaxation parameter ($1 < \omega < 2$).

The calculation steps involved in completing one calculation cycle are as follows:

1. Evaluate the tentative velocity field $\tilde{\mathbf{u}}$ by solving equation (9.2.22) explicitly, based on the present values of $\mathbf{u}^n$ and p^n at time level t. Treat the tentative velocity $\tilde{\mathbf{u}}^n$ and the pressure p^n as $\tilde{\mathbf{u}}^{(k)}$ and $p^{(k)}$ for an iterative evaluation of the velocity and pressure fields at time level $t + \Delta t$.
2. Evaluate the pressure change δp by equation (9.2.29)
3. Adjust the pressure and velocity values using equation (9.2.30) iteratively until the continuity equation is satisfied, i.e., $\tilde{D}^{(k+1)} = 0$.
4. When convergence has been achieved, convert the adjusted velocity and pressure fields $\mathbf{u}^{(k+1)}$ and $p^{(k+1)}$ into the new velocity and pressure fields $\mathbf{u}^{n+1}$ and p^{n+1} at time level $t + \Delta t$, which can be used as the new starting values for the next cycle.
5. Repeat procedures 1–4 for the next calculation cycle.

Different from the MAC and SMAC methods, there is no need to solve the Poisson equation in the SOLA method.

9.2.2.2 The SIMPLE method

The SIMPLE (semi-implicit method for pressure-linked equations) method, which is based on the control volume approach, was originally proposed by Patankar and Spalding [11]. This scheme, which is similar to the MAC method, is based on a guess-and-correct procedure to solve the governing equations. The velocity components are first calculated from the momentum equations using a guessed pressure field, and the pressures and velocities are then corrected, so as to satisfy continuity. This procedure continues until the continuity equation is satisfied. The main difference with the MAC method is in the way in which the pressure and velocity corrections are achieved.

The time integration for solving Navier–Stokes equations in the SIMPLE method takes the fully implicit scheme given by equation (9.2.13).

$$\frac{\mathbf{u}^{(n+1)} - \mathbf{u}^{(n)}}{\Delta t} = -\mathbf{u}^{(n+1)} \cdot \nabla \mathbf{u}^{(n+1)} + \nu \nabla^2 \mathbf{u}^{(n+1)} - \frac{1}{\rho} \nabla p^{(n+1)}. \quad (9.2.13)$$

If a guessed pressure field p^* is introduced to equation (9.2.13), we obtain the following equation which can be solved to calculate an estimated velocity field $\mathbf{u}^*$.

$$\frac{\mathbf{u}^* - \mathbf{u}^{(n)}}{\Delta t} = -(\mathbf{u}^* \cdot \nabla)\mathbf{u}^* + \nu \nabla^2 \mathbf{u}^* - \frac{1}{\rho} \nabla p^*. \quad (9.2.31)$$

However, the velocity field $\mathbf{u}^*$ computed according to equation (9.2.31), which is a guessed velocity field for $\mathbf{u}^{(n+1)}$, will not in general satisfy continuity, i.e., $\nabla \cdot \mathbf{u}^* \neq 0$. It is therefore necessary to improve the guessed pressure p^* such that the estimated velocity field $\mathbf{u}^*$ will progressively get closer to satisfying the continuity equation. Let us define the pressure correction δp as the difference between the correct pressure field and the guessed pressure field.

$$p = p^* + \delta p. \quad (9.2.32)$$

Likewise, the correct velocity field $\mathbf{u}$ can be written as

$$\mathbf{u} = \mathbf{u}^* + \delta \mathbf{u} \quad (9.2.33)$$

where $\mathbf{u}^*$ is the estimated velocity vector and $\delta \mathbf{u}$ is the velocity correction. The pressure correction is related to the velocity correction by an approximate form of the momentum equation as follows.

$$\frac{\partial \delta \mathbf{u}}{\partial t} = -\frac{1}{\rho} \nabla \delta p. \quad (9.2.34)$$

Since the velocity correction can be assumed to be zero at the previous iteration step, the above equation can be expressed as

$$\delta \mathbf{u} = -\frac{\Delta t}{\rho} \nabla \delta p. \quad (9.2.35)$$

By substituting equation (9.2.33) into the continuity equation, equation (9.2.9), we obtain the following Poisson equation about the pressure correction δp.

$$\nabla \cdot \mathbf{u} = \nabla \cdot \mathbf{u}^* - \frac{\Delta t}{\rho} \nabla^2 \delta p = 0$$

or

$$\nabla^2 \delta p = \frac{\rho \nabla \cdot \mathbf{u}^*}{\Delta t}. \qquad (9.2.36)$$

This Poisson equation, which is called the pressure-correction equation, can be solved for the pressure correction. If the estimated velocity vector satisfies continuity at every point, the pressure correction becomes zero at every point.

The solving procedure is briefly summarized as follows:

1. Guess the pressure field p^*.
2. Solve the momentum equations, equation (9.2.31), to obtain the estimated velocity field $\mathbf{u}^*$.
3. Solve the pressure-correction equation, equation (9.2.36), to obtain δp.
4. Correct the pressure using equation (9.2.32).
5. Correct the velocity using equations (9.2.33) and (9.2.35).
6. Treat the corrected pressure p as a new guessed pressure p^* for the next calculation cycle, and return to step 2. Repeat the whole procedure until the solution converges.

As described in this section, the SIMPLE algorithm needs a lot of iterative calculation for obtaining a converged solution since it starts with a guessed pressure field and needs pressure and velocity corrections until a converged solution is obtained. A large number of variations of the SIMPLE scheme have been proposed for the purpose of improving the convergence rate of the scheme, such as SIMPLER, SIMPLEC and PISO. The SIMPLER (SIMPLE Revised) algorithm [7] starts with a guessed velocity field, by which the pressure field can be predicted, resulting in a great decrease of iterative calculation. The SIMPLEC (SIMPLE Consistent) algorithm [12] attempts to approximate the effects of some terms in the momentum equations neglected in the SIMPLE algorithm for δp. The PISO (pressure implicit with splitting of operators) algorithm [13] is a pressure–velocity calculation procedure developed originally for the non-iterative computation of unsteady compressible flows.

9.3 Summary

As described in the previous section, many solution algorithms have been developed and used for modelling fluid flow problems. Each algorithm has its own benefits and is good for simulating a specific type of flow. The vorticity-stream function approach makes use of the vorticity transport equation and the stream function. In the process of deriving the vorticity transport equation, the pressure gradient terms in the momentum equations are eliminated. In cases of two-dimensional flows, only two variables, i.e., vorticity and stream function, need to be solved instead of the velocity and pressure components in the x and y directions. But, since the vorticity in each direction has to be defined together with the velocity potentials, this method becomes much more complicated than the primitive approach. That is the reason why the vorticity-stream function approach is limited to two-dimensional problems.

The MAC, SMAC and SOLA-VOF methods were developed for solving the fluid flow problems. These solving algorithms are based on the explicit scheme which is very useful for the free surface tracking. Especially, the SOLA algorithm coupled with the VOF method has been widely used in the fields of mould filling simulation. In general, the casting process consists of not only mould filling processes, but also solidification processes. When the fluid material is an alloy having a temperature range of solidification, which is very common in practice, flow through the mushy zone consisting of the mixture of solid and liquid phases must be considered. In particular, for semi-solid metal processes such as thixo-forming and rheo-casting, semi-solid state materials, i.e., a mushy state, will be the fluid materials, which do not follow the Newtonian flow characteristics. The application of the explicit scheme might be limited for the simulation of fluid flow in a mushy state due to the complicating phenomena, such as non-Newtonian behavior, double diffusive convection, and damping effect, etc.

The SIMPLE algorithm is an implicit scheme treating the pressure gradient through combining the momentum and continuity equations, and is well known as one of the most accurate algorithms for solving fluid flow problems including two-phase flows. The SIMPLE method is systematically well-organized to adopt various coordinate systems including BFC (Body Fitted Coordinates) and TDC (Time Dependent Coordinates). However, the SIMPLE method has a significant limitation. It takes a large computational time to solve simultaneous equations iteratively when we simulate transient fluid flow problems, such as mould filling with free surfaces. This is due to the fact that the whole algorithm should follow the explicit time steps caused by the free surface motion. This is the reason why the SIMPLE algorithm is hardly applied to the free surface tracking problems. The calculation speed of the SIMPLE algorithm can be improved dramatically, when the Gauss–Seidel method is adopted instead of the line-by-line method based on the TDMA [14], though it is still slower than that of the SOLA-VOF method. Considering the complex casting processes including various physical phenomena and complicated geometry, the SIMPLE algorithm will be widely used in modelling of materials processing in the near future, since it guarantees the effective and accurate calculations in spite of longer calculation time.

References

[1] Tannehill J C, Anderson D A and Pletcher R H 1997 *Computational Fluid Mechanics and Heat Transfer* (Taylor & Francis) p 657
[2] Chorin A J 1967 *J. Comput. Phys.* **2** 12
[3] Harlow F H and Welch J E 1965 *Phys. Fluids* **8** 2182
[4] Amsden A A and Harlow F H 1970 *J. Comput. Phys.* **6** 322
[5] Chorin A J 1968 *Math. Comput.* **22** 745
[6] Hirt C W, Nichols B D and Romero N C 1975 *SOLA-A Numerical Solution Algorithm for Transient Fluid Flows* (Los Alamos, New Mexico: Los Alamos Scientific Lab Report LA-5852)
[7] Patankar S V 1980 *Numerical Heat Transfer and Fluid Flow* (New York: Hemisphere Publishing Co, Taylor & Francis Group)
[8] Issa R I 1986 *J. Comput. Phys.* **62** 40
[9] Harlow F H and Welch J E 1965 *Phys. Fluids* **8** 2182
[10] Amsden A A and Harlow F H 1970 *J. Comput. Phys.* **6** 322
[11] Patankar S V and Spalding D B 1972 *Int. J. Heat Mass Transfer* **15** 1787
[12] Van Doormal J P and Raithby G D 1984 *Numer. Heat Transfer* **7** 147
[13] Issa R I 1986 *J. Comput. Phys.* **62** 40
[14] Hong C P, Lee S Y and Song K 2001 *ISIJ International* **41**(9) 999

Chapter 10

Fluid flow analysis using the SIMPLE method based on the Cartesian coordinate system

In chapter 9, we introduced fundamentals of solution algorithms for the calculation of fluid flow fields. These methods include (i) the vorticity-stream function approach and (ii) the primitive variable approaches such as the MAC, SMAC, SOLA and SIMPLE methods. Among these numerical methods the applications of the SIMPLE method to fluid flow modelling has been increasingly focused on in many materials processing problems. Furthermore, the SIMPLE method has much more potential to be extended for the modelling of various materials processing than other methods. In this chapter we will describe the SIMPLE method for the numerical modelling of incompressible fluid flow problems.

10.1 Governing equations

The governing equations for incompressible fluid flow are the continuity equation and the Navier–Stokes equation, rewritten here as follows.
Continuity equation can be expressed in two different forms.
(integral form)

$$\iint_\Gamma (\rho \mathbf{u}) \cdot \mathbf{n}\, d\Gamma = 0 \qquad (10.1.1a)$$

(differential form)

$$\nabla \cdot (\rho \mathbf{u}) = 0. \qquad (10.1.1b)$$

The momentum balance equation can also be expressed in two different forms.
(integral form)

$$\frac{\partial}{\partial t}\iiint_\Omega (\rho \mathbf{u})\, d\Omega + \iint_\Gamma (\rho \mathbf{u}\mathbf{u}) \cdot \mathbf{n}\, d\Gamma = -\iint_\Gamma (\tau \cdot \mathbf{n})\, d\Gamma + \iiint_\Omega (\mathbf{f_b} - \nabla p)\, d\Omega \qquad (10.1.2a)$$

(differential form)
$$\frac{\partial(\rho\mathbf{u})}{\partial t} + \nabla \cdot (\rho\mathbf{u}\mathbf{u}) = -\nabla \cdot \tau + (\mathbf{f_b} - \nabla p). \tag{10.1.2b}$$

The continuity and Navier–Stokes equations, equations (10.1.1) and (10.1.2) can be summarized as the general forms of the integral and differential equations as follows:

(integral form)
$$\frac{\partial}{\partial t}\iiint_\Omega (\rho\Phi)\,d\Omega + \iint_\Gamma (\rho\mathbf{u}\Phi)\cdot\mathbf{n}\,d\Gamma = -\iint_\Gamma (\mathbf{f}_\Phi\cdot\mathbf{n})\,d\Gamma + \iiint_\Omega S\,d\Omega \tag{10.1.3a}$$

(differential form)
$$\frac{\partial(\rho\Phi)}{\partial t} + \nabla\cdot(\rho\mathbf{u}\Phi) = -\nabla\cdot\mathbf{f}_\Phi + S. \tag{10.1.3b}$$

The transport property Φ indicates $\mathbf{u}$, $C_v T$ or ω_A and $\mathbf{f}_\Phi$ is τ, $\mathbf{q}$ or $\mathbf{j}_A$ for momentum, heat or species transfer, respectively. The definitions of the parameters Φ, $\mathbf{f}_\Phi$ and S are given in table 3.2. The continuity equation is simply obtained by replacing $\Phi = 1$, $\mathbf{f}_\Phi = 0$ and $S = 0$ in equation (10.1.3).

In the above equations the diffusion flux term $(\nabla\cdot\mathbf{f}_\Phi)$ due to the gradient of the general transport property Φ would represent viscous stress, heat flux or species flux, and is expressed by $-K\nabla\Phi$. Since the integral form of the governing equation is the starting point of the finite volume method as described in chapter 4, from now on we will consider the following integral form of the general transport equation

$$\underbrace{\frac{\partial}{\partial t}\iiint_\Omega (\rho\Phi)\,d\Omega}_{(1)} + \underbrace{\iint_\Gamma (\rho\mathbf{u}\Phi)\cdot\mathbf{n}\,d\Gamma}_{(2)} = \underbrace{\iint_\Gamma (K\nabla\Phi)\cdot\mathbf{n}\,d\Gamma}_{(3)} + \underbrace{\iiint_\Omega S\,d\Omega}_{(4)} \tag{10.1.4}$$

where K is the diffusion coefficient. Term (1) on the left-hand side of equation (10.1.4), which is an integral over the control volume, represents the rate of change of transport property Φ in the control volume (Ω), term (2) and (3), the integrals over the faces of the control volume (Γ), indicate the net fluxes of Φ into the control volume by convection and diffusion through the control-volume faces, and term (4), an integral over the control volume, represents the source term within the control volume, respectively.

In steady-state problems the rate of change term in equation (10.1.4) is equal to zero. This leads to the integral form of the steady transport equation.

$$\iint_\Gamma (\rho\mathbf{u}\Phi)\cdot\mathbf{n}\,d\Gamma = \iint_\Gamma (K\nabla\Phi)\cdot\mathbf{n}\,d\Gamma + \iiint_\Omega S\,d\Omega. \tag{10.1.5}$$

150   *Fluid flow analysis based on the Cartesian coordinate system*

In transient problems it is also necessary to integrate with respect to time over a small time interval Δt from t to $t + \Delta t$. This yields the general integrated form of the transport equation as follows.

$$\int_{\Delta t} \frac{\partial}{\partial t} \left\{ \iiint_\Omega (\rho \Phi) \, d\Omega \right\} dt + \int_{\Delta t} \iint_\Gamma (\rho \mathbf{u} \Phi) \cdot \mathbf{n} \, d\Gamma \, dt$$
$$= \int_{\Delta t} \iint_\Gamma (K \nabla \Phi) \cdot \mathbf{n} \, d\Gamma \, dt + \int_{\Delta t} \iiint_\Omega S \, d\Omega \, dt. \quad (10.1.6)$$

10.2 Staggered and non-staggered grids

When we begin to derive the discretized form of the governing equations, we need to construct grids, which fill the domain of interest. In the evaluation of the integral form of governing equations, such as equations (10.1.1) and (10.1.2), over the control volumes, we need to evaluate the mass flow rates across the control-volume faces. The estimation of the mass flow rates across the control-volume faces is dependent upon the type of grids. Here we describe two grid systems, which are generally used in the modelling of fluid flow problems: (i) the non-staggered grid and (ii) the staggered grid.

In the non-staggered grid the velocity components together with other scalar variables such as pressure, temperature and species concentration (p, T, ρ_A) are calculated for the nodal points that lie at the centers of the control volumes, as shown in figure 10.1.

First consider the discretization of momentum equations such as equation (10.1.2) in which the pressure gradient is included as a source term. If we assume that a highly non-uniform 'checker-board' pressure field with a uniform grid is given as shown in figure 10.2, the pressure gradient terms $\partial p / \partial x$ and $\partial p / \partial y$ in the u- and v-momentum equations can be evaluated by linear interpolation as follows.

$$\frac{\partial}{\partial x} = \frac{p_{\Gamma_{i+1/2,j}} - p_{\Gamma_{i-1/2,j}}}{\Delta x} = \frac{(p_{i+1} + p_i)/2 - (p_i + p_{i-1})/2}{\Delta x}$$
$$= \frac{p_{i+1} - p_{i-1}}{2\Delta x} = 0. \quad (10.2.1a)$$

Similarly,

$$\frac{\partial p}{\partial y} = \frac{p_{j+1} - p_{j-1}}{2\Delta y} = 0. \quad (10.2.1b)$$

As seen from the above relations the pressure p_i at the node (i) has no role to play. As a result, a highly non-uniform pressure field would be treated as a uniform pressure field, resulting in the same (zero) momentum source in the x or y direction. This is physically non-realistic.

Let us now consider the discretization of the continuity equation, equation (10.1.1) in which the velocity gradient term is included. If we

Staggered and non-staggered grids   151

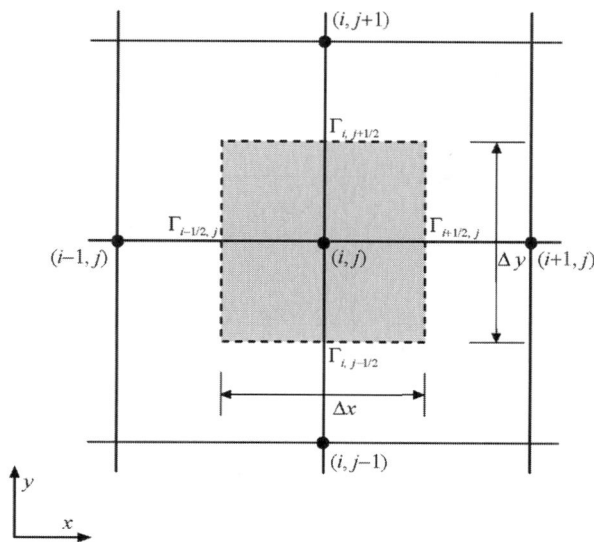

Figure 10.1. Non-staggered grids.

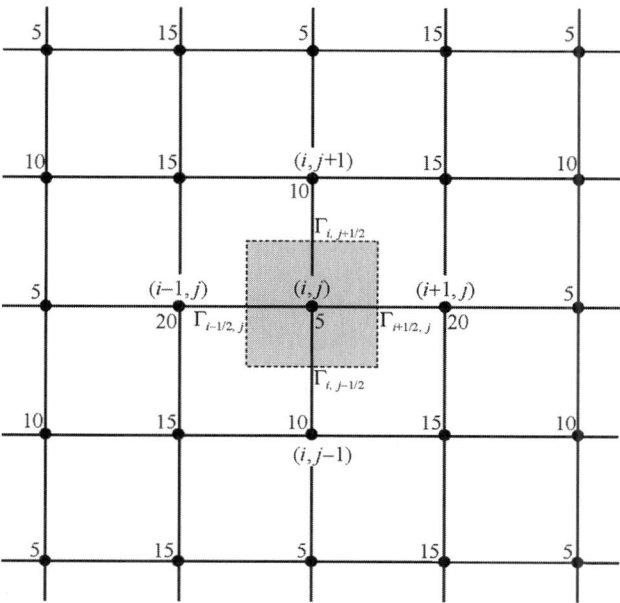

Figure 10.2. Checker-board pressure fields.

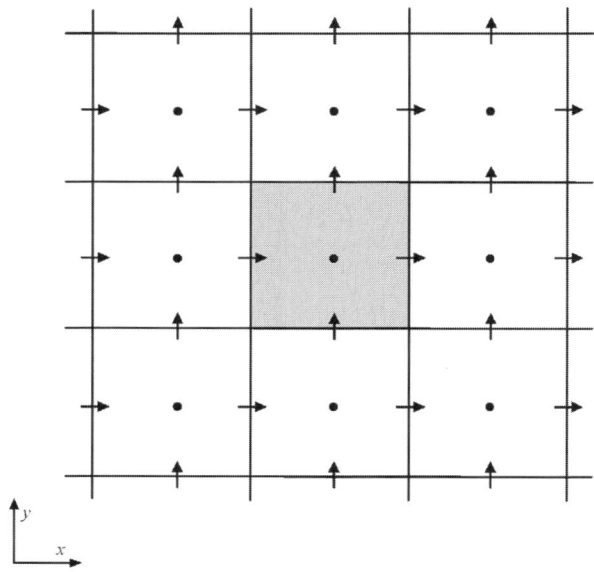

Figure 10.3. Staggered locations for u and v. $\rightarrow = u$; $\uparrow = v$; $\bullet$ = other variables.

assume that a highly non-uniform 'checker-board' velocity field with a uniform grid is also somehow given similar to figure 10.2, a similar kind of problem will arise [1]. Even though the velocity field, which exhibits special oscillations, is not realistic, the continuity condition is satisfied. There are two kinds of method to solve this problem: (i) to use the momentum interpolation [2] to evaluate the velocity field on the control-volume faces and (ii) to use a staggered grid system used in the MAC method [3]. This method has been widely used in other methods, including the SIMPLE procedure [4]. In the staggered grids all scalar variables (p, T, ρ_A) are calculated at the centers of the control volumes, while the velocity components are calculated for the points that lie on the control-volume faces, as shown in figure 10.3. In this case the mass flow rates across the control-volume faces can be evaluated without any interpolation for the relevant velocity components. However, a computer program based on the staggered grids must carry all the indexing and geometric information about the locations of the velocity components and must perform certain rather tiresome interpolations [1].

10.3 Discretization method

10.3.1 Discretization of the integral form of transport equation

In chapter 8, we introduced the typical discretization schemes for convection/diffusion problems. In this section we describe how the general form of transport equation, equation (10.1.6), can be cast into a finite-difference form. Then, it will be simple to discretize the continuity and Navier–Stokes equations using this basic concept. The integral form of the transport equation, which is the starting point for the derivation of finite difference equations in the finite volume approach, is rewritten here as

$$\frac{1}{\Delta t}\int_t^{t+\Delta t}\frac{\partial}{\partial t}\left\{\iiint_\Omega(\rho\Phi)\,d\Omega\right\}dt + \frac{1}{\Delta t}\int_t^{t+\Delta t}\iint_\Gamma(\rho\mathbf{u}\Phi - K\nabla\Phi)\cdot\mathbf{n}\,d\Gamma\,dt$$

$$= \frac{1}{\Delta t}\int_t^{t+\Delta t}\iiint_\Omega S\,d\Omega\,dt. \qquad (10.3.1)$$

For a two-dimensional problem, equation (10.3.1) can be expressed as

$$\frac{1}{\Delta t}\int_t^{t+\Delta t}\frac{\partial}{\partial t}\left\{\iint_\Omega(\rho\Phi)\,dx\,dy\right\}dt$$
(1)

$$+\frac{1}{\Delta t}\int_t^{t+\Delta t}\left\{\int_{\Gamma_{i-1/2}}^{\Gamma_{i+1/2}}\left(\rho u\Phi - K\frac{\partial\Phi}{\partial x}\right)dy + \int_{\Gamma_{j-1/2}}^{\Gamma_{j+1/2}}\left(\rho v\Phi - K\frac{\partial\Phi}{\partial y}\right)dx\right\}dt$$
(2)

$$= \frac{1}{\Delta t}\int_t^{t+\Delta t}\iint_\Omega S\,dx\,dy\,dt. \qquad (10.3.2)$$
(3)

In order to integrate equation (10.3.2), we consider two-dimensional control volumes shown in figure 10.4. In the above equation, $\Gamma_{i\pm 1/2}$ and

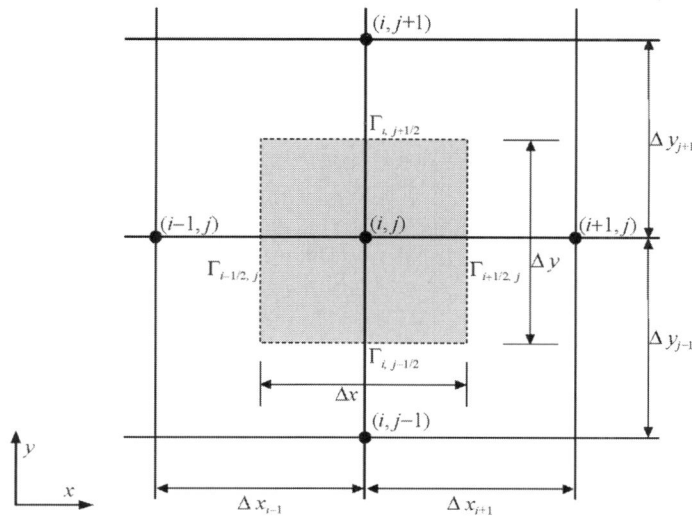

Figure 10.4. Control volume for a transport property Φ.

$\Gamma_{j\pm 1/2}$ indicate the control-volume faces $\Gamma_{i\pm 1/2,j}$ and $\Gamma_{i,j\pm 1/2}$, respectively. This abbreviation will prevail throughout chapters 10 and 11. It is to be noted that equation (10.3.2) is valid for control volumes of any shapes. By applying the conservation law, i.e., the integral form of transport equation, to a fixed region in a space known as the control volumes, we can develop the finite difference equations for two-dimensional problems.

Let us now evaluate the terms in equation (10.3.2) by applying this equation to the control volumes in figure 10.4. If the transport property Φ is assumed to prevail over the control volume (i,j), term (1) can be evaluated as

$$\frac{1}{\Delta t}\int_t^{t+\Delta t}\frac{\partial}{\partial t}\left\{\iint_\Omega (\rho\Phi)\,dx\,dy\right\}dt \equiv \frac{1}{\Delta t}\iint_\Omega\left\{\int_t^{t+\Delta t}\frac{\partial}{\partial t}(\rho\Phi)\,dt\right\}dx\,dy$$

$$= \frac{1}{\Delta t}\iint_\Omega \rho(\Phi^{t+\Delta t}-\Phi^t)\,dx\,dy$$

$$= \frac{\rho}{\Delta t}(\Phi_{i,j}^{t+\Delta t}-\Phi_{i,j}^t)\Delta x\,\Delta y$$

$$= \frac{\rho}{\Delta t}(\Phi_{i,j}^{n+1}-\Phi_{i,j}^n)\Delta x\,\Delta y. \qquad (10.3.3)$$

Let us now consider the integral of term (2) in equation (10.3.2), which represents the total fluxes into the control volume (i,j) through its control-volume faces by convection and diffusion. The convection and diffusion terms can be discretized using the similar schemes described in the previous chapters. If we take the fully implicit scheme for the time integration and the central difference scheme for the spatial derivatives in a uniform grid with a constant flow velocity u, then we obtain

$$\frac{1}{\Delta t}\int_t^{t+\Delta t}\left\{\int_{\Gamma_{i-1/2}}^{\Gamma_{i+1/2}}\left(\rho u\Phi - K\frac{\partial\Phi}{\partial x}\right)dy + \int_{\Gamma_{j-1/2}}^{\Gamma_{j+1/2}}\left(\rho v\Phi - K\frac{\partial\Phi}{\partial y}\right)dx\right\}dt$$

$$= \left\{(\rho u_{\Gamma_{i+1/2}})\Phi\Big|_{\Gamma_{i+1/2}}^{n+1} - K\frac{\Phi_{i+1,j}^{n+1}-\Phi_{i,j}^{n+1}}{\Delta x_{i+1}}\right\}\Delta y$$

$$- \left\{(\rho u_{\Gamma_{i-1/2}})\Phi\Big|_{\Gamma_{i-1/2}}^{n+1} - K\frac{\Phi_{i,j}^{n+1}-\Phi_{i-1,j}^{n+1}}{\Delta x_{i-1}}\right\}\Delta y$$

$$+ \left\{(\rho v_{\Gamma_{j+1/2}})\Phi\Big|_{\Gamma_{j+1/2}}^{n+1} - K\frac{\Phi_{i,j+1}^{n+1}-\Phi_{i,j}^{n+1}}{\Delta y_{j+1}}\right\}\Delta x$$

$$- \left\{(\rho v_{\Gamma_{j-1/2}})\Phi\Big|_{\Gamma_{j-1/2}}^{n+1} - K\frac{\Phi_{i,j}^{n+1}-\Phi_{i,j-1}^{n+1}}{\Delta y_{j-1}}\right\}\Delta x. \qquad (10.3.4)$$

The potential values $\Phi|_{\Gamma_{i\pm 1/2}}^{n+1}$ and $\Phi|_{\Gamma_{j\pm 1/2}}^{n+1}$ on the control-volume faces $\Gamma_{i\pm 1/2}$ and $\Gamma_{j\pm 1/2}$ can be evaluated by taking the simple arithmetic mean

Discretization method 155

between the adjacent nodal values using equation (4.2.28), as described in chapter 4.

$$\Phi\Big|^t_{\Gamma_{i\pm 1/2}} = \tfrac{1}{2}(\Phi^t_{i\pm 1,j} + \Phi^t_{i,j}) = \tfrac{1}{2}(\Phi^n_{i\pm 1,j} + \Phi^n_{i,j}) \qquad (10.3.5a)$$

$$\Phi\Big|^t_{\Gamma_{j\pm 1/2}} = \tfrac{1}{2}(\Phi^t_{i,j\pm 1} + \Phi^t_{i,j}) = \tfrac{1}{2}(\Phi^n_{i,j\pm 1} + \Phi^n_{i,j}). \qquad (10.3.5b)$$

Then, equation (10.3.4) can be expressed as

$$\frac{1}{\Delta t}\int_t^{t+\Delta t}\left\{\int_{\Gamma_{i-1/2}}^{\Gamma_{i+1/2}}\left(\rho u\Phi - K\frac{\partial \Phi}{\partial x}\right)dy + \int_{\Gamma_{j-1/2}}^{\Gamma_{j+1/2}}\left(\rho v\Phi - K\frac{\partial \Phi}{\partial y}\right)dx\right\}dt$$

$$= \left\{(\rho u_{\Gamma_{i+1/2}})\frac{\Phi^{n+1}_{i+1,j}+\Phi^{n+1}_{i,j}}{2} - K\frac{\Phi^{n+1}_{i+1,j}-\Phi^{n+1}_{i,j}}{\Delta x_{i+1}}\right\}\Delta y$$

$$- \left\{(\rho u_{\Gamma_{i-1/2}})\frac{\Phi^{n+1}_{i-1,j}+\Phi^{n+1}_{i,j}}{2} - K\frac{\Phi^{n+1}_{i,j}-\Phi^{n+1}_{i-1,j}}{\Delta x_{i-1}}\right\}\Delta y$$

$$+ \left\{(\rho v_{\Gamma_{j+1/2}})\frac{\Phi^{n+1}_{i,j+1}+\Phi^{n+1}_{i,j}}{2} - K\frac{\Phi^{n+1}_{i,j+1}-\Phi^{n+1}_{i,j}}{\Delta y_{j+1}}\right\}\Delta x$$

$$- \left\{(\rho v_{\Gamma_{j-1/2}})\frac{\Phi^t_{i,j-1}+\Phi^t_{i,j}}{2} - K\frac{\Phi^{n+1}_{i,j}-\Phi^{n+1}_{i,j-1}}{\Delta y_{j-1}}\right\}\Delta x. \qquad (10.3.6)$$

Finally, term (3) in equation (10.3.2), which represents the source term, can be integrated as

$$\frac{1}{\Delta t}\int_t^{t+\Delta t}\iint_\Omega S\,dx\,dy\,dt = S\,\Delta x\,\Delta y. \qquad (10.3.7)$$

In the SIMPLE method the source term is evaluated by the concept of the linearization [1].

By substituting equations (10.3.3), (10.3.6) and (10.3.7) into equation (10.3.2) and rearranging them, we obtain the following discretized equation, similar to the discretized equations mentioned in chapter 8.

$$a_{i,j}\Phi^{n+1}_{i,j} = a_{i+1,j}\Phi^{n+1}_{i+1,j} + a_{i-1,j}\Phi^{n+1}_{i-1,j} + a_{i,j+1}\Phi^{n+1}_{i,j+1} + a_{i,j-1}\Phi^{n+1}_{i,j-1} + b_{i,j}. \qquad (10.3.8)$$

Here the coefficients of the above equation are given by

$$a_{i+1,j} = \left(D_{\Gamma_{i+1}} - \frac{F_{\Gamma_{i+1/2}}}{2}\right)\Delta y \qquad (10.3.9a)$$

$$a_{i-1,j} = \left(D_{\Gamma_{i-1}} + \frac{F_{\Gamma_{i-1/2}}}{2}\right)\Delta y \qquad (10.3.9b)$$

$$a_{i,j+1} = \left(D_{\Gamma_{j+1}} - \frac{F_{\Gamma_{j+1/2}}}{2}\right)\Delta x \tag{10.3.9c}$$

$$a_{i,j-1} = \left(D_{\Gamma_{j-1}} + \frac{F_{\Gamma_{j-1/2}}}{2}\right)\Delta x \tag{10.3.9d}$$

$$a_{i,j} = a_{i+1,j} + a_{i-1,j} + a_{i,j+1} + a_{i,j-1}$$
$$+ (F_{\Gamma_{i+1/2}}\Delta y - F_{\Gamma_{i-1/2}}\Delta y + F_{\Gamma_{j+1/2}}\Delta x - F_{\Gamma_{j-1/2}}\Delta x) + \rho\frac{\Delta x \Delta y}{\Delta t}$$
$$\tag{10.3.10a}$$

$$b_{i,j} = S\Delta x \Delta y + \rho\frac{\Delta x \Delta y}{\Delta t}\Phi_{i,j}^n \tag{10.3.10b}$$

$$F_{\Gamma_{i+1/2}} \equiv \rho u_{\Gamma_{i+1/2}} \qquad D_{\Gamma_{i+1/2}} \equiv \frac{K_{\Gamma_{i+1/2}}}{\Delta x_{i+1}} \tag{10.3.11a}$$

$$F_{\Gamma_{i-1/2}} \equiv \rho u_{\Gamma_{i-1/2}} \qquad D_{\Gamma_{i-1/2}} \equiv \frac{K_{\Gamma_{i-1/2}}}{\Delta x_{i-1}} \tag{10.3.11b}$$

$$F_{\Gamma_{j+1/2}} \equiv \rho v_{\Gamma_{j+1/2}} \qquad D_{\Gamma_{j+1/2}} \equiv \frac{K_{\Gamma_{j+1/2}}}{\Delta y_{j+1}} \tag{10.3.11c}$$

$$F_{\Gamma_{j-1/2}} \equiv \rho v_{\Gamma_{j-1/2}} \qquad D_{\Gamma_{j-1/2}} \equiv \frac{K_{\Gamma_{j-1/2}}}{\Delta y_{j-1}}. \tag{10.3.11d}$$

In the above relation the term $(F_{\Gamma_{i+1/2}}\Delta y - F_{\Gamma_{i-1/2}}\Delta y + F_{\Gamma_{j+1/2}}\Delta x - F_{\Gamma_{j-1/2}}\Delta x)$ in $a_{i,j}$ is zero when the flow field satisfies continuity.

In this section we employed the central difference scheme to calculate the cell face values for the sake of simplicity. As described in chapter 8, the neighbor coefficients for various difference schemes are summarized in table 10.1.

10.3.2 Discretization of momentum equations

In the previous section we described the discretization of the integral form of the general transport equation and obtained the general form of discretized equation, equation (10.3.8). If the pressure field is given, the discretization of the momentum equations can be obtained by employing the similar solution procedure for the general variable Φ. In the momentum equations, Φ represents the relevant velocity components, and K and S are to be given their appropriate meanings. As mentioned earlier, the SIMPLE method adopts the staggered grids in which all scalar variables (p, T, ρ_A) are calculated at the centers of the control volumes, and the velocity components are calculated for the points that lie on the faces of the control volumes. The discretized momentum equations are somewhat different from the discretized equations for the general variable Φ.

Table 10.1. Difference schemes for diffusion and convection terms.

Scheme	Central difference	Upwind difference	Hybrid difference	Power-law
$a_{i-1,j}$	$D_{\Gamma_{i-1/2,j}} + \dfrac{F_{\Gamma_{i-1/2,j}}}{2}$	$D_{\Gamma_{i-1/2,j}} + \max\left\|F_{\Gamma_{i-1/2,j}}, 0\right\|$	$\max\left\|\left(D_{\Gamma_{i-1/2,j}} + \dfrac{F_{\Gamma_{i-1/2,j}}}{2}\right), F_{\Gamma_{i-1/2,j}}, 0\right\|$	$\max\left\|(\beta D_{\Gamma_{i-1/2,j}}), (\beta D_{\Gamma_{i-1/2,j}} + F_{\Gamma_{i-1/2,j}}), F_{\Gamma_{i-1/2,j}}, 0\right\|$ where $\beta = (1 - 0.1 P_e)^5$
$a_{i+1,j}$	$D_{\Gamma_{i+1/2,j}} + \dfrac{F_{\Gamma_{i+1/2,j}}}{2}$	$D_{\Gamma_{i+1/2,j}} + \max\left\|F_{\Gamma_{i+1/2,j}}, 0\right\|$	$\max\left\|\left(D_{\Gamma_{i+1/2,j}} + \dfrac{F_{\Gamma_{i+1/2,j}}}{2}\right), F_{\Gamma_{i+1/2,j}}, 0\right\|$	$\max\left\|(\beta D_{\Gamma_{i+1/2,j}}), (\beta D_{\Gamma_{i+1/2,j}} + F_{\Gamma_{i+1/2,j}}), F_{\Gamma_{i+1/2,j}}, 0\right\|$ where $\beta = (1 - 0.1 P_e)^5$
$a_{i,j-1}$	$D_{\Gamma_{i,j-1/2}} + \dfrac{F_{\Gamma_{i,j-1/2}}}{2}$	$D_{\Gamma_{i,j-1/2}} + \max\left\|F_{\Gamma_{i,j-1/2}}, 0\right\|$	$\max\left\|\left(D_{\Gamma_{i,j-1/2}} + \dfrac{F_{\Gamma_{i,j-1/2}}}{2}\right), F_{\Gamma_{i,j-1/2}}, 0\right\|$	$\max\left\|(\beta D_{\Gamma_{i,j-1/2}}), (\beta D_{\Gamma_{i,j-1/2}} + F_{\Gamma_{i,j-1/2}}), F_{\Gamma_{i,j-1/2}}, 0\right\|$ where $\beta = (1 - 0.1 P_e)^5$
$a_{i,j+1}$	$D_{\Gamma_{i,j+1/2}} + \dfrac{F_{\Gamma_{i,j+1/2}}}{2}$	$D_{\Gamma_{i,j+1/2}} + \max\left\|F_{\Gamma_{i,j+1/2}}, 0\right\|$	$\max\left\|\left(D_{\Gamma_{i,j+1/2}} + \dfrac{F_{\Gamma_{i,j+1/2}}}{2}\right), F_{\Gamma_{i,j+1/2}}, 0\right\|$	$\max\left\|(\beta D_{\Gamma_{i,j+1/2}}), (\beta D_{\Gamma_{i,j+1/2}} + F_{\Gamma_{i,j+1/2}}), F_{\Gamma_{i,j+1/2}}, 0\right\|$ where $\beta = (1 - 0.1 P_e)^5$

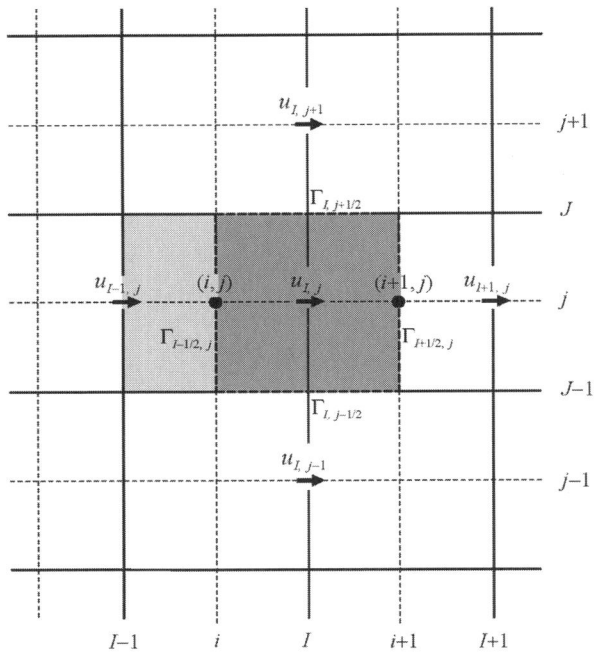

Figure 10.5. Control volume for u. $\bullet$ = nodes for scalar variables and $\rightarrow$ = u.

First, let us consider the discretization of u-momentum equation. A staggered control volume for the x-momentum equation, in which a forward-staggered velocity grid is adopted, is shown in figure 10.5. The control volume for u is staggered in relation to the normal control volume around the main node point (I,j). The staggering is in the x-direction only, such that the control-volume faces normal to that direction pass through the main node points (i,j) and $(i+1,j)$. The source term in the integral form of the general transport equation, such as equation (10.3.1) or (10.3.2) includes the body force and the pressure gradient terms described in equation (10.1.2). For a two-dimensional problem, the integral of the pressure gradient term in the u-momentum equation is given by

$$\frac{1}{\Delta t}\int_t^{t+\Delta t}\iiint_\Omega (-\nabla p)\,\mathrm{d}\Omega\,\mathrm{d}t = -\frac{1}{\Delta t}\int_t^{t+\Delta t}\int_{\Gamma_{I-1/2}}^{\Gamma_{I+1/2}}\int_{\Gamma_{j-1/2}}^{\Gamma_{j+1/2}}\left(\frac{\partial p}{\partial x}\right)\mathrm{d}x\,\mathrm{d}y\,\mathrm{d}t$$

$$= -\frac{1}{\Delta t}\int_t^{t+\Delta t}\left\{A_{I,j}\int_{\Gamma_{I-1/2}}^{\Gamma_{I+1/2}}\frac{\partial p}{\partial x}\mathrm{d}x\right\}\mathrm{d}t \quad (10.3.12)$$

where $A_{I,j}$ is the cell face area ($\Gamma_{I-1/2}$ or $\Gamma_{I+1/2}$) of the u-control-volume on which the pressure difference acts. The u-control-volume faces indicated by $\Gamma_{I\pm 1/2}$ and $\Gamma_{j\pm 1/2}$ indicate the control-volume faces $\Gamma_{I\pm 1/2,j}$ and

$\Gamma_{I,j\pm 1/2}$, respectively, as shown in figure 10.5. Thus, equation (10.3.12) becomes

$$\frac{1}{\Delta t}\int_{t}^{t+\Delta t}\iiint_{\Omega}(-\nabla p)\,d\Omega\,dt = -\frac{1}{\Delta t}\int_{t}^{t+\Delta t}\left\{A_{I,j}\int_{\Gamma_{I-1/2}}^{\Gamma_{I+1/2}}\frac{\partial p}{\partial x}\,dx\right\}dt$$

$$= -A_{I,j}(p_{i+1,j} - p_{i,j}). \qquad (10.3.13)$$

In the case of a two-dimensional problem $A_{I,j}$ is equal to Δy.

The evaluation of the total flux due to convection and diffusion at the u-control-volume faces shown in figure 10.5 would require an appropriate interpolation used for the discretization of the integral form of transport equation. In a similar procedure, we have the following discretized equation for the u-momentum equation, which is similar to equation (10.3.8). For the sake of simplicity, we omit the superscript $(n+1)$.

$$a_{I,j}u_{I,j} = \sum a_{nb}u_{nb} + b_{I,j} + A_{I,j}(p_{i,j} - p_{i+1,j}) \qquad (10.3.14)$$

where the subscript nb indicates the number of neighbors of the u-control-volume: four neighbors for a two-dimensional and six for a three-dimensional problem. The term $b_{I,j}$ is defined in the same manner as in equation (10.3.10b). The term $A_{I,j}(p_{i,j} - p_{i+1,j})$ indicates the pressure force acting on the u-control volume. The neighbor coefficient a_{nb} accounts for the combined convection/diffusion influence on the u-control-volume faces. For a two-dimensional problem, the term $\sum a_{nb}u_{nb}$ is given by

$$\sum a_{nb}u_{nb} = a_{I+1,j}u_{I+1,j} + a_{I-1,j}u_{I-1,j} + a_{I,j+1}u_{I,j+1} + a_{I,j-1}u_{I,j-1}. \qquad (10.3.15)$$

As described in chapter 8, the values of the coefficients $a_{I\pm 1,j\pm 1}$ and a_{nb} may be calculated with any difference scheme (upwind, hybrid or power-law) suitable for convection/diffusion problems.

The v-momentum equation can be handled in a similar manner. Figure 10.6 indicates the control volumes for the v-momentum equation. The control volume for v is staggered in the y direction. The discretized equation for the v-momentum equation is then given by

$$a_{i,J}v_{i,J} = \sum a_{nb}v_{nb} + b_{i,J} + A_{i,J}(p_{i,j} - p_{i,j+1}). \qquad (10.3.16)$$

Similarly, the term $\sum a_{nb}v_{nb}$ for a two-dimensional problem is given by

$$\sum a_{nb}v_{nb} = a_{i+1,J}v_{i+1,J} + a_{i-1,J}v_{i-1,J} + a_{i,J+1}v_{i,J+1} + a_{i,J-1}v_{i,J-1}. \qquad (10.3.17)$$

For three-dimensional problems, a similar discretized equation for the w-momentum equation can also be obtained.

The discretized momentum equations, equations (10.3.14) and (10.3.16), can be solved to obtain the velocity fields when the pressure field is given.

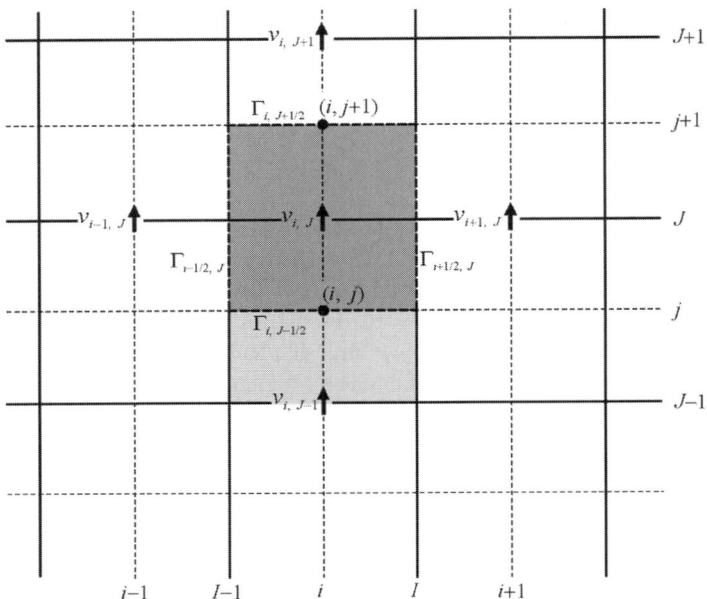

Figure 10.6. Control volume for v. $\bullet$ = nodes for scalar variables and $\uparrow = v$.

Unless the correct pressure field is employed, the resulting velocity field will not satisfy continuity. As the correct pressure field is unknown, we need a method for calculating pressure.

10.3.3 The SIMPLE algorithm

The SIMPLE stands for the Semi-Implicit Method for Pressure-Linked Equations [4]. The algorithm is essentially a guess-and-correct procedure for the calculation of pressure on the staggered grid arrangement.

To initiate the SIMPLE procedure a pressure field p^* needs to be guessed. Then, the discretized momentum equations, equations (10.3.14) and (10.3.16), can be solved using the guessed pressure field to yield the estimated velocity fields denoted by u^* and v^* as follows.

$$a_{I,j} u^*_{I,j} = \sum a_{nb} u^*_{nb} + b_{I,j} + A_{I,j}(p^*_{i,j} - p^*_{i+1,j}) \qquad (10.3.18)$$

$$a_{i,J} v^*_{i,J} = \sum a_{nb} v^*_{nb} + b_{i,J} + A_{i,J}(p^*_{i,j} - p^*_{i,j+1}). \qquad (10.3.19)$$

10.3.3.1 The pressure and velocity corrections

Let us consider finding a way of improving the guessed pressure field p^* such that the resulting estimated velocity fields u^* and v^* will progressively get closer to satisfying continuity. Now define the pressure correction δp as the difference between the correct pressure field p and the guessed pressure field p^*, so that

$$p = p^* + \delta p. \quad (10.3.20)$$

Likewise, the correct velocity field (u, v) can be written as

$$u = u^* + \delta u \quad (10.3.21a)$$
$$v = v^* + \delta v. \quad (10.3.21b)$$

Substitution of the correct pressure field p into the discretized momentum equations yields the correct velocity field (u, v). The discretized momentum equations, equations (10.3.14) and (10.3.16), are related to the correct velocity fields with the correct pressure field.

By subtracting equation (10.3.18) from equation (10.3.14) we obtain

$$a_{I,j}(u_{I,j} - u^*_{I,j}) = \sum a_{nb}(u_{nb} - u^*_{nb}) + A_{I,j}\{(p_{i,j} - p^*_{i,j}) - (p_{i+1,j} - p^*_{i+1,j})\}. \quad (10.3.22)$$

Using the correction formulae, equations (10.3.20)–(10.3.21), the above equation can be rewritten as

$$a_{I,j}\delta u_{I,j} = \sum a_{nb}\delta u_{nb} + A_{I,j}(\delta p_{i,j} - \delta p_{i+1,j}). \quad (10.3.23)$$

At this point, we assume that we can drop the term $\sum a_{nb}\delta u_{nb}$ to simplify the above equation for the velocity correction. This is the main approximation of the SIMPLE algorithm, referred to be a semi-implicit method [1]. Then, we obtain

$$\delta u_{I,j} = d_{I,j}(\delta p_{i,j} - \delta p_{i+1,j}) \quad (10.3.24)$$

where

$$d_{I,j} = \frac{A_{I,j}}{a_{I,j}}. \quad (10.3.25)$$

Equation (10.3.24) is called the velocity-correction formula. Substitution of equation (10.3.24) into equation (10.3.21a) gives

$$u_{I,j} = u^*_{I,j} + d_{I,j}(\delta p_{i,j} - \delta p_{i+1,j}). \quad (10.3.26)$$

This procedure shows how the guessed velocity field u^* is to be corrected in response to the pressure corrections to produce u.

Similarly, we have the correction formula for the velocity component v.

$$v_{i,J} = v^*_{i,J} + d_{i,J}(\delta p_{i,j} - \delta p_{i,j+1}). \quad (10.3.27)$$

10.3.3.2 The pressure-correction equation

Thus far we have considered the momentum equations. Let us now consider obtaining a discretized equation for the pressure correction δp. As described

162   Fluid flow analysis based on the Cartesian coordinate system

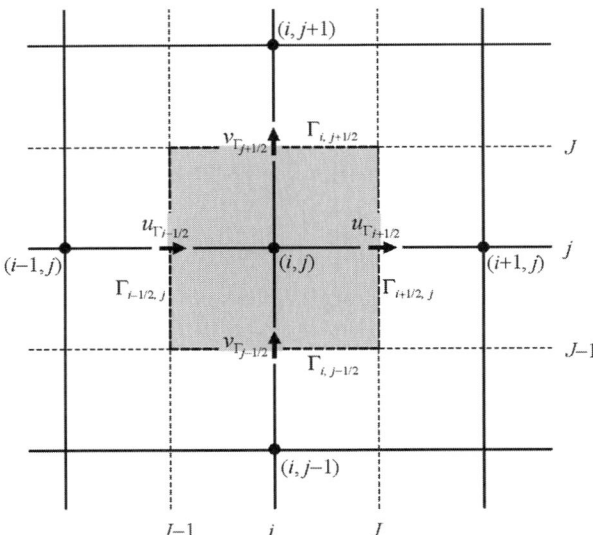

Figure 10.7. Control volume for δp. $u_{\Gamma_{i-1/2,j}} = u_{I-1,j}$; $u_{\Gamma_{i+1/2,j}} = u_{I,j}$; $v_{\Gamma_{i,j-1/2}} = v_{i,J-1}$; $v_{\Gamma_{i,j+1/2}} = v_{i,J}$.

in section 10.3.1, the discretization of the continuity equation, equation (10.1.1), can be simply made in a similar way as in the discretization of the general transport equation, such as equation (10.3.2).

If we assume that the density is constant, equation (10.1.1a) is simply expressed as

$$\iint_\Gamma \mathbf{u} \cdot \mathbf{n}\, d\Gamma = 0. \tag{10.3.28}$$

Let us evaluate the integral of equation (10.3.28) over the shaded control volume shown in figure 10.7, which is the same control volume used for deriving the discretized equation for the general variable Φ.

$$\begin{aligned}
\iint_\Gamma \mathbf{u}\cdot\mathbf{n}\,d\Gamma &= \int_{\Gamma_{i-1/2}}^{\Gamma_{i+1/2}} u\,dy + \int_{\Gamma_{j-1/2}}^{\Gamma_{j+1/2}} v\,dx \\
&= \{u_{\Gamma_{i+1/2}}A_{\Gamma_{i+1/2}} - u_{\Gamma_{i-1/2}}A_{\Gamma_{i-1/2}}\} + \{v_{\Gamma_{j+1/2}}A_{\Gamma_{j+1/2}} - v_{\Gamma_{j-1/2}}A_{\Gamma_{j-1/2}}\} \\
&= \{u_{\Gamma_{i+1/2}} - u_{\Gamma_{i-1/2}}\}\Delta y + \{v_{\Gamma_{j+1/2}} - v_{\Gamma_{j-1/2}}\}\Delta x.
\end{aligned}$$

In the above equation, $\Gamma_{i\pm 1/2}$ and $\Gamma_{j\pm 1/2}$ indicate the control-volume faces $\Gamma_{i\pm 1/2,j}$ and $\Gamma_{i,j\pm 1/2}$, respectively, as shown in figure 10.7. Then, we obtain

$$\{u_{\Gamma_{i+1/2}} - u_{\Gamma_{i-1/2}}\}\Delta y + \{v_{\Gamma_{j+1/2}} - v_{\Gamma_{j-1/2}}\}\Delta x = 0. \tag{10.3.29}$$

By replacing the velocity notations shown in figure 10.7, equation (10.3.29) becomes

$$\{u_{I,j} - u_{I-1,j}\}\Delta y + \{v_{i,J} - v_{i,J-1}\}\Delta x = 0. \tag{10.3.30}$$

If we substitute the velocity-correction formulas, equations (10.3.26) and (10.3.27), together with the equations for $u_{I-1,j}$ and $v_{i,J-1}$ into the discretized continuity equation, equation (10.3.30), and rearrange it, we obtain the following discretized equation for δp, which is similar to equation (10.3.8).

$$a_{i,j}\delta p_{i,j} = a_{i+1,j}\delta p_{i+1,j} + a_{i-1,j}\delta p_{i-1,j} + a_{i,j+1}\delta p_{i,j+1} + a_{i,j-1}\delta p_{i,j-1} + b \tag{10.3.31}$$

where the coefficients are given by

$$a_{i+1,j} = \Delta y \frac{A_{\Gamma_{i+1/2}}}{a_{\Gamma_{i+1/2}}} = \Delta y \, d_{\Gamma_{i+1/2}} \tag{10.3.32a}$$

$$a_{i-1,j} = \Delta y \frac{A_{\Gamma_{i-1/2}}}{a_{\Gamma_{i-1/2}}} = \Delta y \, d_{\Gamma_{i-1/2}} \tag{10.3.32b}$$

$$a_{i,j+1} = \Delta x \frac{A_{\Gamma_{j+1/2}}}{a_{\Gamma_{j+1/2}}} = \Delta x \, d_{\Gamma_{j+1/2}} \tag{10.3.32c}$$

$$a_{i,j-1} = \Delta x \frac{A_{\Gamma_{j-1/2}}}{a_{\Gamma_{j-1/2}}} = \Delta x \, d_{\Gamma_{j-1/2}} \tag{10.3.32d}$$

$$a_{i,j} = a_{i+1,j} + a_{i-1,j} + a_{i,j+1} + a_{i,j-1} \tag{10.3.32e}$$

$$b = \{u^*_{I,j} - u^*_{I-1,j}\}\Delta y + \{v^*_{i,J} - v^*_{i,J-1}\}\Delta x. \tag{10.3.32f}$$

Equation (10.3.21) indicates the discretized continuity equation as an equation for the pressure correction δp. The source term b stands for the continuity imbalance arising from the incorrect velocity field u^* and v^*. By solving equation (10.3.31), the pressure correction field δp can be obtained at all nodal points. Then, the correct pressure field p can be calculated using equation (10.3.20) and the correct velocity fields using equations (10.3.21a) and (10.3.21b). It is to be noted that the omission of the term $\sum a_{nb}\delta u_{nb}$ in the derivation process does not affect the final solution because the pressure correction and the velocity corrections will be zero in converged solutions giving $p^* = p$, $u^* = u$ and $v^* = v$.

10.3.3.3 Solution scheme

We have now formulated all the equations needed for calculating the velocity and pressure fields. The discretized equations for the velocity and pressure fields, equations (10.3.14), (10.3.16) and (10.3.31), must be set up at each of the nodal points including the boundary regions in order to solve the problems. The resultant system of linear algebraic equations is then solved to obtain the distribution of the transport property Φ at nodal points. The transport property Φ in the SIMPLE algorithm represents u, v or δp.

In the iterative solutions of the algebraic equations which frequently have non-linearity, it is often desirable to accelerate the iterative procedure. This process is called over-relaxation or under-relaxation. The over-relaxation is often used in conjunction with the Gauss–Seidel iteration method, in which the relaxation parameter is in the range $1 < W < 2$. For non-linear problems, the under-relaxation $(0 < W < 1)$ is often employed to avoid the divergence in the iterative solution.

It is to be noted that the pressure correct equation, equation (10.3.31), is susceptible to divergence unless some under-relaxation is used during the iterative process. Then, equation (10.3.20) can be expressed as

$$p^{\text{new}} = p^* + W_p \delta p \tag{10.3.33}$$

where p^{new} represents the new, improved pressure field and W_p is the pressure under-relaxation parameter.

Similarly, the iteratively improved velocity components u^{new} and v^{new} are also given by

$$u^{\text{new}} = W_u u^* + (1 - W_u) u^{(n-1)} \tag{10.3.34}$$

$$v^{\text{new}} = W_v v^* + (1 - W_v) v^{(n-1)} \tag{10.3.35}$$

where u and v are the corrected velocity components without relaxation and $u^{(n-1)}$ and $v^{(n-1)}$ represent the values obtained in the previous iteration, and W_u and W_v are the u- and v-velocity under-relaxation parameters. In a large number of fluid-flow computations [1], the pressure under-relaxation parameter W_p is about 0.8, and the velocity under-relaxation parameters W_u and W_v are about 0.5. However, it is not implied that these values are the optimum ones. A correct choice of under-relaxation parameters is essential for cost-effective computations. The optimum values of under-relaxation parameters are usually problem-dependent. It is to be noted that the relaxation parameters must be small enough to avoid divergence, but large enough to speed up the iterative process.

After rearrangement we have the discretized u- and v-momentum equations with under-relaxation as follows.

$$\frac{a_{I,j}}{W_u} u_{I,j} = \sum a_{nb} u_{nb} + b_{I,j} + A_{I,j}(p_{i,j} - p_{i+1,j}) + \left\{ (1 - W_u) \frac{a_{I,j}}{W_u} \right\} u_{I,j}^{(n-1)}. \tag{10.3.36}$$

$$\frac{a_{i,J}}{W_v} v_{i,J} = \sum a_{nb} v_{nb} + b_{i,J} + A_{i,J}(p_{i,j} - p_{i,j+1}) + \left\{ (1 - W_v) \frac{a_{i,J}}{W_v} \right\} v_{i,J}^{(n-1)}. \tag{10.3.37}$$

The important sequences of solution procedure can be summarized as follows:

1. Guess the pressure, velocity and other transport property fields, p^*, u^*, v^* and Φ^*.
2. Solve the discretized momentum equations, equations (10.3.14) and (10.3.16), to obtain u^* and v^*.
3. Solve the pressure correction equation, equation (10.3.31), to obtain δp.
4. Correct the pressure field by equation (10.3.33).
5. Correct the velocity fields using the velocity-correction formulas, equations (10.3.34) and (10.3.35).
6. Solve the discretized equations for other Φ (such as temperature and species concentration etc.) using the general form of the discretized equation, equation (10.3.8), if they influence the flow field through fluid properties, source terms, such as in the problems including natural convection, etc. If a particular Φ does not affect the flow field, it is better to calculate it after a converged solution for the flow field has been obtained.
7. Set the corrected values as the new guessed values, $p^* = p$, $u^* = u$, $v^* = v$ and $\Phi^* = \Phi$, and return to step 2, and repeat the whole procedure until a converged solution is obtained.

10.4 Treatment of free surfaces

Fluid flow having a free surface often appears in materials processing, such as flow in ladles or tundishs, mould filling in casting processes, welding process, Czochralski crystal growth, zone melting, continuous casting, and strip casting, etc. It is important to model flow problems involved in materials processing since fluid flow phenomena have major effects not only on the quality of products, but also on the optimization of manufacturing processes. In order to model fluid flow phenomena, we need to couple the flow fields with the variation of free surfaces. The important characteristics in solving fluid flow having a free surface is in treating boundary conditions on free surfaces. The boundary conditions on free surfaces need to be described by the normal and tangential stress conditions whilst the solid wall is usually described by the velocity condition on the wall. Thus, the shape of a free surface varies with time according to the flow patterns.

There are several computational techniques for solving fluid flow with a free surface, such as MAC [3], SMAC [5], SOLA-VOF (Solution Algorithm–Volume of Fluid) [6], and BFC (Boundary Fitted Coordinate) [7]. In this section, we will briefly describe the MAC and VOF techniques for tracking free surfaces.

10.4.1 The MAC method

The MAC (Marker And Cell) method was primarily developed in the Los Alamos Scientific Laboratory for nuclear and civil engineering applications [3], and first applied to simulate the fluid flow in mould filling having a free surface by Hwang et al [8].

The MAC technique is distinguished by four features: the form of the primitive-variable equations, the finite difference scheme, the cell structure, and the use of marker particles. The form of the primitive-variable equations, i.e., the Navier–Stokes equations in their primitive form including all non-linear terms, are given by equation (9.2.10). The finite differencing scheme has also been described in chapter 9.

In this section, we will briefly decribe the basic feature of the marker and cell technique for treating a free surface problem, i.e., the cell structure and the use of marker particles. The primary dependent variables are the pressure and velocity components. It combines the best features of the two classical computational schemes of fluid dynamics, i.e., the ideal Eulerian approach and the ideal Lagrangian approach. The cavity is divided into a large number of Eulerian cells which remains at rest while a Lagrangian meshes of particles ('markers') represent the elements of the fluid which move through the meshes of cells. The Eulerian mesh is used to characterize the fluid variables (such as velocity and pressure) while the markers are used to represent the fluid itself and define the location of the free surface. It is to be noted that the marker particles used in the MAC method are 'massless particles' which move with the fluid flow.

The positions, or the Lagrangian coordinates, of each marker particle (x_p^n, y_p^n) are determined by numerical integration from an initial position (x_p^0, y_p^0) at time $t = 0$ as follows.

$$x_p^n = x_p^0 + \int_0^t u_p \, dt \tag{10.4.1a}$$

$$y_p^n = y_p^0 + \int_0^t v_p \, dt \tag{10.4.1b}$$

where u_p and v_p are the velocities in the Eulerian mesh at the time-dependent location of the particle.

The velocities of marker particles are evaluated from the nearest four mesh velocities in each coordinate direction using a standard bilinear interpolation.

$$u_p = \frac{A_1 u_1 + A_2 u_2 + A_3 u_3 + A_4 u_4}{\Delta x \, \Delta y}. \tag{10.4.2}$$

The areas A_1 through A_4 are defined by the location (x_p^n, y_p^n), as shown in figure 10.8.

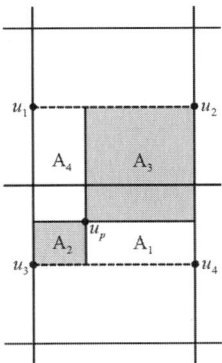

Figure 10.8. Bilinear interpolation for u_p.

The new position of a marker particle after a time interval Δt can be determined sequentially by the forward time integration of equation (10.4.1).

$$x_p^{n+1} = x_p^n + u_p \Delta t \qquad (10.4.3\text{a})$$

$$y_p^{n+1} = y_p^n + v_p \Delta t. \qquad (10.4.3\text{b})$$

As mentioned above, the marker particles move with the fluid flow field. The cells within the computational domain can be defined into five types, depending on whether they contain marker particles or not: (i) an empty (E) cell contains no fluid, (ii) a full (F) cell contains fluid and has no empty cells for neighbors on any of its four faces, (iii) a free-surface (S) cell contains fluid, but has at least one empty cell among the neighbors on its four faces, (iv) a boundary (B) cell indicating the outer boundary of the computational domain is defined for handling boundary conditions, and (v) an obstacle (OB) cell is defined to handle an obstacle inside the computational domain. Figure 10.9 indicates the free surface tracking by a distribution of marker particles and the classification of cells.

In the case of mould filling, all the cells in the computational domain are initially empty. However, the types of cells change from E to F when the liquid metal flows into the domain through its gate. So, the marker particles need to be generated continuously according to the inlet velocity at the gate. The number of marker particles generated at the gate are usually determined by experience (i.e., 3×3 or 4×4 markers per cell) in two-dimensional cases.

Important improvements for obtaining solution accuracy and computational efficiency were made by adopting pressure interpolation on the free surface [9, 10]. The MAC method is also readily extendable to three-dimensional problems, provided the increased storage requirements can be tolerated [11, 12]. The SMAC method is a simplified MAC method in

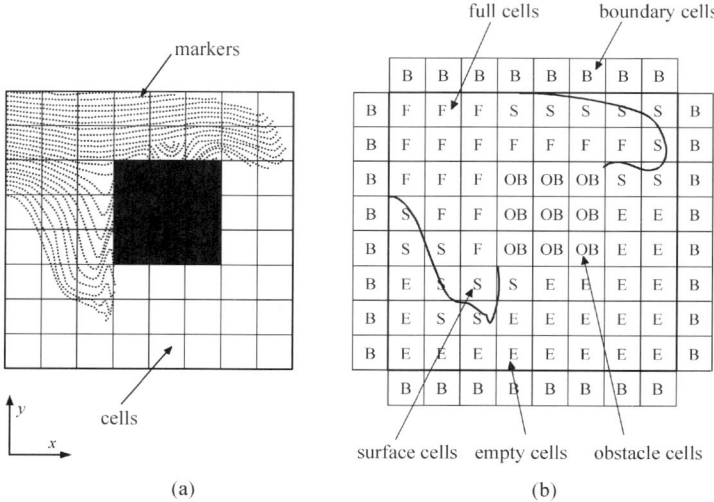

Figure 10.9. Marker particles and classification of cells in the MAC technique.

which pressure is not solved, but the divergence (continuity) condition on velocity fields is satisfied directly [5].

10.4.2 The VOF (volume of fluid) method

The basis of the SOLA-VOF method [11] is the fractional volume of fluid scheme for tracking free surfaces. In this technique, a function $F(x, y, t)$ is defined to represent the fractional volume of the cells occupied by fluid. Free surfaces are reconstructed by means of a scalar field $F(x, y, t)$, where

$$F(x, y, t) = \begin{cases} 1, & \text{a cell full of fluid} \\ 0 < F < 1, & \text{a cell at the free surface} \\ 0, & \text{an empty cell.} \end{cases} \quad (10.4.4)$$

For example, the free surface can be represented by $F(x, y, t)$ value as shown in figure 10.10. Thus, the VOF method [13] provides the same coarse interface information available to the MAC method. In addition, it requires only one storage word for each cell, which is consistent with the storage requirements for all other dependent variables.

The fractional volume of fluid in each cell can be determined by solving the following equation.

$$\frac{\partial F}{\partial t} + u \frac{\partial F}{\partial x} + v \frac{\partial F}{\partial x} = 0 \quad (10.4.5)$$

where (u, v) are fluid velocities in (x, y) coordinates. This equation represents that F moves with the fluid.

Treatment of free surfaces

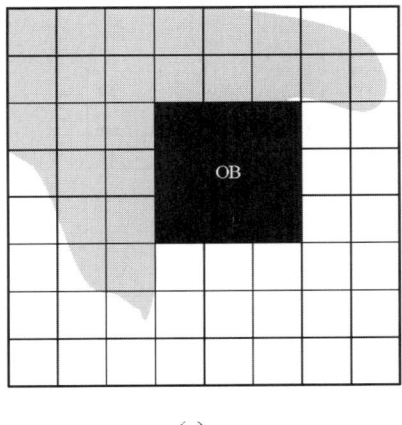

1.0	1.0	1.0	1.0	0.98	0.85	0.53	0.22
1.0	1.0	1.0	1.0	1.0	0.99	0.99	0.61
1.0	1.0	1.0	OB			0.03	0.14
0.4	1.0	1.0				0.0	0.0
0.03	0.92	1.0				0.0	0.0
0.0	0.51	1.0	0.0	0.0	0.0	0.0	0.0
0.0	0.0	0.21	0.0	0.0	0.0	0.0	0.0
0.0	0.0	0.0	0.0	0.0	0.0	0.0	0.0

(a)   (b)

Figure 10.10. Expression of a free surface shape by F value.

The basic advection method can be understood by considering the amount of F to be fluxed through the interface of neighboring cells during a time step of Δt. Fluxes across other cell faces in two- and three-dimensional problems are completely analogous. The total flux of fluid volume and void volume crossing the cell face is determined by the normal velocity at the cell face, u. The sign of u determines the donor and acceptor cells, i.e., the cells losing and gaining fluid volume, respectively. If u is positive, the upstream or left cell is the donor and the downstream or right cell the acceptor. The total flux depends not only on the velocity, but also on the orientation of a free surface.

The amount of F on the free surface can be evaluated for three cases, as shown in figure 10.11 [14]. When the orientation of a free surface is perpendicular to the direction of flow as in case (a) of figure 10.11, the air or void first flows from the left cell (donor cell) into the right cell (acceptor cell) and then the fluid flows. When the orientation of a free surface is parallel to the direction of flow as in case (b), the fluid and void flow simultaneously. However, when the free surface is inclined to the direction of flow as in case (c), we may combine the concepts of (a) and (b) to evaluate F, as shown in figure 10.12. The amount of δF fluxed across the cell face in a time step can be evaluated as follows.

$$\delta F = \frac{u}{|u|} \min \| F_{AD} \times Q + CF, \quad F_D \times \delta x_D \| \tag{10.4.6a}$$

$$CF = \max \|(1 - F_{AD})Q - (1 - F_D)\delta x_D, \quad 0.0\| + F_{AD} \times Q \tag{10.4.6b}$$

$$Q = |u| \times \Delta t \tag{10.4.6c}$$

where δx_D is the distance between the center points of the two adjacent cells (donor and acceptor), u is the velocity of flow at the cell face, Δt is

170   *Fluid flow analysis based on the Cartesian coordinate system*

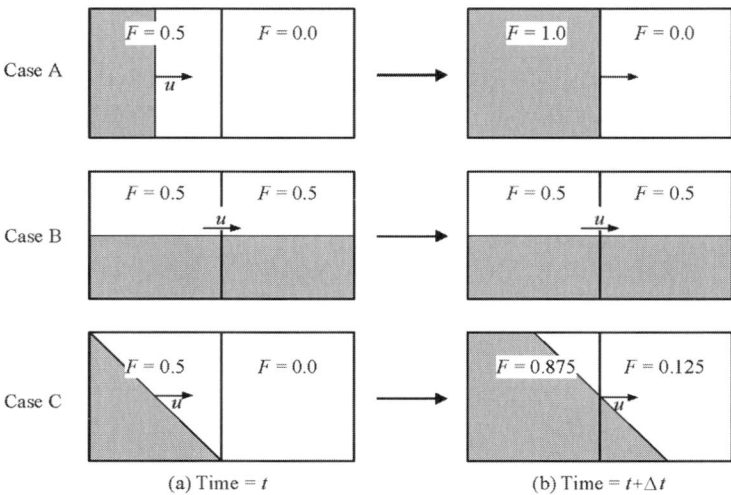

Figure 10.11. Examples of free surface shapes used in the advection of F value [14].

the time step, and Q is the total flux of fluid volume and void volume per unit cross-sectional area. The min $\|\ \|$ feature prevents the fluxing of more fluid from the donor cell than it has to give, while the max $\|\ \|$ feature accounts for an additional fluid flux, CF, if the amount of void to be fluxed exceeds

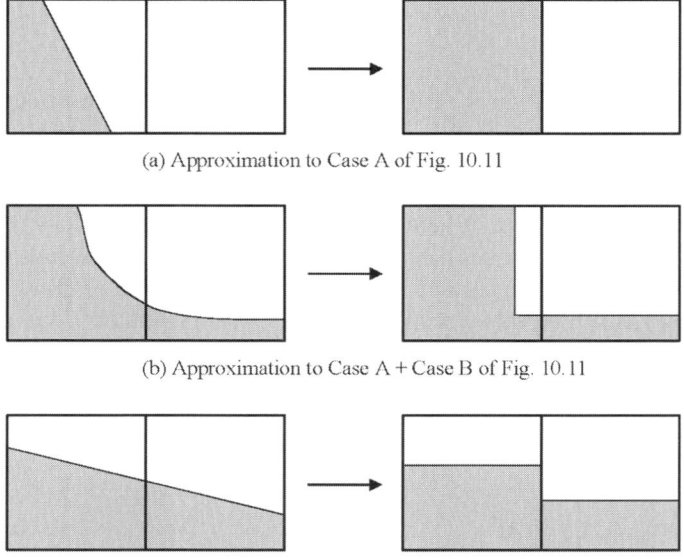

Figure 10.12. Approximation of inclined free surfaces [14].

the amount available. F_A and F_D denote F values for the acceptor and donor cells, respectively. F_{AD} refers to either F_A or F_D, depending on the orientation of the free surface relative to the direction of flow as follow.

$$F_{AD} = \begin{cases} F_A & \text{for figure 10.12(a) and (b)} \\ F_D & \text{for figure 10.12(c).} \end{cases} \quad (10.4.7)$$

The calculated F values may occasionally have values slightly less than zero or slightly greater than unity. It is therefore necessary to reset values of F less than zero back to zero and values of F greater than unity back to unity.

10.5 Boundary conditions

The governing equations of momentum transfer can be solved using appropriate initial and boundary conditions. Initial conditions refer to the values of all the flow variables to be specified in the computational domain at time $t = 0$, for transient problems. The most common boundary conditions appearing in confined fluid flow problems described in chapter 2 are such as the inlet and outlet boundary conditions, the wall boundary condition, the prescribed pressure boundary condition, the symmetry condition, etc. The details of implementation of boundary conditions in the discretized equations of the finite volume method can be referred to in the literature [15].

In this section we will briefly describe the boundary conditions for fluid flow having a free surface in mould filling. In the free surface region, the continuity condition given by equation (10.1.1) is not valid because the mass within the cells of the free surface region changes with time even though the momentum balance conditions are still satisfied. The surface region contains the interface between the liquid and the atmosphere surrounding it, and the normal and tangential stresses on a free surface should balance with the sum of the applied pressure and the surface tension as follows [13].

(Normal stress condition)

$$2\mu \left[n_x n_x \frac{\partial u}{\partial x} + n_x n_y \left(\frac{\partial u}{\partial y} + \frac{\partial v}{\partial x} \right) + n_y n_y \frac{\partial v}{\partial y} \right] = p_a + p_s. \quad (10.5.1)$$

(Tangential stress condition)

$$\mu \left[2 n_x m_x \frac{\partial u}{\partial x} + (n_x m_y + n_y m_x) \left(\frac{\partial u}{\partial y} + \frac{\partial v}{\partial x} \right) + 2 n_y m_y \frac{\partial v}{\partial y} \right] = 0 \quad (10.5.2)$$

where u and v are the velocity components in the x- and y-directions, and n_x, n_y, m_x and m_y are the x- and y-components of the unit vectors which refer to the local normal and tangential directions to the free surface, as shown in figure 10.13. p_a is the applied gas pressure in the empty region and p_s is the

172   *Fluid flow analysis based on the Cartesian coordinate system*

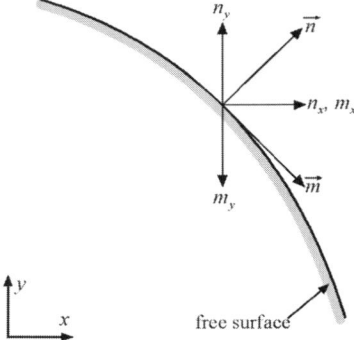

Figure 10.13. Distribution of the normal and tangential vectors at the cell face.

surface tension pressure. The flow field in the surface regions is calculated by these principles.

10.6 Turbulent flow

The laminar flow is limited to a finite value of Reynolds number, i.e. a critical Reynolds number. As the Reynolds number increases, the turbulent flow is inevitable. Thus, in most flow problems appearing in materials processing, turbulence modelling based on numerical analyses is essential for simulating high Reynolds number flow.

In this section, the Reynolds stress equation is introduced together with its physical meaning. In order to simplify the long and complex conservation equations for a single Reynolds stress, the tensor form of Navier–Stokes equation is expressed based on the tensor notation as

$$\frac{\partial u_i}{\partial t} + u_j \frac{\partial u_i}{\partial x_j} = -\frac{1}{\rho}\frac{\partial p}{\partial x_i} + \frac{1}{\rho}\frac{\partial}{\partial x_j}\left(\mu \frac{\partial u_i}{\partial x_j}\right) \qquad (10.6.1)$$

where i corresponds to the direction of the coordinate and j indicates the summation of each coordinate component according to the common rule of tensor notation. In the same manner the continuity equation for the incompressible flow is

$$\frac{\partial u_i}{\partial x_i} = 0. \qquad (10.6.2)$$

The velocity u_i or the pressure p in the incompressible turbulent flow is resolved into the mean value ($\bar{u}_i$ or $\bar{p}$) and the fluctuation value (u_i' or p') as shown in figure 10.14 as follows.

$$u_i = \bar{u}_i + u_i', \qquad p = \bar{p} + p'. \qquad (10.6.3)$$

Turbulent flow 173

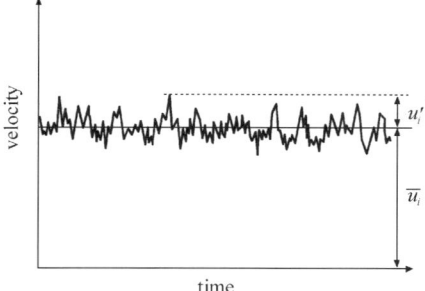

Figure 10.14. Mean value and fluctuation in a turbulent flow.

The rules of time averaging are applicable to equations (10.6.1) and (10.6.2), leading to the following equations [16].

$$\frac{\partial \overline{u_i}}{\partial x_i} = 0 \tag{10.6.4}$$

$$\frac{\partial \overline{u_i}}{\partial t} + \overline{u_j}\frac{\partial \overline{u_i}}{\partial x_j} = -\frac{1}{\rho}\frac{\partial \overline{p}}{\partial x_i} + \frac{1}{\rho}\frac{\partial}{\partial x_j}\left(\mu \frac{\partial \overline{u_i}}{\partial x_j} - \rho\overline{u'_i u'_j}\right). \tag{10.6.5}$$

The newly created term $-\rho\overline{u'_i u'_j}$ is the Reynolds or turbulent stress. Applying the time averaging rule yields the Reynolds stress equation as follows.

$$\frac{\partial \overline{u'_i u'_j}}{\partial t} + u_k \frac{\partial \overline{u'_i u'_j}}{\partial x_k} = -\frac{\partial}{\partial x_k}(\overline{u'_i u'_j u'_k}) - \frac{\partial}{\partial x_k}\left(\nu \frac{\partial \overline{u'_i u'_j}}{\partial x_k}\right)$$

$$- \frac{1}{\rho}\left(\overline{\frac{\partial u'_i p'}{\partial x_j}} + \overline{\frac{\partial u'_j p'}{\partial x_i}}\right)$$

$$- \left(\overline{u'_j u'_k}\frac{\partial \overline{u_i}}{\partial x_k} + \overline{u'_i u'_k}\frac{\partial \overline{u_j}}{\partial x_k}\right)$$

$$- 2\nu \overline{\frac{\partial u'_i}{\partial x_k}\frac{\partial u'_j}{\partial x_k}} + \frac{\overline{p'}}{\rho}\left(\frac{\partial u'_i}{\partial x_j} + \frac{\partial u'_j}{\partial x_i}\right). \tag{10.6.6}$$

Equation (10.6.6) indicates that the rate of change of Reynolds stress in the left-hand side is balanced with the diffusion of Reynolds stresses (the first three terms), the generation of stresses (the forth and fifth terms), dissipation (the sixth term), and the pressure strain effects (the last term) in the right-hand side. There are several turbulent models being used: (i) zero-equation model, (ii) one-equation model, (iii) two-equation model and (iv) Reynolds stress model. The detailed information on turbulence models are referred to in the literature [16].

174   *Fluid flow analysis based on the Cartesian coordinate system*

10.7 Case studies

10.7.1 Flow over a semi-circular core between two plates

10.7.1.1 Description of the problem

Now consider a two-dimensional fluid flow problem between two plates having a semi-circular core on the bottom plate. Figures 10.15(a) and (b) indicate a schematic drawing of the calculation domain and a grid system having 1200 grids used in the present simulation. The working fluid is water and the flow rate is 10^{-4} m/s.

10.7.1.2 Material properties and boundary conditions

The material properties and boundary conditions used in the present simulation are shown in tables 10.2 and 10.3.

10.7.1.3 Simulation

Let us simulate the flow sequences and flow patterns using the execution file [flow1.exe].

Data input
```
> Input inlet flow rate (m/s) : 0.0001
> Input density of fluid (kg/m³) : 1000
> Input dynamic viscosity of fluid (kg/m s) : 0.001
> number of iterations : 1000
```

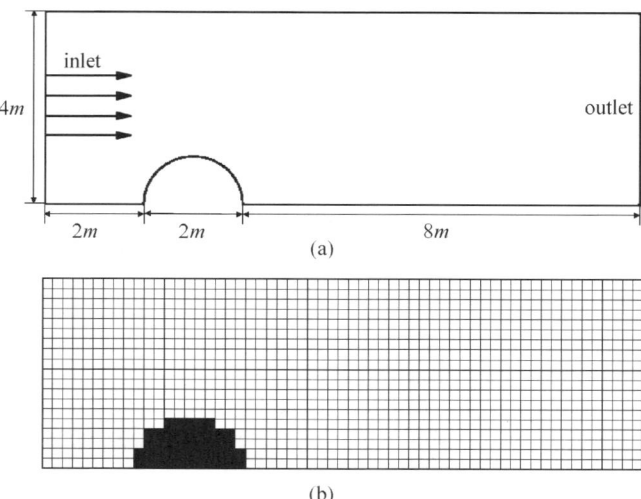

Figure 10.15. A schematic drawing of (a) the geometry and dimension of a computational domain and (b) a grid configuration.

Case studies

Table 10.2. Material properties of water.

Material property	Value
ρ density (kg/m^3)	1000
μ dynamic viscosity (kg/m s)	1×10^{-3}

Table 10.3. Boundary conditions.

Inlet	Constant flow rate (10^{-4} m/s)
Outlet	Outflow
Upper plate	No slip wall
Lower plate	No slip wall

Results

Figure 10.16 indicates the simulated flow patterns. As the flow passes over the semi-circular core located on the bottom plate, a secondary circulating flow is formed in the rear zone of the core. There are a few dominant properties that affect the flow pattern, such as the boundary conditions and the material properties.

10.7.2 Internal flow in a U-tube

10.7.2.1 Description of the problem

Now consider another example of internal flow in a U-tube. Figures 10.17(a) and (b) indicate a schematic drawing of the computational domain and a grid

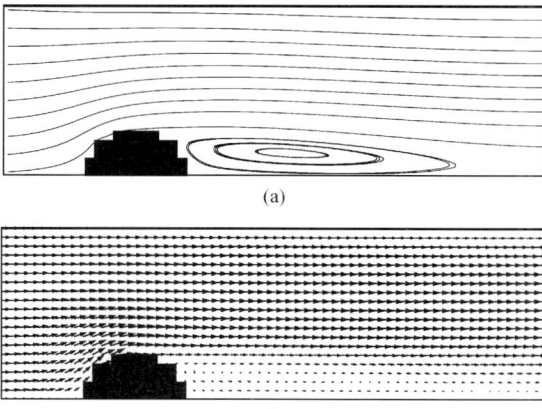

Figure 10.16. Flow between two plates having a semi-circular core on the bottom plate: (a) a stream function and (b) a flow velocity profile.

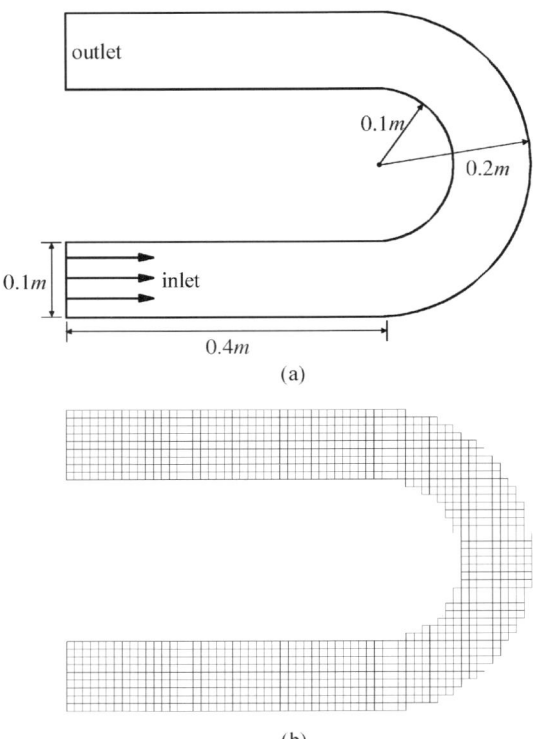

Figure 10.17. A schematic drawing of (a) the geometry and dimension of a computational domain and (b) a grid configuration.

system having 2400 grids used in the present simulation. The working fluid is water and the flow rate is 10^{-3} m/s.

10.7.2.2 Material properties and boundary conditions

The material properties and boundary conditions used in the present simulation are shown in tables 10.2 and 10.3.

10.7.2.3 Simulation

Let us simulate the flow sequences and flow patterns using the execution file [flow2.exe].

Data input
```
> Input inlet flow rate (m/s) : 0.001
> Input density of fluid (kg/m³) : 1000
> Input dynamic viscosity of fluid (kg/m s) : 0.001
> number of iterations : 1000
```

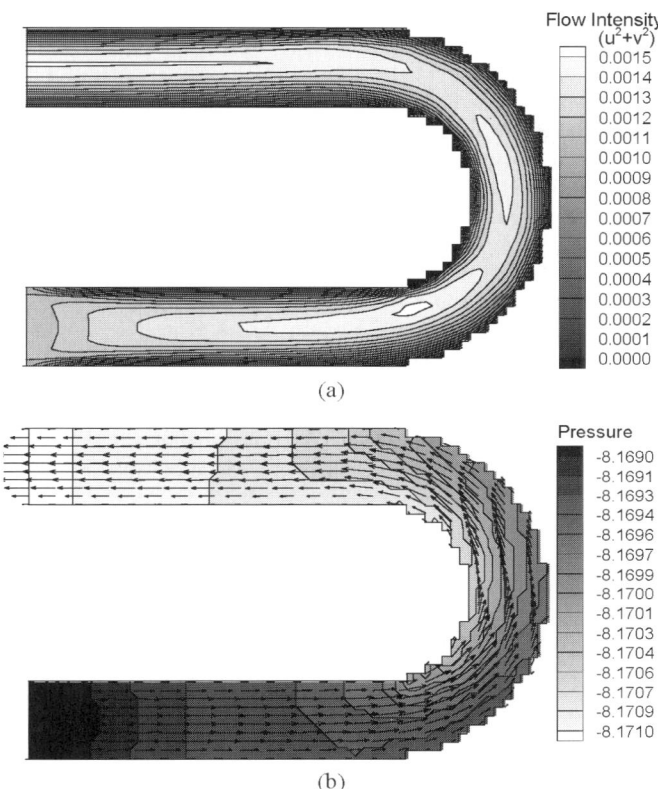

Figure 10.18. Internal flow in a U-tube: (a) a flow intensity profile and (b) a pressure distribution.

Results

Figure 10.18 indicates the simulated flow patterns. In cases of modelling the fluid flow in a curved shape cavity, such as the fluid flow in a U-tube, the orthogonal grid system based on the Cartesian coordinate as shown in figure 10.17 might cause a large solution error because of its stair-like grid shapes, especially in the case with a large mesh size. If we adopt the body-fitted coordinate (BFC) system to solve this problem, the solution errors caused by the stair-like grids can be eliminated, as will be discussed in chapter 12.

References

[1] Patankar S V 1980 *Numerical Heat Transfer and Fluid Flow* (New York: Hemisphere Publishing Co, Taylor & Francis Group)

[2] Rhie C M and Chow W L 1983 *AIAA J.* **21** 1525
[3] Harlow F H and Welch J E 1965 *Phys. Fluids* **8** 2182
[4] Patankar S V and Spalding D B 1972 *Int. J. Heat Mass Transfer* **15** 1787
[5] Amsden A A and Harlow F H 1970 *J. Comput. Phys.* **6** 322
[6] Nichols B D, Hirt C W and Welch J E 1980 *Technical Report LA-8355* Los Alamos Scientific Laboratory
[7] Sarraf S, Kahawita R and Camareo R 1987 *Int. J. Num. Meth. Fluids* **7** 465
[8] Hwang W S and Stoehr R A 1983 *J. Metals* **35** 22
[9] Chan R K-C and Street R L 1970 *J. Comput. Phys.* **6** 68
[10] Nichols B D and Hirt C W 1971 *J. Comput. Phys.* **8** 434
[11] Harlow F H, Amsden A A and Nix J R 1976 *J. Comput. Phys.* **20** 119
[12] Kim S B 1996 PhD thesis, Yonsei University, Seoul, Korea
[13] Hirt C W and Nichols B D 1981 *J. Comput. Phys.* **39** 201
[14] Anzai K 1992 *J. Japan Foundry Engng. Soc.* **64** 410
[15] Versteeg H K and Malalasekera W 1995 *An Introduction to Computational Fluid Dynamics—The Finite Volume Method* (Harlow: Longman Scientific & Technical)
[16] White F M 1991 *Viscous Fluid Flow* (Singapore: McGraw-Hill)

Chapter 11

Fluid flow analysis using the SIMPLE method based on the body-fitted coordinate system

In chapter 10, we described the SIMPLE method for modelling fluid flow problems using orthogonal grids in the Cartesian coordinate system. The standard SIMPLE method based on the Cartesian coordinates is useful for solving problems of simple geometries. However, it may encounter some difficulties and limitations in dealing with complex geometries, such as thin-walled and curved-shape geometries, since an orthogonal grid system might cause stair-like boundaries along the surface of a computational domain. Recently, a numerical approach based on the curvilinear coordinate system has been developed to overcome this limitation. In this chapter we will describe the fundamentals of the SIMPLE method based on non-orthogonal grids and also introduce the transformed VOF method for free surface tracking in the body-fitted coordinate system.

11.1 Introduction

When numerical models based on the Cartesian coordinate system are applied to complex geometries, an orthogonal grid system cannot exactly represent curved-shape surfaces of a computational domain. In this case the circular or curved-shape surfaces may be approximated by stair-like grids, as shown in figure 11.1. This approximation has considerable disadvantages since the stair-like boundary approximation has a low computational accuracy causing some numerical errors. In addition, further disadvantages including excessive computer memory and resources will arise due to (i) the blocking of the grids in blank regions and (ii) the introduction of fine orthogonal grids in a particular region of interest causing unnecessary refinement of grids in another region of minimal interest. Algorithms based on the body-fitted coordinate system using non-orthogonal grids have been

180   *Fluid flow analysis using the body-fitted coordinate system*

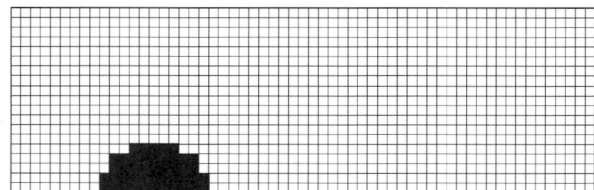

Figure 11.1. Cartesian grid arrangements for modelling fluid flow between two plates having a semi-circular obstacle on the bottom plate.

developed to overcome the limitations mentioned above and increasingly applied to model fluid flow problems [1–5]. Recently, the SIMPLE algorithm coupled with the VOF method based on the body-fitted coordinate (BFC) system has been developed for modelling the mould filling process for thin-walled and curved-shape castings [6].

Figure 11.2 shows a body-fitted grid system for the same domain used in figure 11.1. The geometrical flexibility offered by the body-fitted grid technique is useful in modelling practical problems having complex and curved-shape physical domains. The main differences between the two formulations are in the grid arrangement and in the choice of dependent variables in momentum equations. In fluid flow modelling based on the body-fitted coordinate system, the use of non-staggered or collocated grids for velocity components becomes more popular than the use of staggered grids. The use of non-staggered grids requires a special procedure to ensure proper velocity and pressure coupling and to avoid the unrealistic pressure fields, but need small storage allocations compared to that of staggered grids.

Let us now compare a simple example of flow field simulated using the two-grid systems. Figure 11.3 indicates the results obtained from the Cartesian coordinate system: (a) an orthogonal grid configuration with a stair-like arrangement and (b) a simulated velocity field. Figure 11.4 indicates the results obtained from the body-fitted coordinate system: (a) a non-orthogonal grid configuration and (b) a simulated velocity field. This example clearly demonstrates the advantages of the body-fitted coordinate system for treating curved-shape geometries. It can be seen from the

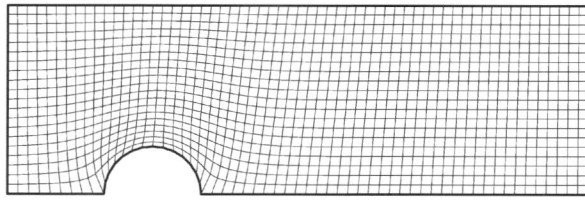

Figure 11.2. Body-fitted grid arrangements for modelling fluid flow between two plates having a semi-circular obstacle on the bottom plate.

Introduction 181

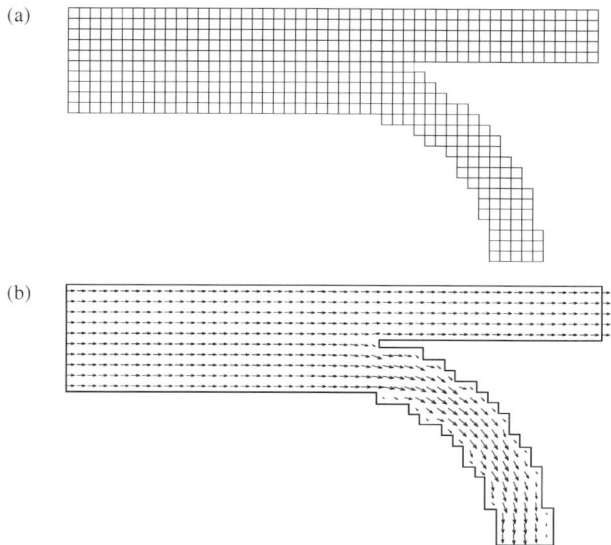

Figure 11.3. An example of flow field simulation: (a) an orthogonal Cartesian grid and (b) a simulated flow pattern.

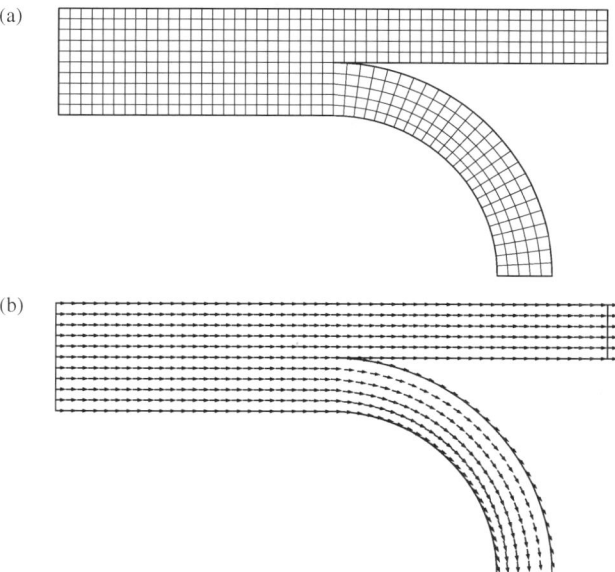

Figure 11.4. An example of flow field simulation: (a) a non-orthogonal body-fitted grid and (b) a simulated flow pattern.

figures that the model based on the body-fitted coordinate system provides grid independent results compared to those based on the Cartesian coordinate system [7].

11.2 Transformation of coordinate systems

Most of physical domains of interest are complex and curved-shape. A physical domain with a complex geometry in the Cartesian coordinate system (x, y) can be transformed into a computational domain with an orthogonal geometry in a generalized coordinate system (ξ, η), as shown in figure 11.5. Then, non-orthogonal grids in the physical domain are transformed into their corresponding orthogonal grids in the computational domain.

Figure 11.5 describes the relationship between the physical domain and its transformed computational domain. The derivatives in the ξ-direction can be expressed by the following relation, i.e., *chain rule*.

$$\frac{\partial}{\partial \xi} = \frac{\partial}{\partial x}\frac{\partial x}{\partial \xi} + \frac{\partial}{\partial y}\frac{\partial y}{\partial \xi}. \qquad (11.2.1)$$

Similarly, the derivatives in the η-direction can be given. Therefore, the first derivatives can be expressed as a matrix form as follows.

$$\begin{bmatrix} \dfrac{\partial}{\partial \xi} \\ \dfrac{\partial}{\partial \eta} \end{bmatrix} = \begin{bmatrix} \dfrac{\partial x}{\partial \xi} & \dfrac{\partial y}{\partial \xi} \\ \dfrac{\partial x}{\partial \eta} & \dfrac{\partial y}{\partial \eta} \end{bmatrix} \begin{bmatrix} \dfrac{\partial}{\partial x} \\ \dfrac{\partial}{\partial y} \end{bmatrix}. \qquad (11.2.2)$$

In consequence, the partial derivatives of the variable f can be evaluated as

$$\begin{bmatrix} \dfrac{\partial f}{\partial \xi} \\ \dfrac{\partial f}{\partial \eta} \end{bmatrix} = \begin{bmatrix} \dfrac{\partial x}{\partial \xi} & \dfrac{\partial y}{\partial \xi} \\ \dfrac{\partial x}{\partial \eta} & \dfrac{\partial y}{\partial \eta} \end{bmatrix} \begin{bmatrix} \dfrac{\partial f}{\partial x} \\ \dfrac{\partial f}{\partial y} \end{bmatrix}. \qquad (11.2.3)$$

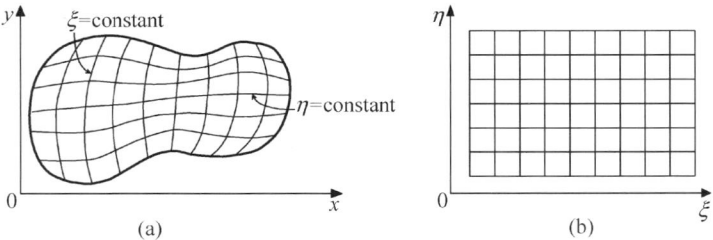

Figure 11.5. A schematic drawing: (a) a physical domain and (b) a computational domain.

By rearranging equation (11.2.3), we can obtain other derivatives for $\partial f/\partial x$ and $\partial f/\partial y$:

$$\frac{\partial f}{\partial x} = \frac{1}{J}\left(\frac{\partial y}{\partial \eta}\frac{\partial f}{\partial \xi} - \frac{\partial y}{\partial \xi}\frac{\partial f}{\partial \eta}\right) \tag{11.2.4a}$$

$$\frac{\partial f}{\partial y} = \frac{1}{J}\left(-\frac{\partial x}{\partial \eta}\frac{\partial f}{\partial \xi} + \frac{\partial x}{\partial \xi}\frac{\partial f}{\partial \eta}\right) \tag{11.2.4b}$$

where J is the Jacobian of the transformation given by

$$J = \frac{\partial x}{\partial \xi}\frac{\partial y}{\partial \eta} - \frac{\partial x}{\partial \eta}\frac{\partial y}{\partial \xi}. \tag{11.2.5}$$

By exchanging (x, y) with (ξ, η), equation (11.2.3) can be expressed as

$$\begin{bmatrix}\dfrac{\partial f}{\partial x}\\[4pt]\dfrac{\partial f}{\partial y}\end{bmatrix} = \begin{bmatrix}\dfrac{\partial \xi}{\partial x} & \dfrac{\partial \eta}{\partial x}\\[4pt]\dfrac{\partial \xi}{\partial y} & \dfrac{\partial \eta}{\partial y}\end{bmatrix}\begin{bmatrix}\dfrac{\partial f}{\partial \xi}\\[4pt]\dfrac{\partial f}{\partial \eta}\end{bmatrix}. \tag{11.2.6}$$

Geometric variables $\partial\xi/\partial x$, $\partial\xi/\partial y$, $\partial\eta/\partial x$ and $\partial\eta/\partial y$ can be obtained from equation (11.2.4) and equation (11.2.6) as follows.

$$\frac{\partial \xi}{\partial x} = \frac{1}{J}\frac{\partial y}{\partial \eta},\quad \frac{\partial \xi}{\partial y} = -\frac{1}{J}\frac{\partial x}{\partial \eta},\quad \frac{\partial \eta}{\partial x} = -\frac{1}{J}\frac{\partial y}{\partial \xi},\quad \frac{\partial \eta}{\partial y} = \frac{1}{J}\frac{\partial x}{\partial \xi}. \tag{11.2.7}$$

11.3 Transformation of basic equations

As described in chapter 10, numerical modelling of transient fluid flow problems using the SIMPLE method starts from the integral form of transport equation given by equation (10.1.6).

$$\int_{\Delta t}\frac{\partial}{\partial t}\left\{\iiint_\Omega (\rho\Phi)\,d\Omega\right\}dt + \int_{\Delta t}\iint_\Gamma (\rho\mathbf{u}\Phi)\cdot\mathbf{n}\,d\Gamma\,dt$$

$$= \int_{\Delta t}\iint_\Gamma (K\nabla\Phi)\cdot\mathbf{n}\,d\Gamma\,dt + \int_{\Delta t}\iiint_\Omega S\,d\Omega\,dt. \tag{10.1.6}$$

In order to apply the body-fitted coordinate system to the SIMPLE method, the transformation of the integral form of transport equation also needs to be carried out corresponding to the transformation of a physical domain to a computational domain. Transformation can be made using the algorithms, given by equations (11.2.6) and (11.2.7). Then, the general form of the

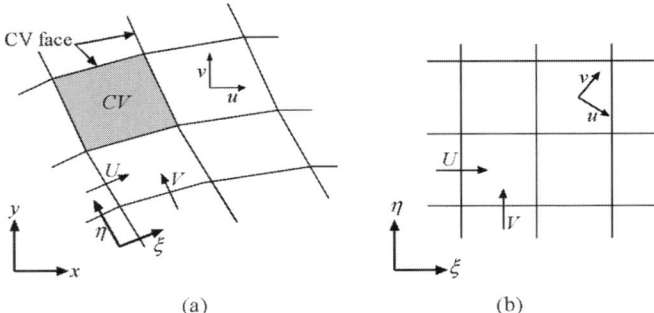

Figure 11.6. Grid configurations and velocity components in (a) the physical domain and (b) the computational domain. u and v are the velocity components in the Cartesian coordinates, and U and V are the contravariant velocities.

transport equation, equation (10.1.6), can be rearranged for the body-fitted coordinate system as follows [8].

$$\int_{\Delta t} \frac{\partial}{\partial t} \left\{ \iiint_{\Omega_{\xi\eta}} (\rho J \Phi) \, d\Omega \right\} dt + \int_{\Delta t} \iint_{\Gamma_{\xi\eta}} (\rho \mathbf{U} \Phi) \cdot \mathbf{n} \, d\Gamma \, dt$$

$$= \int_{\Delta t} \iint_{\Gamma_{\xi\eta}} \left(\frac{K}{J} D_i^j \nabla \Phi \right) \cdot \mathbf{n} \, d\Gamma \, dt + \int_{\Delta t} \iiint_{\Omega_{\xi\eta}} J S \, d\Omega \, dt \qquad (11.3.1)$$

where $\Gamma_{\xi\eta}$ and $\Omega_{\xi\eta}$ indicate the control-volume surface and the control volume for the body-fitted coordinate system. It is noted that equation (11.3.1) indicates the integral form of transport equation in the body-fitted coordinate system whilst equation (10.1.6) represents that in the Cartesian coordinate system. D_i^j is a geometric coefficient, J is a Jacobian and $\mathbf{U}(U, V)$ is a contravariant velocity at the cell face shown in figure 11.6. A detailed explanation will be given in section 11.4.

Equation (11.3.1) can be discretized by the control volume approach based on the non-staggered grid system, as shown in figure 11.7. In the non-staggered grid system, the velocity and the pressure are calculated at the center of a control volume, and the contravariant velocities at the cell faces are evaluated by the momentum interpolation method [9]. This interpolation method prevents the zig-zag problem, known to be a serious problem in non-staggered grids, as described in chapter 10.

11.4 Discretization method

11.4.1 Discretization of the integral form of transport equation

The computational domain is divided into quadrilateral control volumes, and all the Cartesian velocity components and scalar variables are solved

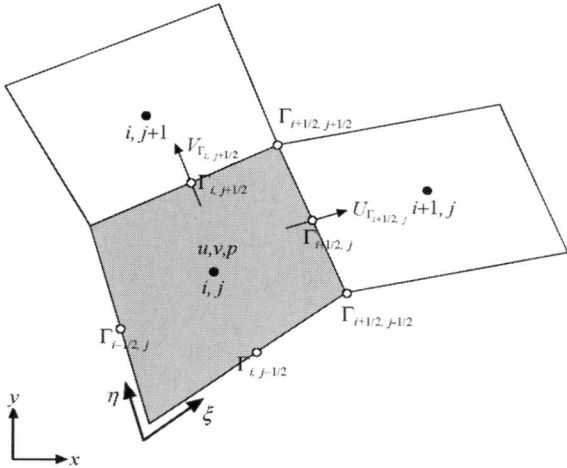

Figure 11.7. Notations of the calculation points, the cell-face points, the contravariant velocities, and the cell-centered velocities.

at the geometric center of each control volume, as shown in figure 11.8. The discretization of the transport equation in the physical domain is performed using the control volume approach.

Equation (11.3.1) can be rearranged as

$$\int_{\Delta t} \frac{\partial}{\partial t} \left\{ \iiint_{\Omega_{\xi\eta}} (\rho J \Phi) \, d\Omega \right\} dt + \int_{\Delta t} \iint_{\Gamma_{\xi\eta}} \left(\rho \mathbf{U} \Phi - \frac{K}{J} D_i^j \nabla \Phi \right) \cdot \mathbf{n} \, d\Gamma \, dt$$

$$= \int_{\Delta t} \iiint_{\Omega_{\xi\eta}} JS \, d\Omega \, dt. \qquad (11.4.1)$$

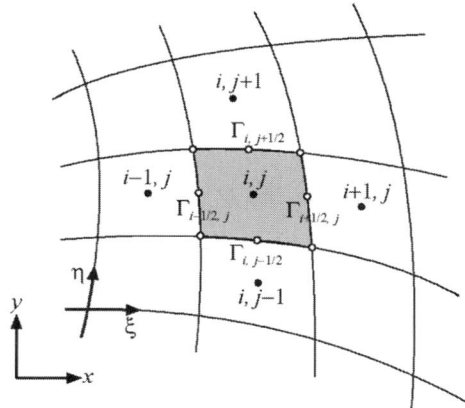

Figure 11.8. A typical control volume in the ξ and η coordinates.

For a two-dimensional problem, equation (11.4.1) can be expressed as

$$\frac{1}{\Delta t} \int_t^{t+\Delta t} \frac{\partial}{\partial t} \left\{ \iint_{\Omega_{\xi\eta}} (\rho J \Phi) \, d\xi \, d\eta \right\} dt$$
(1)

$$+ \frac{1}{\Delta t} \int_t^{t+\Delta t} \left\{ \int_{\Gamma_{i-1/2}}^{\Gamma_{i+1/2}} \left(\left(\rho U \Phi - \frac{K}{J} D_1^1 \frac{\partial \Phi}{\partial \xi} - \frac{K}{J} D_2^1 \frac{\partial \Phi}{\partial \eta} \right) d\eta \right) \right\} dt$$
(2) \quad (3) \quad (4)

$$+ \frac{1}{\Delta t} \int_t^{t+\Delta t} \left\{ \int_{\Gamma_{j-1/2}}^{\Gamma_{j+1/2}} \left(\left(\rho V \Phi - \frac{K}{J} D_1^2 \frac{\partial \Phi}{\partial \xi} - \frac{K}{J} D_2^2 \frac{\partial \Phi}{\partial \eta} \right) d\xi \right) \right\} dt$$
(2)' \quad (3)' \quad (4)'

$$= \frac{1}{\Delta t} \int_t^{t+\Delta t} JS \, d\xi \, d\eta \, dt \qquad (11.4.2)$$
(5)

where

$$U = \beta_1^1 u + \beta_2^1 v \qquad (11.4.3a)$$

$$V = \beta_1^2 u + \beta_2^2 v \qquad (11.4.3b)$$

$$D_1^1 = \left(\frac{\partial x}{\partial \eta}\right)^2 + \left(\frac{\partial y}{\partial \eta}\right)^2 = \beta_1^1 \beta_1^1 + \beta_2^1 \beta_2^1 \qquad (11.4.3c)$$

$$D_2^2 = \left(\frac{\partial x}{\partial \xi}\right)^2 + \left(\frac{\partial y}{\partial \xi}\right)^2 = \beta_1^2 \beta_1^2 + \beta_2^2 \beta_2^2 \qquad (11.4.3d)$$

$$D_2^1 = -\left(\frac{\partial x}{\partial \xi}\frac{\partial x}{\partial \eta} + \frac{\partial y}{\partial \xi}\frac{\partial y}{\partial \eta}\right) = \beta_1^2 \beta_1^1 + \beta_2^2 \beta_2^1 \qquad (11.4.3e)$$

$$D_1^2 = D_2^1 \qquad (11.4.3f)$$

$$\beta_1^1 = \frac{\partial y}{\partial \eta}, \quad \beta_2^1 = -\frac{\partial x}{\partial \eta}, \quad \beta_1^2 = -\frac{\partial y}{\partial \xi}, \quad \beta_2^2 = \frac{\partial x}{\partial \xi}. \qquad (11.4.3g)$$

The diffusion terms including D_1^1 and D_2^2 in equation (11.4.2) are the primary diffusion terms. They consist of the normal gradient forms, as described in chapter 10. However, the diffusion terms including D_2^1 and D_1^2 are expressed as $\partial^2 \Phi / (\partial \xi \, \partial \eta)$ and called as the cross diffusion terms or the secondary diffusion terms, which originate from the cross derivatives in the non-orthogonal coordinate system. When a perfect square cell is used, these values become zero.

Let us now evaluate the terms in equation (11.4.2) by applying this equation to the control volumes. If the transport property Φ is assumed to prevail over the control volume (i,j), term (1) can be evaluated as

$$\frac{1}{\Delta t}\int_{t}^{t+\Delta t}\frac{\partial}{\partial t}\left\{\iint_{\Omega_{\xi\eta}}(\rho J\Phi)\,d\xi\,d\eta\right\}dt \equiv \frac{1}{\Delta t}\iint_{\Omega_{\xi\eta}}\left\{\int_{t}^{t+\Delta t}\frac{\partial}{\partial t}(\rho J\Phi)\,dt\right\}d\xi\,d\eta$$

$$= \frac{1}{\Delta t}\iint_{\Omega_{\xi\eta}}\rho J(\Phi^{t+\Delta t}-\Phi^{t})\,d\xi\,d\eta$$

$$= \frac{\rho}{\Delta t}J_{i,j}(\Phi_{i,j}^{t+\Delta t}-\Phi_{i,j}^{t})$$

$$= \frac{\rho}{\Delta t}J_{i,j}(\Phi_{i,j}^{n+1}-\Phi_{i,j}^{n}). \qquad (11.4.4)$$

Let us now consider the evaluation of term (2) in equation (11.4.2), which represents the first convection term. If we take the fully implicit scheme for the time integration and the central difference scheme for the spatial derivatives, then we obtain

$$\frac{1}{\Delta t}\int_{t}^{t+\Delta t}\left\{\int_{\Gamma_{i-1/2}}^{\Gamma_{i+1/2}}(\rho U\Phi)\,d\eta\right\}dt = (\rho U_{\Gamma_{i+1/2}})\Phi\bigg|_{\Gamma_{i+1/2}}^{n+1} - (\rho U_{\Gamma_{i-1/2}})\Phi\bigg|_{\Gamma_{i-1/2}}^{n+1}.$$

$$(11.4.5)$$

By applying equation (10.2.12a), equation (11.4.5) can be rewritten as

$$\frac{1}{\Delta t}\int_{t}^{t+\Delta t}\left\{\int_{\Gamma_{i-1/2}}^{\Gamma_{i+1/2}}(\rho U\Phi)\,d\eta\right\}dt$$

$$= (\rho U_{\Gamma_{i+1/2}})\frac{\Phi_{i+1,j}^{n+1}+\Phi_{i,j}^{n+1}}{2} - (\rho U_{\Gamma_{i-1/2}})\frac{\Phi_{i-1,j}^{n+1}+\Phi_{i,j}^{n+1}}{2}. \qquad (11.4.6)$$

Term (3) in equation (11.4.2), which represents the primary diffusion term, can be expressed similarly to equation (10.3.6), as follows.

$$-\frac{1}{\Delta t}\int_{t}^{t+\Delta t}\left\{\int_{\Gamma_{i-1/2}}^{\Gamma_{i+1/2}}\left(\frac{K}{J}D_{1}^{1}\frac{\partial\Phi}{\partial\xi}\right)d\eta\right\}dt$$

$$= -\left\{\frac{K}{J_{\Gamma_{i+1/2}}}D_{1\Gamma_{i+1/2}}^{1}(\Phi_{i+1,j}^{n+1}-\Phi_{i,j}^{n+1}) - \frac{K}{J_{\Gamma_{i-1/2}}}D_{1\Gamma_{i-1/2}}^{1}(\Phi_{i,j}^{n+1}-\Phi_{i-1,j}^{n+1})\right\}.$$

$$(11.4.7)$$

Term (4), which represents the secondary diffusion term originated from the non-orthogonality of grids, can be integrated as

$$-\frac{1}{\Delta t}\int_{t}^{t+\Delta t}\left\{\int_{\Gamma_{i-1/2}}^{\Gamma_{i+1/2}}\left(\frac{K}{J}D_{2}^{1}\frac{\partial\Phi}{\partial\eta}\right)d\eta\right\}dt$$

$$= -K\left\{\left(\frac{D_{2}^{1}}{J}J\frac{\partial\Phi}{\partial\eta}\right)_{\Gamma_{i+1/2}}^{n+1} - \left(\frac{D_{2}^{1}}{J}J\frac{\partial\Phi}{\partial\eta}\right)_{\Gamma_{i-1/2}}^{n+1}\right\}$$

$$= -\frac{K}{2}\left\{\left(\frac{D_2^1|_{i+1}}{J_{i+1}}\frac{\partial\Phi}{\partial\eta}\bigg|_{i+1}^{n+1} + \frac{D_2^1|_i}{J_i}\frac{\partial\Phi}{\partial\eta}\bigg|_i^{n+1}\right)\right.$$

$$\left. -\left(\frac{D_2^1|_{i-1}}{J_{i-1}}\frac{\partial\Phi}{\partial\eta}\bigg|_{i-1}^{n+1} + \frac{D_2^1|_i}{J_i}\frac{\partial\Phi}{\partial\eta}\bigg|_i^{n+1}\right)\right\}. \quad (11.4.8)$$

Here,

$$\frac{\partial\Phi}{\partial\eta}\bigg|_{i\pm 1}^{n+1}\left(\equiv \frac{\partial\Phi}{\partial\eta}\bigg|_{i\pm 1,j}^{n+1}\right)$$

can be evaluated as follows.

$$\frac{\partial\Phi}{\partial\eta}\bigg|_{i\pm 1,j}^{n+1} = \left(\frac{\Phi_{j+1}^{n+1} - \Phi_{j-1}^{n+1}}{2\Delta\eta}\right)_{i\pm 1} = \frac{\Phi_{i\pm 1,j+1}^{n+1} - \Phi_{i\pm 1,j-1}^{n+1}}{2}.$$

Then, equation (11.4.8) can be rewritten as

$$-\frac{1}{\Delta t}\int_t^{t+\Delta t}\left\{\int_{\Gamma_{i-1/2}}^{\Gamma_{i+1/2}}\left(\frac{K}{J}D_2^1\frac{\partial\Phi}{\partial\eta}\right)d\eta\right\}dt$$

$$= -\left\{\frac{K}{2}\left(\frac{D_2^1|_{i+1}}{J_{i+1}}\frac{\Phi_{i+1,j+1}^{n+1} - \Phi_{i+1,j-1}^{n+1}}{2} - \frac{D_2^1|_{i-1}}{J_{i-1}}\frac{\Phi_{i-1,j+1}^{n+1} - \Phi_{i-1,j-1}^{n+1}}{2}\right)\right\}.$$

$$(11.4.9)$$

In equation (11.4.9), besides the usual neighboring four nodes, i.e., $(i+1, j)$, $(i-1, j)$, $(i, j+1)$ and $(i, j-1)$, extra Φ quantities are also considered such as for the nodes $(i+1, j+1)$, $(i+1, j-1)$, $(i-1, j+1)$ and $(i-1, j-1)$. These terms are generated from the cross derivatives in equation (11.4.2), resulting from the non-orthogonal coordinate system. They are usually very small, and thus can be combined with the source term and treated as known quantities. Similarly, terms (2)', (3)' and (4)' in equation (11.4.2) can also be evaluated.

Finally, term (5) in equation (11.4.2), which represents the source term, can be integrated as follows.

$$\frac{1}{\Delta t}\int_t^{t+\Delta t} JS\,d\xi\,d\eta\,dt = J_{i,j}S. \quad (11.4.10)$$

The resultant algebraic equation for a variable Φ can be written in the following general form.

$$a_{i,j}\Phi_{i,j}^{n+1} = a_{i+1,j}\Phi_{i+1,j}^{n+1} + a_{i-1,j}\Phi_{i-1,j}^{n+1} + a_{i,j+1}\Phi_{i,j+1}^{n+1} + a_{i,j-1}\Phi_{i,j-1}^{n+1} + b_{i,j}.$$

$$(11.4.11)$$

where

$$a_{i+1,j} = \left(D_{\Gamma_{i+1}} - \frac{F_{\Gamma_{i+1/2}}}{2}\right) \quad (11.4.12a)$$

$$a_{i-1,j} = \left(D_{\Gamma_{i-1}} + \frac{F_{\Gamma_{i-1/2}}}{2}\right) \tag{11.4.12b}$$

$$a_{i,j+1} = \left(D_{\Gamma_{j+1}} - \frac{F_{\Gamma_{j+1/2}}}{2}\right) \tag{11.4.12c}$$

$$a_{i,j-1} = \left(D_{\Gamma_{j-1}} + \frac{F_{\Gamma_{j-1/2}}}{2}\right) \tag{11.4.12d}$$

$$a_{i,j} = a_{i+1,j} + a_{i-1,j} + a_{i,j+1} + a_{i,j-1}$$
$$+ (F_{\Gamma_{i+1/2}} - F_{\Gamma_{i-1/2}} + F_{\Gamma_{j+1/2}} - F_{\Gamma_{j-1/2}}) + J_{i,j}\frac{\rho}{\Delta t} \tag{11.4.12e}$$

$$b_{i,j} = J_{i,j} S + J_{i,j}\frac{\rho}{\Delta t}\Phi_{i,j}^n + S^c \tag{11.4.12f}$$

$$F_{\Gamma_{i+1/2}} \equiv \rho U_{\Gamma_{i+1/2}} \quad D_{\Gamma_{i+1/2}} \equiv \mathrm{K}\left(\frac{D_1^1}{J}\right)_{\Gamma_{i+1/2}} \tag{11.4.13a}$$

$$F_{\Gamma_{i-1/2}} \equiv \rho U_{\Gamma_{i-1/2}} \quad D_{\Gamma_{i-1/2}} \equiv \mathrm{K}\left(\frac{D_1^1}{J}\right)_{\Gamma_{i-1/2}} \tag{11.4.13b}$$

$$F_{\Gamma_{j+1/2}} \equiv \rho V_{\Gamma_{j+1/2}} \quad D_{\Gamma_{j+1/2}} \equiv \mathrm{K}\left(\frac{D_2^2}{J}\right)_{\Gamma_{j+1/2}} \tag{11.4.13c}$$

$$F_{\Gamma_{j-1/2}} \equiv \rho V_{\Gamma_{j-1/2}} \quad D_{\Gamma_{j-1/2}} \equiv \mathrm{K}\left(\frac{D_2^2}{J}\right)_{\Gamma_{j-1/2}} \tag{11.4.13d}$$

$$S^c = \mathrm{K}\left\{\frac{1}{4}\frac{D_2^1|_{i+1,j}}{J_{i+1,j}}(\Phi_{i+1,j+1}^n - \Phi_{i+1,j-1}^n) - \frac{1}{4}\frac{D_2^1|_{i-1,j}}{J_{i-1,j}}(\Phi_{i-1,j+1}^n - \Phi_{i-1,j-1}^n)\right.$$
$$+ \frac{1}{4}\frac{D_1^2|_{i,j+1}}{J_{i,j+1}}(\Phi_{i+1,j+1}^n - \Phi_{i-1,j+1}^n)$$
$$\left. - \frac{1}{4}\frac{D_1^2|_{i,j-1}}{J_{i,j-1}}(\Phi_{i+1,j-1}^n - \Phi_{i-1,j-1}^n)\right\} \tag{11.4.13e}$$

where S^c is the source term caused by the non-orthogonality of the grids.

11.4.2 Discretization of momentum equations

Similar to the discretization of momentum equations described in chapter 10 (section 10.3.2), in the body-fitted coordinate system, the integral form of the pressure gradient term in the u-momentum equation in equation (11.4.2) can be evaluated as

$$\frac{1}{\Delta t}\int_t^{t+\Delta t}\iint_{\Omega_{\xi\eta}} JS\,d\xi\,d\eta\,dt$$

$$= -\frac{1}{\Delta t}\int_t^{t+\Delta t}\iint_{\Omega_{\xi\eta}} J\left(\frac{\partial p}{\partial x}\right)d\xi\,d\eta\,dt$$

$$= -\frac{1}{\Delta t}\int_t^{t+\Delta t}\iint_{\Omega_{\xi\eta}} J\left(\frac{\partial p}{\partial \xi}\frac{\partial \xi}{\partial x}+\frac{\partial p}{\partial \eta}\frac{\partial \eta}{\partial x}\right)d\xi\,d\eta\,dt$$

$$= -\frac{1}{\Delta t}\int_t^{t+\Delta t}\iint_{\Omega_{\xi\eta}} \left(\frac{\partial p}{\partial \xi}\frac{\partial y}{\partial \eta}-\frac{\partial p}{\partial \eta}\frac{\partial y}{\partial \xi}\right)d\xi\,d\eta\,dt$$

$$= -\frac{1}{\Delta t}\int_t^{t+\Delta t}\iint_{\Omega_{\xi\eta}} \left(\frac{\partial p}{\partial \xi}\frac{\partial y}{\partial \eta}\right)d\xi\,d\eta\,dt+\frac{1}{\Delta t}\int_t^{t+\Delta t}\iint_{\Omega_{\xi\eta}} \left(\frac{\partial p}{\partial \eta}\frac{\partial y}{\partial \xi}\right)d\xi\,d\eta\,dt$$

$$= -\left(\frac{p_{i+1,j}-p_{i-1,j}}{2}\right)\left(\frac{\partial y}{\partial \eta}\right)_{i,j}+\left(\frac{p_{i,j+1}-p_{i,j-1}}{2}\right)\left(\frac{\partial y}{\partial \xi}\right)_{i,j}. \quad (11.4.14)$$

In equation (11.4.14), the terms $(\partial y/\partial \eta)_{i,j}$ and $(\partial y/\partial \xi)_{i,j}$ become unity and zero, respectively, in a particular case such as when the (x, y) coordinates coincide with the (ξ, η) coordinates. Then, equation (11.4.14) simplifies to equation (10.3.13).

The momentum equations can be solved implicitly at the cell-centered locations [1, 8]. The discretized forms of momentum equations for the cell-centered Cartesian velocity components can be written as follows.

$$u_{i,j}^{n+1} = (H_u)_{i,j} + (D_u^1)_{i,j}(p_{\Gamma_{i-1}} - p_{\Gamma_{i+1}}) + (D_u^2)_{i,j}(p_{\Gamma_{j-1}} - p_{\Gamma_{j+1}}) + (1-\alpha)u_{i,j}^n$$
(11.4.15)

$$v_{i,j}^{n+1} = (H_v)_{i,j} + (D_v^1)_{i,j}(p_{\Gamma_{i-1}} - p_{\Gamma_{i+1}}) + (D_v^2)_{i,j}(p_{\Gamma_{j-1}} - p_{\Gamma_{j+1}}) + (1-\alpha)v_{i,j}^n$$
(11.4.16)

where

$$H_u = \frac{\alpha(\sum_{nb} a_{nb}^u u_{nb} + b_u)}{a_{i,j}^u} \quad (11.4.17a)$$

$$H_v = \frac{\alpha(\sum_{nb} a_{nb}^v v_{nb} + b_v)}{a_{i,j}^v} \quad (11.4.17b)$$

$$D_u^1 = \frac{\alpha\beta_1^1}{a_{i,j}^u}, \qquad D_u^2 = \frac{\alpha\beta_1^2}{a_{i,j}^u} \quad (11.4.17c)$$

$$D_v^1 = \frac{\alpha\beta_2^1}{a_{i,j}^v}, \qquad D_v^2 = \frac{\alpha\beta_2^2}{a_{i,j}^v} \quad (11.4.17d)$$

where α is the under-relaxation factor and the subscript nb indicates the neighboring node.

The momentum equations for the Cartesian velocity components at the cell-face locations can also be written as follows.

$$u^{n+1}_{\Gamma_{i+1/2,j}} = (H_u)_{\Gamma_{i+1/2,j}} + (D^1_u)_{\Gamma_{i+1/2,j}}(p_{i,j} - p_{i+1,j})$$
$$+ (D^2_u)_{\Gamma_{i+1/2,j}}(p_{\Gamma_{i+1/2,j-1/2}} - p_{\Gamma_{i+1/2,j+1/2}}) + (1-\alpha)u^n_{\Gamma_{i+1/2,j}} \quad (11.4.18a)$$

$$v^{n+1}_{\Gamma_{i+1/2,j}} = (H_v)_{\Gamma_{i+1/2,j}} + (D^1_v)_{\Gamma_{i+1/2,j}}(p_{i,j} - p_{i+1,j})$$
$$+ (D^2_v)_{\Gamma_{i+1/2,j}}(p_{\Gamma_{i+1/2,j-1/2}} - p_{\Gamma_{i+1/2,j+1/2}}) + (1-\alpha)v^n_{\Gamma_{i+1/2,j}} \quad (11.4.18b)$$

$$u^{n+1}_{\Gamma_{i-1/2,j}} = (H_u)_{\Gamma_{i-1/2,j}} + (D^1_u)_{\Gamma_{i-1/2,j}}(p_{i-1,j} - p_{i,j})$$
$$+ (D^2_u)_{\Gamma_{i-1/2,j}}(p_{\Gamma_{i-1/2,j-1/2}} - p_{\Gamma_{i-1/2,j+1/2}}) + (1-\alpha)u^n_{\Gamma_{i-1/2,j}} \quad (11.4.18c)$$

$$v^{n+1}_{\Gamma_{i-1/2,j}} = (H_v)_{\Gamma_{i-1/2,j}} + (D^1_v)_{\Gamma_{i-1/2,j}}(p_{i-1,j} - p_{i,j})$$
$$+ (D^2_v)_{\Gamma_{i-1/2,j}}(p_{\Gamma_{i-1/2,j-1/2}} - p_{\Gamma_{i-1/2,j+1/2}}) + (1-\alpha)v^n_{\Gamma_{i-1/2,j}} \quad (11.4.18d)$$

$$u^{n+1}_{\Gamma_{i,j+1/2}} = (H_u)_{\Gamma_{i,j+1/2}} + (D^1_u)_{\Gamma_{i,j+1/2}}(p_{\Gamma_{i-1/2,j+1/2}} - p_{\Gamma_{i+1/2,j+1/2}})$$
$$+ (D^2_u)_{\Gamma_{i,j+1/2}}(p_{i,j} - p_{i,j+1}) + (1-\alpha)u^n_{\Gamma_{i,j+1/2}} \quad (11.4.18e)$$

$$v^{n+1}_{\Gamma_{i,j+1/2}} = (H_v)_{\Gamma_{i,j+1/2}} + (D^1_v)_{\Gamma_{i,j+1/2}}(p_{\Gamma_{i-1/2,j+1/2}} - p_{\Gamma_{i+1/2,j+1/2}})$$
$$+ (D^2_v)_{\Gamma_{i,j+1/2}}(p_{i,j} - p_{i,j+1}) + (1-\alpha)v^n_{\Gamma_{i,j+1/2}} \quad (11.4.18f)$$

$$u^{n+1}_{\Gamma_{i,j-1/2}} = (H_u)_{\Gamma_{i,j-1/2}} + (D^1_u)_{\Gamma_{i,j-1/2}}(p_{\Gamma_{i-1/2,j-1/2}} - p_{\Gamma_{i+1/2,j-1/2}})$$
$$+ (D^2_u)_{\Gamma_{i,j-1/2}}(p_{i,j-1} - p_{i,j}) + (1-\alpha)u^n_{\Gamma_{i,j-1/2}} \quad (11.4.18g)$$

$$v^{n+1}_{\Gamma_{i,j-1/2}} = (H_v)_{\Gamma_{i,j-1/2}} + (D^1_v)_{\Gamma_{i,j-1/2}}(p_{\Gamma_{i-1/2,j-1/2}} - p_{\Gamma_{i+1/2,j-1/2}})$$
$$+ (D^2_v)_{\Gamma_{i,j-1/2}}(p_{i,j-1} - p_{i,j}) + (1-\alpha)v^n_{\Gamma_{i,j-1/2}} \quad (11.4.18h)$$

The cell-face Cartesian velocity components can be evaluated through the interpolation of the momentum equations for the neighboring cell-centered Cartesian velocity components. The detailed procedures are referred to in the literature [1, 2, 7, 8, 10].

11.4.3 The SIMPLE algorithm

The SIMPLE method was employed to solve the discretized continuity and momentum equations [11]. Following are the details of the pressure-correction associated with employing the contravariant velocity components,

11.4.3.1 The pressure and velocity corrections

The integrated continuity equation for a control volume around a grid point (i,j) shown in figure 11.8 can be written as

$$(\rho U)_{\Gamma_{i+1/2,j}} - (\rho U)_{\Gamma_{i-1/2,j}} + (\rho V)_{\Gamma_{i,j+1/2}} - (\rho V)_{\Gamma_{i,j-1/2}} = 0. \quad (11.4.19)$$

The momentum equations for the contravariant velocity components can be derived by substituting the momentum equations for the cell-face Cartesian velocity components, equations (11.4.18a) through (11.4.18h), into equations (11.4.3a) and (11.4.3b). Then, we have the following equations for the contravariant velocity components.

$$U^{n+1}_{\Gamma_{i+1/2,j}} = (H_U)_{\Gamma_{i+1/2,j}} + (D^1_U)_{\Gamma_{i+1/2,j}}(p_{i,j} - p_{i+1,j})$$
$$+ (D^2_U)_{\Gamma_{i+1/2,j}}(p_{\Gamma_{i+1/2,j-1/2}} - p_{\Gamma_{i+1/2,j+1/2}}) + (1-\alpha)U^n_{\Gamma_{i+1/2,j}}$$
$$(11.4.20a)$$

$$V^{n+1}_{\Gamma_{i+1/2,j}} = (H_V)_{\Gamma_{i+1/2,j}} + (D^1_V)_{\Gamma_{i+1/2,j}}(p_{i,j} - p_{i+1,j})$$
$$+ (D^2_V)_{\Gamma_{i+1/2,j}}(p_{\Gamma_{i+1/2,j-1/2}} - p_{\Gamma_{i+1/2,j+1/2}}) + (1-\alpha)V^n_{\Gamma_{i+1/2,j}}$$
$$(11.4.20b)$$

$$U^{n+1}_{\Gamma_{i-1/2,j}} = (H_U)_{\Gamma_{i-1/2,j}} + (D^1_U)_{\Gamma_{i-1/2,j}}(p_{i-1,j} - p_{i,j})$$
$$+ (D^2_U)_{\Gamma_{i-1/2,j}}(p_{\Gamma_{i-1/2,j-1/2}} - p_{\Gamma_{i-1/2,j+1/2}}) + (1-\alpha)U^n_{\Gamma_{i-1/2,j}}$$
$$(11.4.20c)$$

$$V^{n+1}_{\Gamma_{i-1/2,j}} = (H_V)_{\Gamma_{i-1/2,j}} + (D^1_V)_{\Gamma_{i-1/2,j}}(p_{i-1,j} - p_{i,j})$$
$$+ (D^2_V)_{\Gamma_{i-1/2,j}}(p_{\Gamma_{i-1/2,j-1/2}} - p_{\Gamma_{i-1/2,j+1/2}}) + (1-\alpha)V^n_{\Gamma_{i-1/2,j}}$$
$$(11.4.20d)$$

$$U^{n+1}_{\Gamma_{i,j+1/2}} = (H_U)_{\Gamma_{i,j+1/2}} + (D^1_U)_{\Gamma_{i,j+1/2}}(p_{\Gamma_{i-1/2,j+1/2}} - p_{\Gamma_{i+1/2,j+1/2}})$$
$$+ (D^2_U)_{\Gamma_{i,j+1/2}}(p_{i,j} - p_{i,j+1}) + (1-\alpha)U^n_{\Gamma_{i,j+1/2}} \quad (11.4.20e)$$

$$V^{n+1}_{\Gamma_{i,j+1/2}} = (H_V)_{\Gamma_{i,j+1/2}} + (D^1_V)_{\Gamma_{i,j+1/2}}(p_{\Gamma_{i-1/2,j+1/2}} - p_{\Gamma_{i+1/2,j+1/2}})$$
$$+ (D^2_V)_{\Gamma_{i,j+1/2}}(p_{i,j} - p_{i,j+1}) + (1-\alpha)V^n_{\Gamma_{i,j+1/2}} \quad (11.4.20f)$$

Discretization method 193

$$U^{n+1}_{\Gamma_{i,j-1/2}} = (H_U)_{\Gamma_{i,j-1/2}} + (D^1_U)_{\Gamma_{i,j-1/2}}(p_{\Gamma_{i-1/2,j-1/2}} - p_{\Gamma_{i+1/2,j-1/2}})$$
$$+ (D^2_U)_{\Gamma_{i,j-1/2}}(p_{i,j-1} - p_{i,j}) + (1-\alpha)U^n_{\Gamma_{i,j-1/2}} \quad (11.4.20\text{g})$$

$$V^{n+1}_{\Gamma_{i,j-1/2}} = (H_V)_{\Gamma_{i,j-1/2}} + (D^1_V)_{\Gamma_{i,j-1/2}}(p_{\Gamma_{i-1/2,j-1/2}} - p_{\Gamma_{i+1/2,j-1/2}})$$
$$+ (D^2_V)_{\Gamma_{i,j-1/2}}(p_{i,j-1} - p_{i,j}) + (1-\alpha)V^n_{\Gamma_{i,j-1/2}} \quad (11.4.20\text{h})$$

where

$$H_U = \beta^1_1 H_u + \beta^1_2 H_v \quad (11.4.21\text{a})$$

$$H_V = \beta^2_1 H_u + \beta^2_2 H_v \quad (11.4.21\text{b})$$

$$D^1_U = \beta^1_1 D^1_u + \beta^1_2 D^1_v, \qquad D^2_U = \beta^1_1 D^2_u + \beta^1_2 D^2_v \quad (11.4.21\text{c})$$

$$D^1_V = \beta^2_1 D^1_u + \beta^2_2 D^1_v, \qquad D^2_V = \beta^2_1 D^2_u + \beta^2_2 D^2_v. \quad (11.4.21\text{d})$$

The contravariant velocity components, given by equations (11.4.20a) through (11.4.20h), do not generally satisfy the mass conservation unless the pressure field is correct. The velocity components are corrected to satisfy the continuity equation by the following velocity correction formulae.

$$U_{\Gamma_{i+1/2,j}} = U^*_{\Gamma_{i+1/2,j}} + (D^1_U)_{\Gamma_{i+1/2,j}}(\delta p_{i,j} - \delta p_{i+1,j}) \quad (11.4.22\text{a})$$

$$V_{\Gamma_{i+1/2,j}} = V^*_{\Gamma_{i+1/2,j}} + (D^1_V)_{\Gamma_{i+1/2,j}}(\delta p_{i,j} - \delta p_{i+1,j}) \quad (11.4.22\text{b})$$

$$U_{\Gamma_{i-1/2,j}} = U^*_{\Gamma_{i-1/2,j}} + (D^1_U)_{\Gamma_{i-1/2,j}}(\delta p_{i-1,j} - \delta p_{i,j}) \quad (11.4.22\text{c})$$

$$V_{\Gamma_{i-1/2,j}} = V^*_{\Gamma_{i-1/2,j}} + (D^1_V)_{\Gamma_{i-1/2,j}}(\delta p_{i-1,j} - \delta p_{i,j}) \quad (11.4.22\text{d})$$

$$U_{\Gamma_{i,j+1/2}} = U^*_{\Gamma_{i,j+1/2}} + (D^2_U)_{\Gamma_{i,j+1/2}}(\delta p_{i,j} - \delta p_{i,j+1}) \quad (11.4.22\text{e})$$

$$V_{\Gamma_{i,j+1/2}} = V^*_{\Gamma_{i,j+1/2}} + (D^2_V)_{\Gamma_{i,j+1/2}}(\delta p_{i,j} - \delta p_{i,j+1}) \quad (11.4.22\text{f})$$

$$U_{\Gamma_{i,j-1/2}} = U^*_{\Gamma_{i,j-1/2}} + (D^2_U)_{\Gamma_{i,j-1/2}}(\delta p_{i,j-1} - \delta p_{i,j}) \quad (11.4.22\text{g})$$

$$V_{\Gamma_{i,j-1/2}} = V^*_{\Gamma_{i,j-1/2}} + (D^2_V)_{\Gamma_{i,j-1/2}}(\delta p_{i,j-1} - \delta p_{i,j}) \quad (11.4.22\text{h})$$

where δp is the pressure correction. The pressure correction terms on the corners of the grids are neglected, resulting in a simple and diagonal dominant pressure correction equation.

Substitution of the velocity-correction formulae into the continuity equation, equation (11.4.19), yields an equation for the pressure correction as follows.

$$a_{i,j}\delta p_{i,j} = a_{i+1,j}\delta p_{i+1,j} + a_{i-1,j}\delta p_{i-1,j} + a_{i,j+1}\delta p_{i,j+1} + a_{i,j-1}\delta p_{i,j-1} + b_{i,j}$$
$$(11.4.23)$$

where

$$a_{i+1,j} \equiv \rho(D_U^1)_{\Gamma_{i+1/2,j}} \qquad (11.4.24\text{a})$$

$$a_{i-1,j} \equiv \rho(D_U^1)_{\Gamma_{i-1/2,j}} \qquad (11.4.24\text{b})$$

$$a_{i,j+1} \equiv \rho(D_V^2)_{\Gamma_{i,j+1/2}} \qquad (11.4.24\text{c})$$

$$a_{i,j-1} \equiv \rho(D_V^2)_{\Gamma_{i,j-1/2}} \qquad (11.4.24\text{d})$$

$$a_{i,j} = a_{i+1,j} + a_{i-1,j} + a_{i,j+1} + a_{i,j-1} \qquad (11.4.24\text{e})$$

$$b_{i,j} = (\rho U^*)_{\Gamma_{i-1/2,j}} - (\rho U^*)_{\Gamma_{i+1/2,j}} + (\rho V^*)_{\Gamma_{i,j-1/2}} - (\rho V^*)_{\Gamma_{i,j+1/2}}. \qquad (11.4.24\text{f})$$

After solving the above pressure-correction equation, the pressure and the cell-centered Cartesian velocity components are updated by the following formulae.

$$p = p^* + \alpha \delta p \qquad (11.4.25)$$

$$u_{i,j} = u_{i,j}^* + (D_u^1)_{i,j}(\delta p_{\Gamma_{i-1/2,j}} - \delta p_{\Gamma_{i+1/2,j}}) + (D_u^2)_{i,j}(\delta p_{\Gamma_{i,j-1/2}} - \delta p_{\Gamma_{i,j+1/2}})$$
$$(11.4.26)$$

$$v_{i,j} = v_{i,j}^* + (D_v^1)_{i,j}(\delta p_{\Gamma_{i-1/2,j}} - \delta p_{\Gamma_{i+1/2,j}}) + (D_v^2)_{i,j}(\delta p_{\Gamma_{i,j-1/2}} - \delta p_{\Gamma_{i,j+1/2}})$$
$$(11.4.27)$$

where p^* is the guessed pressure field. The above velocity-correction equations for the cell-centered Cartesian velocity components are derived under the assumption of a linearly varying pressure field. The details of the pressure-correction scheme are referred to in the literature [3, 7, 12, 13].

11.4.3.2 Solution scheme

The overall solution procedure is similar to that of the SIMPLE procedure mentioned in chapter 10. It can be summarized as follows:

1. Guess the pressure, velocity and other transport property fields, p^*, u^*, v^* and Φ^*.
2. Solve the discretized momentum equations, equations (11.4.15) and (11.4.16), to obtain u^* and v^*.
3. Solve the discretized momentum equations for the contravariant velocity components, equations (11.4.20), to obtain U^* and V^*.
4. Solve the pressure correction equation, equation (11.4.23), to obtain δp.
5. Correct the pressure field by equation (11.4.25).
6. Correct the velocity fields using the velocity-correction formulae, equations (11.4.26) and (11.4.27).
7. Solve the discretized equations for other Φ (such as temperature and species concentration, etc.) using the general form of the discretized equation, equation (11.4.11) if they influence the flow field through fluid

properties, source terms, such as in the problems including natural convection, etc. If a particular Φ does not affect the flow field, it is better to calculate it after a converged solution for the flow field has been obtained.
8. Set the corrected values as the new guessed values, $p^* = p$, $u^* = u$, $v^* = v$ and $\Phi^* = \Phi$, return to step 2, and repeat the whole procedure until a converged solution is obtained.

11.5 The VOF method in the body-fitted coordinate system

The standard VOF method has been known as more effective than the marker and cell method in the case of tracing the free surface in mould filling because it is based on the mass conservation equation and uses less memory size. However, the standard VOF method cannot be directly applied to the body-fitted coordinate system. The standard VOF method was modified for the body-fitted coordinate system as follows.

In the VOF method, the fluid function F is calculated by

$$\frac{\partial F}{\partial t} + u \frac{\partial F}{\partial x} + v \frac{\partial F}{\partial y} = 0. \tag{11.5.1}$$

Similar to the case for the transformation of equation (11.4.1), equation (11.5.1) can also be transformed as follows.

$$J \frac{\partial F}{\partial t} + U \frac{\partial F}{\partial \xi} + V \frac{\partial F}{\partial \eta} = 0. \tag{11.5.2}$$

The discretized equation of equation (11.5.2) is given by

$$F^{n+1} = F^n - \frac{\Delta t}{J}(U\,\delta F + V\,\delta F). \tag{11.5.3}$$

In order to preserve the sharp definition of a free surface, special care must be taken in computing the cell flux, $U\,\delta F$. The Donor and Acceptor Flux Approximation (DAFA), which is usually used in the SOLA-VOF method, is adopted in the present method. However, the standard DAFA, which is proposed for an orthogonal grid system, cannot be used directly. The modified DAFA for a non-orthogonal grid system is given as follows.

$$U\,\delta F \cdot \Delta t = \text{sgn}(U)\min \|F_{\text{AD}} \cdot |U \cdot \Delta t| + CF_\xi, F_{\text{D}} L_\xi\| \tag{11.5.4a}$$

$$V\,\delta F \cdot \Delta t = \text{sgn}(V)\min \|F_{\text{AD}} \cdot |V \cdot \Delta t| + CF_\eta, F_{\text{D}} L_\eta\|. \tag{11.5.4b}$$

Here,

$$CF_\xi = \max \|(\langle F \rangle - F_{\text{AD}}) \cdot |U \cdot \Delta t| - (\langle F \rangle - F_{\text{AD}}) L_\xi, 0.0\| \tag{11.5.5a}$$

$$CF_\eta = \max \|(\langle F \rangle - F_{\text{AD}}) \cdot |V \cdot \Delta t| - (\langle F \rangle - F_{\text{AD}}) L_\eta, 0.0\|. \tag{11.5.5b}$$

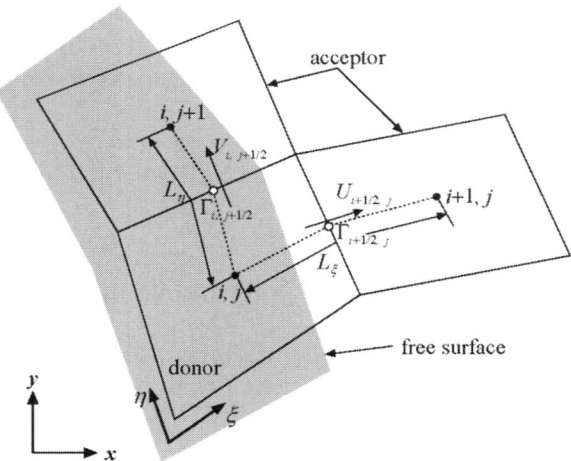

Figure 11.9. Notation of the donors and the acceptors in the body-fitted coordinate system.

and

$$\langle F \rangle = \max \|F_{\mathrm{D}}, F_{\mathrm{DM}}, 0.1\|. \quad (11.5.6)$$

The selection of the subscripts D, DM and AD is similar to those described in chapter 10. The distances L_ξ and L_η, and the contravariant velocity components are shown in figure 11.9.

11.6 Treatment of a surface cell

11.6.1 Momentum equations

In order to calculate the momentum on a surface cell in figure 11.10, the following tangential and normal stress conditions are applied.

The normal stress condition is given by

$$\sigma_{\xi\xi}|_{\Gamma_{i+1/2}} = 0 \quad (11.6.1)$$

and the tangential stress condition is

$$\tau_{\xi\eta}|_{\Gamma_{i+1/2}} = 0. \quad (11.6.2)$$

In order to explain the procedure for applying these conditions, the momentum equations for a two-dimensional problem are represented with respect to the normal and tangential stresses as follows.

$$\rho \frac{\mathrm{D}u}{\mathrm{D}t} = \frac{\partial \sigma_{xx}}{\partial x} + \frac{\partial \tau_{xy}}{\partial y} + \rho \cdot f_{bx} \quad (11.6.3\mathrm{a})$$

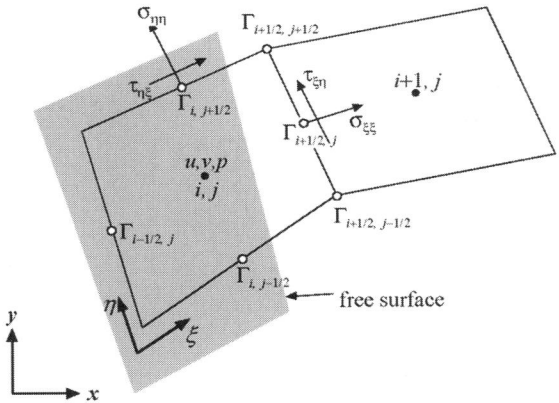

Figure 11.10. Distribution of the normal and shear stresses at the cell faces.

$$\rho \frac{Dv}{Dt} = \frac{\partial \tau_{yx}}{\partial x} + \frac{\partial \sigma_{yy}}{\partial y} + \rho \cdot f_{by}. \qquad (11.6.3b)$$

The above equations are transformed as follows.

$$\rho \frac{Du}{Dt} = \frac{\partial}{\partial \xi}(\sigma_{xx}\xi_x + \tau_{xy}\xi_y) + \frac{\partial}{\partial \eta}(\sigma_{xx}\eta_x + \tau_{xy}\eta_y) + \rho \cdot f_{bx} \qquad (11.6.4a)$$

$$\rho \frac{Dv}{Dt} = \frac{\partial}{\partial \xi}(\tau_{yx}\xi_x + \sigma_{yy}\xi_y) + \frac{\partial}{\partial \eta}(\tau_{yx}\eta_x + \sigma_{yy}\eta_y) + \rho \cdot f_{by}. \qquad (11.6.4b)$$

According to the rule of transformation, the stresses in the body-fitted coordinates can be written as

$$\sigma_{\xi\xi} = \sigma_{xx}\xi_x + \tau_{xy}\xi_y \qquad (11.6.5a)$$
$$\sigma_{\eta\eta} = \sigma_{yy}\eta_x + \tau_{yx}\eta_y \qquad (11.6.5b)$$
$$\tau_{\xi\eta} = \sigma_{xx}\eta_x + \tau_{xy}\eta_y \qquad (11.6.5c)$$
$$\tau_{\eta\xi} = \tau_{yx}\xi_x + \sigma_{yy}\xi_y. \qquad (11.6.5d)$$

By substituting equation (11.6.5) into equation (11.6.4a), the discretized form is obtained as follows.

$$\rho \frac{\partial u}{\partial t} + (U \cdot u)_{\Gamma_{i+1/2}} - (U \cdot u)_{\Gamma_{i-1/2}} + (V \cdot u)_{\Gamma_{j+1/2}} - (V \cdot u)_{\Gamma_{j-1/2}}$$
$$= \sigma_{\xi\xi}|_{\Gamma_{i+1/2}} - \sigma_{\xi\xi}|_{\Gamma_{i-1/2}} + \tau_{\xi\eta}|_{\Gamma_{j+1/2}} - \tau_{\xi\eta}|_{\Gamma_{j-1/2}} + \rho \cdot f_{bx}. \qquad (11.6.6)$$

If the stress condition is applied to equation (11.6.6) on a surface cell in figure 11.10, the equation can be simplified as follows.

$$\rho \frac{\partial u}{\partial t} + (U \cdot u)_{\Gamma_{i+1/2}} - (U \cdot u)_{\Gamma_{i-1/2}} + (V \cdot u)_{\Gamma_{j+1/2}} - (V \cdot u)_{\Gamma_{j-1/2}}$$
$$= -\sigma_{\xi\xi}|_{\Gamma_{i-1/2}} + \tau_{\xi\eta}|_{\Gamma_{j+1/2}} - \tau_{\xi\eta}|_{\Gamma_{j-1/2}} + \rho \cdot f_{bx} \qquad (11.6.7a)$$

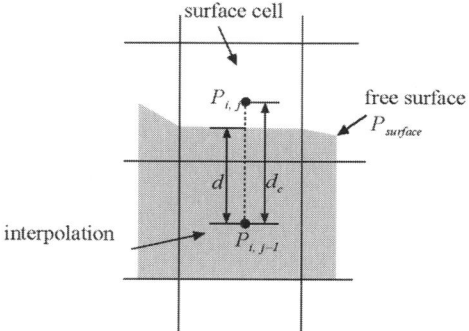

Figure 11.11. Calculation of the pressure of a surface cell according to the standard SOLA-VOF method.

and the equation for the v-component is similarly given by

$$\rho \frac{\partial v}{\partial t} + (U \cdot v)_{\Gamma_{i+1/2}} - (U \cdot v)_{\Gamma_{i-1/2}} + (V \cdot v)_{\Gamma_{j+1/2}} - (V \cdot v)_{\Gamma_{j-1/2}}$$
$$= -\sigma_{\xi\xi}|_{\Gamma_{i-1/2}} + \tau_{\xi\eta}|_{\Gamma_{j+1/2}} - \tau_{\xi\eta}|_{\Gamma_{j-1/2}} + \rho \cdot f_{by}. \qquad (11.6.7b)$$

11.6.2. Pressure at a surface cell

In the SOLA-VOF method, the pressure $P_{i,j}$ of a surface cell is evaluated by interpolating from its neighbor cells. The interpolation function is given by

$$P_{i,j} = \left(1 - \frac{d_c}{d}\right) P_{i,j-1} + \frac{d_c}{d} P_{\text{surface}}. \qquad (11.6.8)$$

The nomenclatures of equation (11.6.8) are shown in figure 11.11. The pressure $P_{i,j}$ is rapidly changed from a negative to a positive value during a small time interval because it depends on the distance d between the neighbor cell and the free surface, giving an unstable result and an increase in computational time. In order to solve this problem, the distance d in figure 11.11 is redefined as a distance from the neighbor cell to the across surface of the control volume of a surface cell, as shown in figure 11.12.

It is also to be noted that entrapped air in a closed loop is one of the principal factors affecting the evolution of a free surface in mould filling. Entrapped air may float due to the buoyancy force or deform because of its back pressure, affecting the geometry of the free surface. However, the effects of buoyancy force and a rise in the pressure of entrapped air or air bubbles are not considered in the present numerical model.

Treatment of a surface cell 199

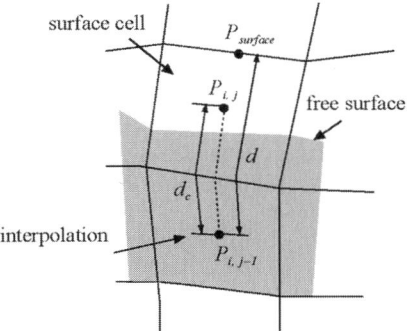

Figure 11.12. Evaluation of the pressure of a surface cell.

11.6.3 Contravariant velocity

The contravariant velocity on the surface of a control volume can be calculated by the momentum interpolation method [9] when two neighbor cells are not empty cells. Therefore, if an empty cell exists as in figure 11.10, the contravariant velocity is obtained not by the momentum interpolation method, but by the mass conservation equation suggested in the SMAC method [14].

For a two-dimensional problem, the mass conservation equation is given by

$$U_{\Gamma_{i+1/2}} - U_{\Gamma_{i-1/2}} + V_{\Gamma_{j+1/2}} - V_{\Gamma_{j-1/2}} = 0. \quad (11.6.9)$$

The contravariant velocity $U_{\Gamma_{i+1/2}}$ in figure 11.7 is calculated by equation (11.6.9) using the known velocities, $U_{\Gamma_{i-1/2}}$, $V_{\Gamma_{j+1/2}}$ and $V_{\Gamma_{j-1/2}}$. According to the scheme used in the SMAC method, 16 cases for a two-dimensional problem and 64 cases for a three-dimensional problem must be considered.

In addition, when an empty cell becomes a surface cell, the initial velocity u_{inlet} at the center point of a control volume can be estimated by averaging the values of its neighbor cells.

11.6.4 Determination of the direction normal to a free surface

It is important to find the direction normal to a free surface cell for calculating the volume of fluid and the pressure. In the SOLA-VOF method [15] the slope of a free surface is first calculated and then the direction normal to a free surface is selected. This method is very simple to use for two-dimensional mould filling problems, but not easy for three-dimensional problems since the definition of the slope for three-dimensional surfaces is very complicated. A new approach has been developed to determine the normal direction to a free surface.

The procedure is as follows:

1. Check neighbor cells and find the locations of the full fluid cells.
2. If there is no full fluid cell in the first step, check flow fluxes through all surfaces of a control volume and find the direction of the maximum flux.
3. If there is no flux through the cell surface, check the value of fluid function (F) of the neighbor cells and find the location of the cell having the maximum value of F.

11.7 Case studies

11.7.1 Flow over a semi-circular core between two plates

11.7.1.1 Description of the problem

Let us consider a two-dimensional fluid flow problem between two plates having a semi-circular core on the bottom plate based on the body-fitted coordinate system. This problem was also examined in chapter 10 based on the Cartesian coordinate system. Figures 11.13(a) and (b) indicate a schematic drawing of the computational domain and a grid system having 66×40 grids. As shown in figure 11.13(b), the body-fitted grid system allows fine grids to be arranged around the bottom plate especially near the semi-circular core. The working fluid is water and the flow rate is 10^{-4} m/s.

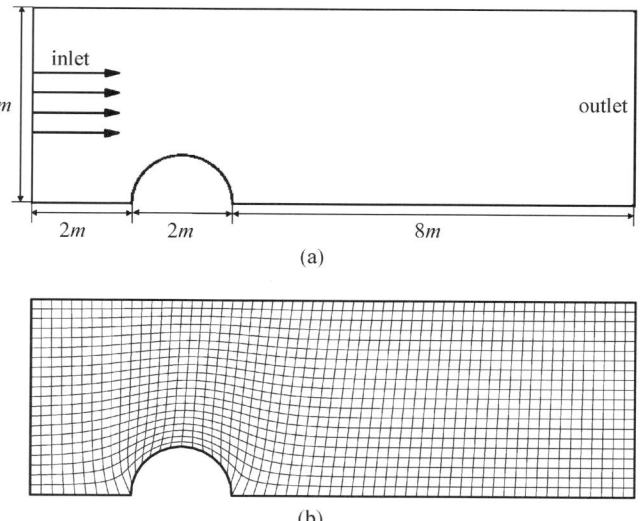

Figure 11.13. A schematic drawing of (a) the geometry and dimension of a computational domain and (b) a grid configuration.

Table 11.1. Material properties of water.

Material property	Value
ρ density (kg/m^3)	1000
μ dynamic viscosity (kg/m s)	1×10^{-3}

Table 11.2. Boundary conditions.

Inlet	Constant flow rate (10^{-4} m/s)
Outlet	Outflow
Upper plate	No slip wall
Lower plate	No slip wall

11.7.1.2 Material properties and boundary conditions

The material properties and boundary conditions used in the present simulation are shown in tables 11.1 and 11.2.

11.7.1.3 Simulation

Let us simulate the flow sequences and patterns using the execution file [bfc1.exe]. And compare the simulated results with those obtained based on the orthogonal grid system in chapter 10.

Data input
```
> Input inlet flow rate (m/s): 0.0001
> Input density of fluid (kg/m³): 1000
> Input dynamic viscosity of fluid (kg/m s): 0.001
> number of iterations: 1000
```

Results
Figure 11.14 shows the simulated flow patterns. As the flow passes over a semi-circular core, a secondary rotational flow is formed in the rear zone of the core. The flow sequences and patterns are similar to the results of figure 10.16. However, it is noted that the simulation based on the body-fitted coordinate system can describe the flow sequences and patterns more quantitatively compared to those by the Cartesian coordinate system using the orthogonal grids, especially around the semi-circular core and the rear zone of the core.

11.7.2 Internal flow in a U-tube

11.7.2.1 Description of the problem

Now consider another example of internal flow in a U-tube, which was also examined in chapter 10. Figures 11.15(a) and (b) indicate a schematic

202 *Fluid flow analysis using the body-fitted coordinate system*

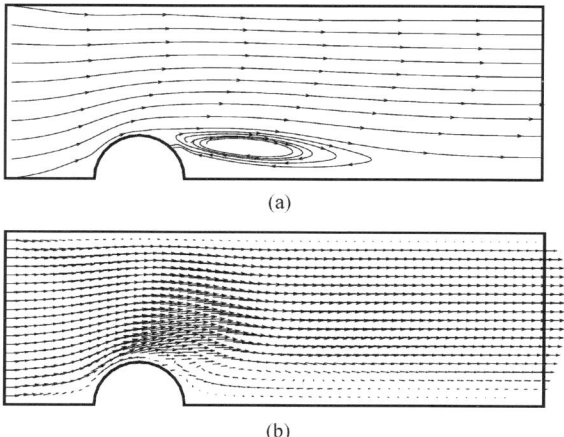

Figure 11.14. Flow between two plates having a semi-circular core on the bottom plate: (a) a stream function and (b) a flow velocity profile.

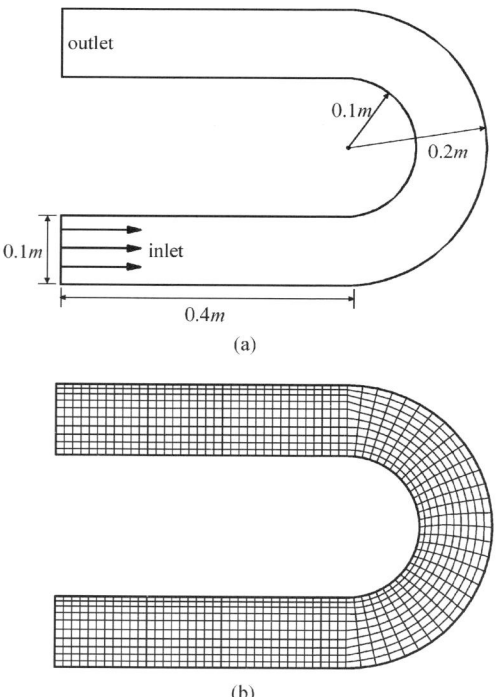

Figure 11.15. A schematic drawing of (a) the geometry and dimension of a computational domain and (b) a grid configuration.

Case studies 203

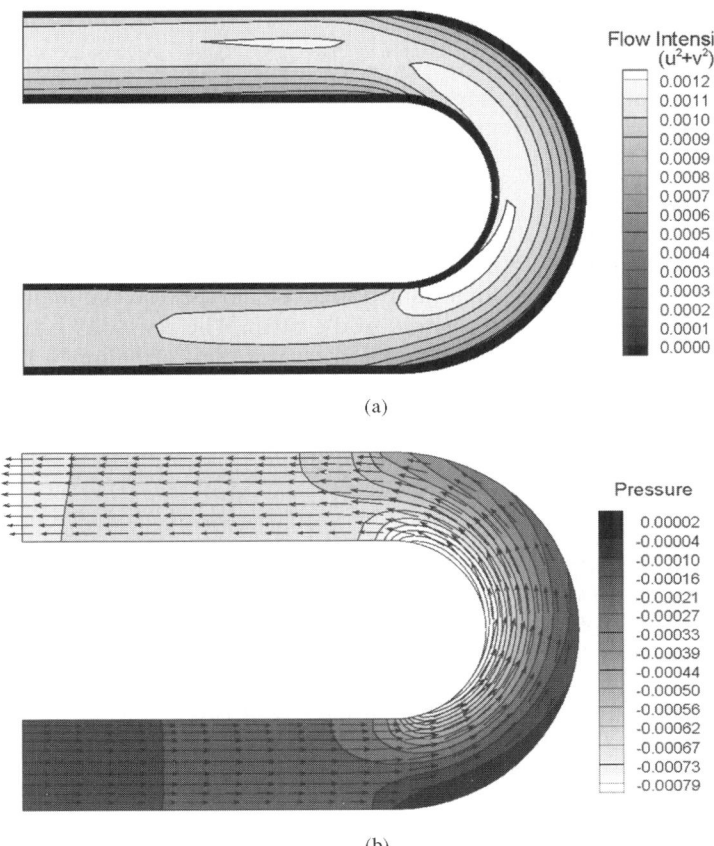

Figure 11.16. Internal flow in a U-tube: (a) a flow intensity profile and (b) a pressure distribution.

drawing of the computational domain and a grid system having 130×50 grids. Fine grids are arranged near the plate wall especially around the curved boundaries. The working fluid is water and the flow rate is 10^{-4} m/s.

11.7.2.2 Material properties and boundary conditions

The material properties and boundary conditions used in the present simulation are shown in tables 11.1 and 11.2.

11.7.2.3 Simulation

Let us simulate the flow sequences and patterns using the execution file [bfc2.exe]. And compare the simulation results with those of chapter 10.

Data input
```
> Input inlet flow rate (m/s): 0.0001
> Input density of fluid (kg/m³): 1000
> Input dynamic viscosity of fluid (kg/m s): 0.001
> number of iterations: 1000
```

Results

Figure 11.16 shows the simulated flow patterns. The velocity profiles are well developed near the inlet and outlet regions. The highest velocity is predicted not in the center region of a tube, but in the region slightly closer to the inside curved boundary. This result is quite different from the results of figure 10.18, which are based on an orthogonal grid system. The pressure distribution, shown in figure 11.16(b), is much different from that of figure 10.18(b). As mentioned in chapter 10, the orthogonal grid system might cause a large error by the stair-like grids along the curved boundary.

References

[1] Rhie C M and Chow W L 1983 *AIAA J* **10** 131
[2] Peric M 1985 *Finite Volume Method for the Prediction of Three-Dimensional Fluid Flow in Complex Ducts* PhD thesis, Imperial College, London
[3] Karki K C and Patankar S V 1988 *Numer. Heat Transfer* **14** 295
[4] Demirdzic I 1982 *A Finite Volume Method for Computation of Fluid Flow in Complex Geometries* PhD Thesis, Imperial College, London
[5] Shyy W and Vu T G 1991 *J. Comput. Phys.* **58** 97
[6] Hong C P, Lee S Y and Song K 2001 *ISIJ International* **41**(9) 999
[7] Rodi W, Majumdar S and Schonung B 1989 *Comput. Meth. Appl. Mech. Eng.* **75** 369
[8] Arakawa C 1994 *Computational Fluid Dynamics for Engineering* (University of Tokyo Press)
[9] Rhie C M and Chow W L 1983 *AIAA J.* **21** 1525
[10] Choi S K, Nam H Y and Cho M 1993 *Numer. Heat Transfer* **23** 21
[11] Patankar S V 1980 *Numerical Heat Transfer and Fluid Flow* (Washington, DC: Hemisphere)
[12] Peric M 1985 *A Finite-Volume Method for the Prediction of Three-Dimensional Fluid Flow in Complex Ducts* PhD thesis, University of London
[13] Kobayashi M H and Pereira J C F 1991 *Numer. Heat Transfer* **19** 243
[14] Amsden A A and Harlow F H 1970 *Report L A 4370* (Los Alamos, New Mexico: Los Alamos Scientific Lab)
[15] Nichols B D, Hirt C W and Hotchkiss R S 1980 *Tech. Report L A 8355* (Los Alamos, New Mexico: Los Alamos Scientific Lab)

Chapter 12

Modelling of mould filling

In chapters 10 and 11, the SIMPLE methods based on both the Cartesian and the body-fitted coordinate systems were described for modelling fluid flow. The algorithms for modelling free surface tracking were also introduced. In this chapter we will describe the application of the SIMPLE methods for the simulation of mould filling. The SIMPLE methods based on the two coordinate systems, i.e., the Cartesian and body-fitted coordinates, will be examined for mould filling simulation in various types of mould cavities using orthogonal and non-orthogonal grid systems.

12.1 Introduction

Most casting processes consist of the filling stage of molten metals and their solidification stage in the mould cavity, including complicated physical phenomena. The methodology of casting is various. The gravity castings make use of the gravitational force in feeding molten metals into a mould cavity through the gating and running system, such as sand mould castings, gravity die castings, etc. The low-pressure die castings are usually made by the up-hill filling against the gravitational force. The high-pressure die castings have some advantages such as high productivity and relatively low production cost, but do not guarantee high quality of products because of high-speed filling of molten metals into the mould cavities. Casting defects, such as porosity, entrapment of gas or oxide film, cold shut and misrun etc., which are caused by improper mould filling, are closely related to surface turbulence [1]. It is well known that qualities of final casting products are closely related to the stages of mould filling and solidification. It is therefore very important to predict and control the mould filling and solidification sequences for obtaining high quality casting products.

Since the MAC and SOLA-VOF methods were first applied to simulate fluid flow of molten metals into a die cavity having a free surface [2–4] extensive efforts have been made to develop numerical models for simulating mould filling in various types of casting processes. Several models have been suggested during the last several decades at a series of conferences on Modelling of Casting, Welding and Advanced Solidification Processes (MCWASP) [5], including a benchmark model to verify the models for the simulation of mould filling and solidification processes [6]. Most models for solving fluid flow having a free surface in mould filling, which have been developed in recent years, are based on the VOF method because of its effective computational efficiency compared to the MAC method. The SOLA algorithm for solving fluid flow, which is generally performed by the finite volume method based on the Cartesian coordinate system with orthogonal grids, is limited to mould filling simulations in simply shaped and relatively thick mould cavities. The finite element method is considered effective to generate a grid system for a complex geometry. However, it is known to be unsuitable for predicting incompressible flow and has a relatively low computational efficiency compared to the finite volume method [7].

Recently, there is an increasing demand to produce thin-walled and curved-shape casting products. Casting processes consisting of mould filling and solidification stages include various physical phenomena: (i) flow with a free surface, (ii) turbulent flow in the mould filling stage, (iii) evolution of latent heat and solute redistribution at the solid/liquid interface and in the mushy zone during phase change, and (iv) various zones of different characteristics such as fluid, mushy, and solidified zones. The SIMPLE algorithm is generally considered to have more potential for modelling various physical phenomena based on transport phenomena, transformation kinetics and continuum mechanics appearing in materials processing, compared to the SOLA algorithm. A series of studies has reported on the development of numerical models for mould filling simulation based on the SIMPLE algorithm coupled with the VOF method [8, 9]. However, the SIMPLE-VOF method is based on the Cartesian coordinate system using orthogonal grids, which is not still unsuitable for modelling mould filling of thin-walled and curved-shape cavities. Recently, the SIMPLE algorithm coupled with the VOF method based on the body-fitted coordinate (BFC) system has been developed for modelling the mould filling process for thin-walled and curved-shape castings [10].

12.2 Numerical analysis of filling process

12.2.1 Governing equations

As described in chapter 10, the integral forms of the governing equations for solving mould filling are as follows.

The continuity equation is given by

$$\iint_\Gamma (\rho \mathbf{u}) \cdot \mathbf{n} \, d\Gamma = 0. \qquad (10.1.1a)$$

The momentum balance equation is given by

$$\frac{\partial}{\partial t}\iiint_\Omega (\rho\mathbf{u})\,d\Omega + \iint_\Gamma (\rho\mathbf{uu})\cdot\mathbf{n}\,d\Gamma = -\iint_\Gamma (\tau\cdot\mathbf{n})\,d\Gamma + \iiint_\Omega (\mathbf{f_b}-\nabla p)\,d\Omega.$$
(10.1.2a)

As described in chapter 9, the above mentioned governing equations can be discretized using the MAC, SMAC, SOLA or SIMPLE methods. The discretized equations combined with the marker and cell method or the volume of fluid (VOF) method are solved to analyze physical phenomena including free surface tracking in mould filling.

In this chapter, we will adopt the SIMPLE scheme to discretize the above mentioned governing equations.

12.2.2 Free surface tracking in mould filling

Most castings are made by pouring the molten metal into a mould cavity through its gating system under the action of gravity or using other types of pressure. Figure 12.1 indicates an example of a mould assembly with its gating system. The molten metal is supplied to the pouring basin, and fills the mould cavity through the sprue, runner and gates. The boundary conditions used for the numerical analysis of mould filling are the normal and tangential stress conditions on the free surface, no-slip condition on the wall, and the mass source from the inlet. For the solidification analysis, the heat transfer through the running system during and after the filling stage and the contact heat transfer on the mould/cast and cast/air interfaces are also to be taken into consideration. In order to analyze the mould filling

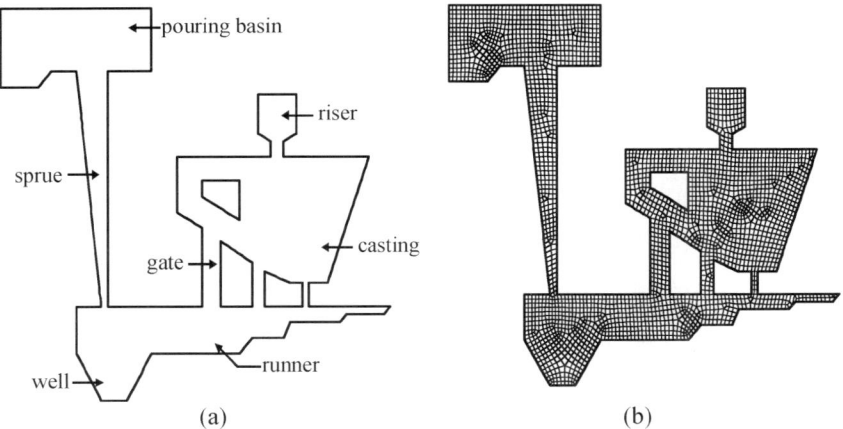

Figure 12.1. An example of a mould assembly with its gating system: (a) a physical model and (b) a grid configuration.

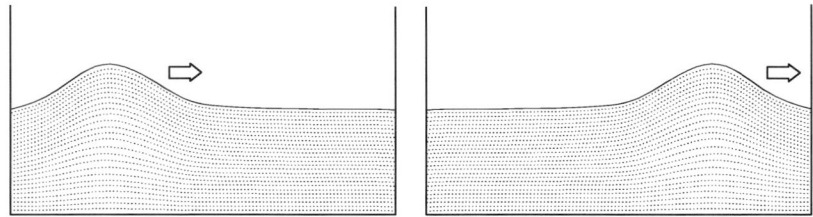

Figure 12.2. Tracking of a free surface by the MAC method.

process, the discretized equations and the boundary conditions need to be applied to the grid system shown in figure 12.1(b).

The numerical techniques for free surface tracking are introduced in detail in chapter 10. In this section we will briefly describe how the MAC and VOF techniques are applied for fluid flow simulation having free surfaces in a mould cavity.

12.2.2.1 MAC method

Figure 12.2 indicates the marker particles representing the free surface of a liquid zone. As an initial condition, the markers are distributed in each control volume with an arbitrary number of density. The number of markers in a finite control volume needs to be large enough to indicate that the control volume is a liquid cell or a free surface cell. When a cell that contains markers in it is open to a cell or cells of empty state, the cell is considered as a free surface cell. Figure 12.3 shows the markers on the inlet boundary at an initial stage of mould filling simulation. The markers need to be generated repeatedly at every time interval of simulation according to the flow rate through the inlet boundary and move into the mould cavity. This procedure is repeated until the end of mould filling stage. Thus, an enormous number of markers is necessary in three-dimensional analysis, causing a large increase of the computer memory size. Recently,

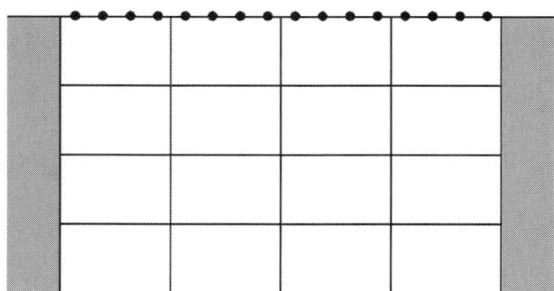

Figure 12.3. Distribution of marker particles along an inlet boundary.

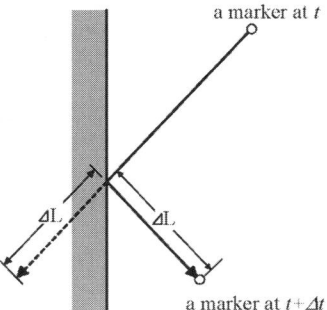

Figure 12.4. Behavior of a marker particle reflected from a solid wall.

the MAC technique is generally used for the purpose of particles trajectory rather than free surface tracking.

Figure 12.4 shows the motion of a particle near the mould wall. When a marker collides with the wall, the incidence and reflection angles will obey the perfect elastic behavior of a massless marker. The travel length, ΔL, of the reflected marker during Δt is equal to that of a marker without collision.

12.2.2.2 VOF method

The VOF method classifies the liquid zone based on the quantitative mathematical value, which is different from the MAC method in which a qualitative decision is made based on the existence of markers. As mentioned in chapter 10, the transport equation for the volume of fluid is given by

$$\frac{\partial F}{\partial t} + u\frac{\partial F}{\partial x} + v\frac{\partial F}{\partial y} = 0. \tag{10.4.5}$$

F has a value between 0 and 1 ($0 \leq F \leq 1$). The above equation can be discretized according to the conditions of the surrounding cells, as shown in figure 12.5(a).

$$F_{i,j}^{\text{new}} = F_{i,j} + \frac{u_{I,j}\Delta t\, F_{I,j} - u_{I-1,j}\Delta t\, F_{I-1,j}}{\Delta x_i} + \frac{v_{i,J}\Delta t\, F_{i,J} - v_{i,J-1}\Delta t\, F_{i,J-1}}{\Delta y_j}. \tag{12.2.1}$$

In the case of a cell (i, J) having an inlet boundary shown in figure 12.5(b), the velocity of the cell (i, J) is given by $v_{i,J} = v_{\text{inlet}}$.

The value of F that transfers through the face of a cell can be evaluated by equation (10.4.6) and the calculated F value on the face is to be substituted for each facial value in equation (12.2.1). $F = 1$ indicates that the cell is a fully fluid cell, while $F = 0$ represents that the cell is an empty cell. When F has a value greater than 0 and less than unity, the cell represents a free surface cell. The VOF method classifies the free surface zones according to

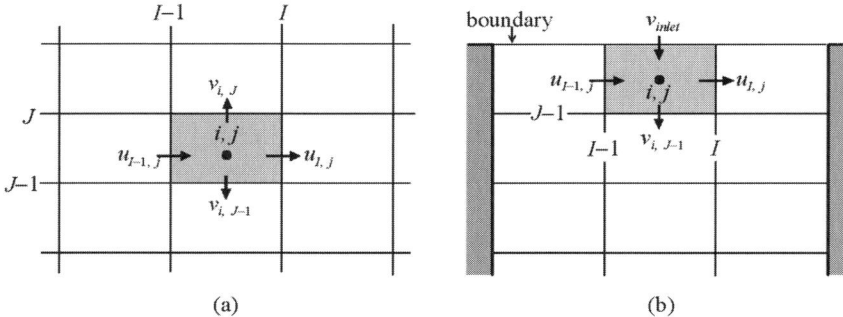

Figure 12.5. A finite control volume near an inlet boundary.

the exact value of the volume of fluid. The necessary memory to trace the free surface in the VOF method is identical to the number of grids, whereas in the MAC method the necessary memory size is proportional to the total number of markers which is much greater than that of grids. This is the reason why the VOF method is much more beneficial in three-dimensional simulations than the MAC method. Different from the free surface motion of the fixed amount of volume of fluid, the free surface motion of mould filling should include mass source in the computational domain matching the following relation.

$$\sum_{i=1}^{n_{\text{inlet}}} v_{\text{inlet}} \Delta t \Delta A_i = \sum_{j=1}^{n_{\text{free surface}}} \Delta V_{\text{at free surface}, j}. \quad (12.2.2)$$

Equation (12.2.2) indicates that the increased liquid volume (ΔV) in the free surface zone needs to be balanced with the supplied liquid volume from the inlet.

12.2.3 Algorithms for free surface tracking in the SIMPLE method

It is generally considered that the SOLA-VOF method based on the coupling of the SOLA algorithm and the VOF method is more useful than the MAC or the SMAC method in modelling the fluid flow problems with free surfaces. However, the SOLA-VOF method has some limitations to be applied to simulate various physical phenomena appearing in mould filling and solidification processes, such as the evolution of solidification microstructures, the formation of macro- and micro-segregation, the evolution of porosities and the entrapment of air or oxides, etc. Furthermore, the algorithm is based on the Cartesian coordinate system using the orthogonal grids and thus is not available to simulate mould filling in thin-walled and curved-shape cavities. On the other hand, the SIMPLE algorithm can effectively treat the damping effect, the double diffusive convection and the turbulent properties in the mushy zone in solidification of alloys [11, 12]. As mentioned in section 12.1, the SIMPLE algorithm is generally considered to have more potential for modelling various physical phenomena based on transport phenomena, transformation kinetics and continuum mechanics appearing in materials processing, compared to the SOLA algorithm.

Coupling methods of the SIMPLE algorithm with the MAC method [8] and with the VOF method [9] are developed to analyze mould filling and solidification processes. The SIMPLE algorithm coupled with the VOF method based on the body-fitted-coordinate (BFC) [10] has also been developed for the simulation of mould filling in thin-walled and curved-shape mould cavities.

The calculation steps of the SIMPLE-VOF method are as follows:

1. Input initial conditions.
2. Calculate the fractional volume of fluid transferred through the interfaces between cells using equation (10.4.6).
3. Calculate the fractional volume of fluid in each cell using equation (12.2.1).
4. Classify the fully fluid cells and the free surface cells by the value of fractional volume of fluid.
5. Apply the boundary conditions (normal and tangential stress conditions, no-slip wall boundary condition and inlet boundary condition) for the calculation of the flow field at $t + \Delta t$, and solve the discretized momentum transfer equations.
6. Solve the other discretized equations for various dependent variables.
7. Calculate the new properties depending on temperature or concentration, and go back to step 5 to calculate the new velocity field based on the new properties.
8. Repeat procedures 2–7 until the desired state of filling or solidification is reached.

12.3 Examples of mould filling simulation

12.3.1 Filling in a mould cavity with a straight and tapered gating system using the SIMPLE-VOF method

As a simple example of mould filling simulation using the SIMPLE-VOF method, we consider a mould assembly having a straight and tapered running system, as shown in figure 12.6: (a) a grid configuration and (b) simulated filling sequences.

If the grids are uniform based on the orthogonal grid system, a small number of grids might cause inaccurate calculation. It is therefore necessary to have a large number of grids in order to obtain accurate simulation results. In the present simulation, 142 200 grids are used for the calculation even

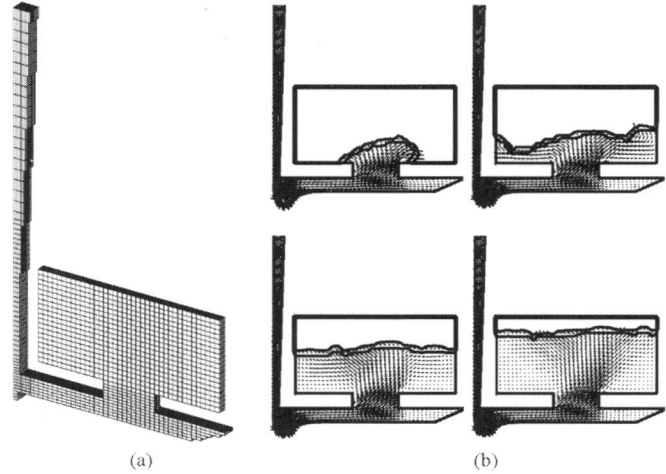

Figure 12.6. (a) A grid configuration and (b) simulated filling sequences.

though the casting shape is very simple. The experimental observation via x-ray shooting is available in reference [6].

12.3.2 Filling in a mould cavity with a curved gating system using the SIMPLE-BFC-VOF method

As mentioned in the previous example, the standard SIMPLE-VOF method based on the Cartesian coordinates using an orthogonal grid system is not suitable for the simulation of mould filling with a curved gating system. An enormous number of grids needs to be distributed along the boundary in order to minimize the calculation errors caused by inaccurate boundary fitting.

Numerical models based on the body-fitted coordinate system are available for the simulation of mould filling in thin-walled and curved-shape mould cavities with a curved gating system. The SIMPLE-BFC-VOF method based on the body-fitted coordinate system described in chapter 11 was applied to simulate the filling sequences in a mould cavity with a curved gate, as shown in figure 12.7: (a) a grid configuration, (b) experimental results [13] and (c) simulation results. As shown in figure 12.7(b), the experimental observation indicates that the melt through a curved gate is ejected into the mould cavity with a slight slope against the vertical axis and reaches up to the top plane of the mould cavity until the end of filling. The simulation can exactly predict the same results as obtained from the experimental observation even though a relatively small number of grids, 5736, are used.

Examples of mould filling simulation   213

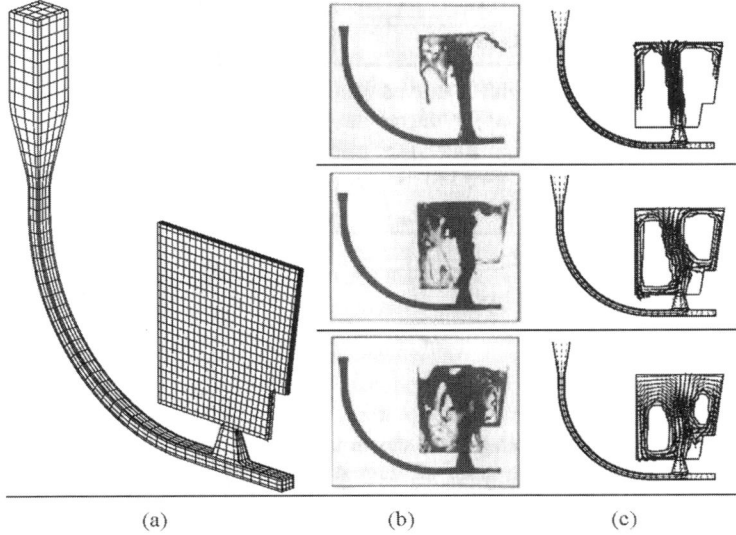

Figure 12.7. (a) A grid configuration, (b) experimental results [13], and (c) simulation results.

12.3.3 Comparison of the standard SIMPLE-VOF and the SIMPLE-BFC-VOF methods

In order to compare the two simulation methods in more detail, we adopt a mould assembly shown in figure 12.8(a). The gating system consists of three arms of gate: one arm is horizontal or perpendicular to the axial direction, and the other two arms are at 30° to the axial direction. Figure 12.8(b) shows an orthogonal grid configuration in the Cartesian coordinate system and figure 12.8(c) a non-orthogonal grid configuration in the body-fitted coordinate system. As shown in figure 12.8(b), in the case of an orthogonal grid configuration, the three arms have different types of grids: one arm has uniform grids, but the other two have stair-like grids. However, in the case of the body-fitted coordinate system shown in figure 12.8(c), the three arms have the same grids of uniform shapes, well fitted to the boundaries of three arms.

Figure 12.9 indicates the simulated filling sequences obtained from (a) the SIMPLE-VOF method with an orthogonal grid system based on the Cartesian coordinate and (b) the SIMPLE-BFC-VOF method with a non-orthogonal grid system based on the body-fitted coordinate system. In the Cartesian coordinate system shown in figure 12.9(a), the stair-like grids of the two arms cause the over-evaluated momentum energy loss along the boundaries in the two arms, compared to the other arm. As a result, the

214 *Modelling of mould filling*

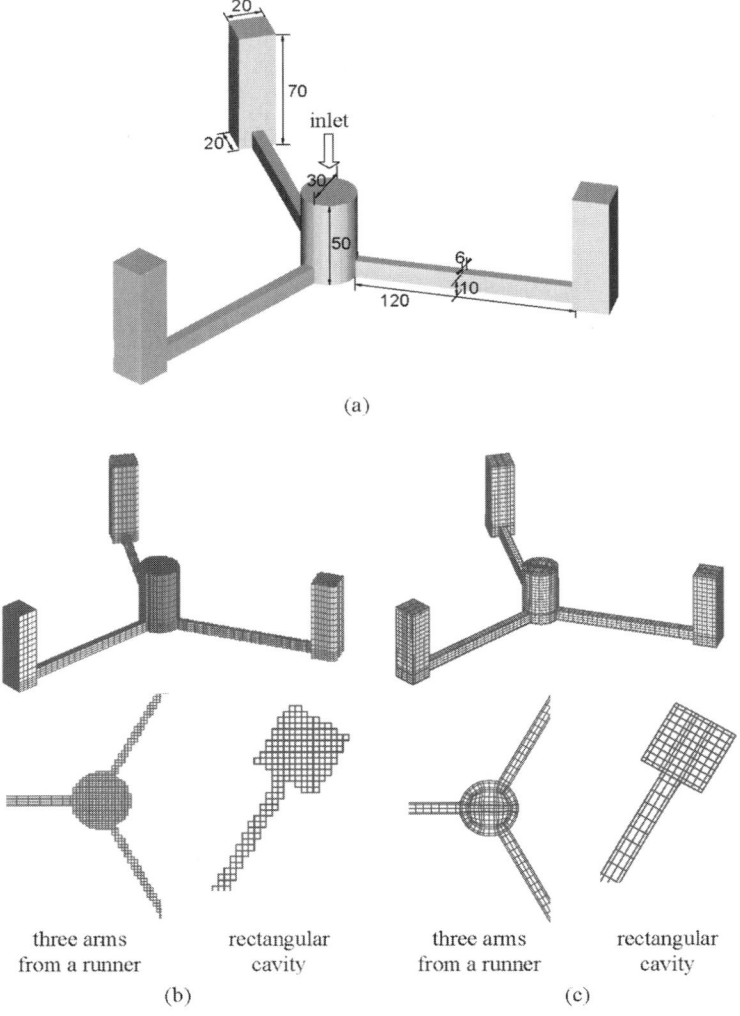

Figure 12.8. A mould assembly with three arms of gate having the same filling conditions: (a) a physical model, (b) an orthogonal grid configuration in the Cartesian coordinate system, and (c) a non-orthogonal grid configuration in the body-fitted coordinate system.

simulated filling patterns and sequences through the gates of two arms are not identical to the other one, which is not reasonable in practical casting processes. However, in the case of the body-fitted coordinate system shown in figure 12.9(b), the simulated filling patterns and sequences through the gates of three arms are identical to each other, which is reasonable for this problem in practice.

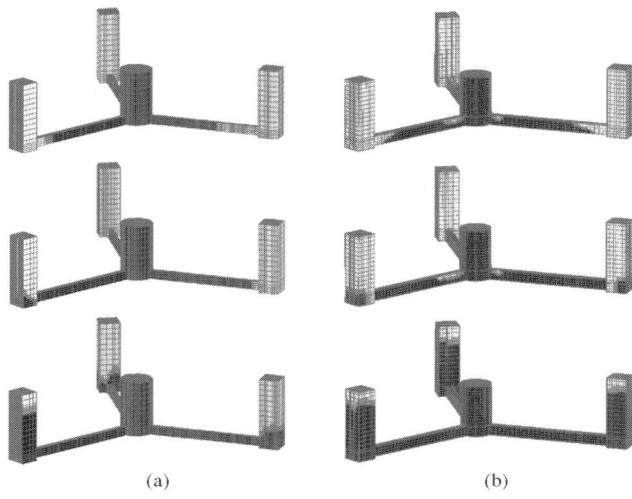

Figure 12.9. Simulated filling sequences in a mould cavity with three arms of gate obtained from: (a) the SIMPLE-VOF method and (b) the SIMPLE-BFC-VOF method.

12.4 Case studies

12.4.1 Filling in a rectangular cavity with a semicircular core on the bottom plate

12.4.1.1 Description of the problem

Now, examine the above mentioned two simulation methods to simulate a two-dimensional mould filling problem in a rectangular cavity with a semicircular core on the bottom plate. Figure 12.10 indicates a schematic

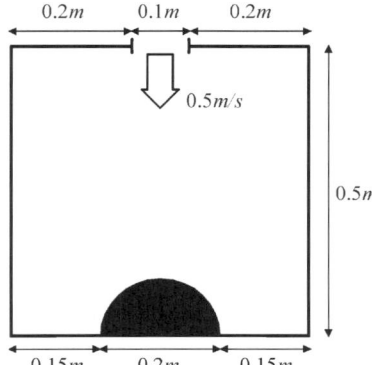

Figure 12.10. A schematic diagram of a mould assembly used in the simulation.

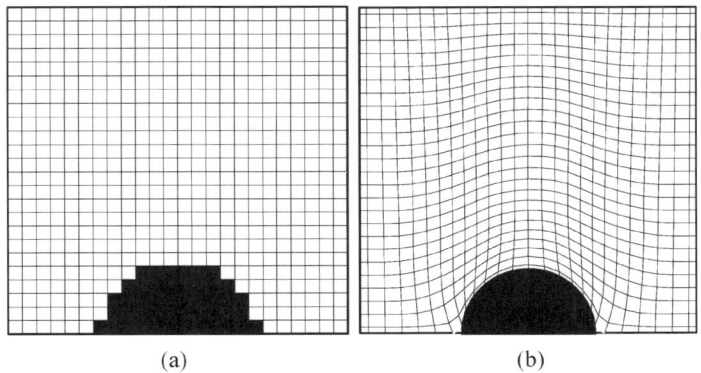

Figure 12.11. Grid configurations: (a) an orthogonal grid system based on the Cartesian coordinate system and (b) a non-orthogonal grid system based on the body-fitted coordinate system.

drawing of the computational domain. The dimension of the cavity is 0.5 m × 0.5 m. The working fluid is pure aluminum. The inlet velocity through the gate is 0.5 m/s.

12.4.1.2 Grids used in the simulation

Figure 12.11 indicates the grid configurations used in the present simulation for two cases: (a) the SIMPLE-VOF method with orthogonal grids based on the Cartesian coordinate system and (b) the SIMPLE-BFC-VOF method with non-orthogonal grids based on the body-fitted coordinate system. The numbers of grids are 625 for (a) and (b).

12.4.1.3 Material properties and boundary conditions

Table 12.1. Material properties of pure aluminum.

Material property	Variable	Value
Density (kg/m^3)	RHO	2380
Dynamic viscosity (kg/m s)	AMU	0.05474

Table 12.2. Boundary conditions.

Inlet	Constant flow rate (0.5 m/s)
Free surface	Normal and tangential stress conditions
Cavity wall	No slip wall

12.4.1.4 Simulation

Let us simulate the flow sequences and flow patterns using the execution files: [flow3.exe] for the SIMPLE-VOF method and [bfc3.exe] for the SIMPLE-BFC-VOF method, respectively.

Data input
```
> Input inlet flow rate (m/s): 0.5
> Input density of fluid (kg/m³): 2380
> Input dynamic viscosity of fluid (kg/m s): 0.05474
> Every number of iterations for data acquisition: 100
```

Results

Figure 12.12 indicates the simulated filling patterns and sequences obtained from (a) the standard SIMPLE-VOF method using orthogonal grids in the

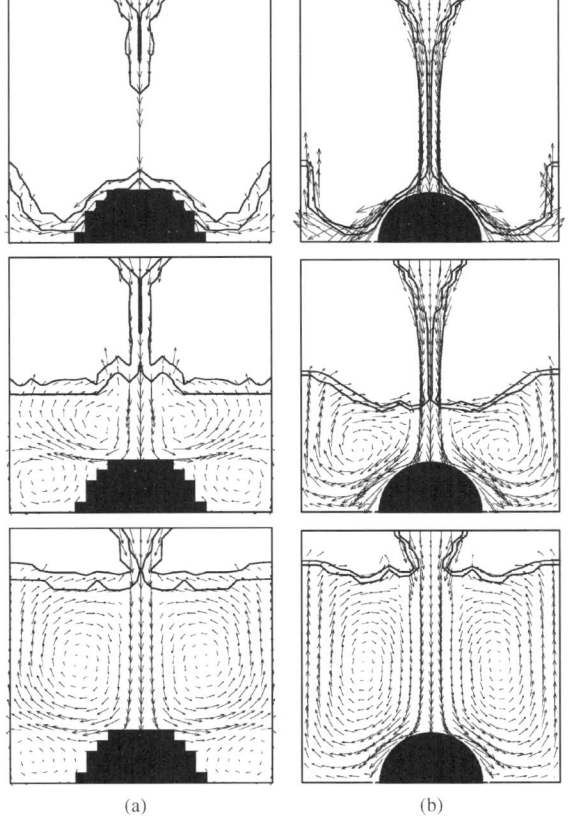

Figure 12.12. Simulated filling sequences in a rectangular cavity with a semi-circular core on the bottom plate obtained from (a) the SIMPLE-VOF method and (b) the SIMPLE-BFC-VOF method.

Cartesian coordinate system and (b) the SIMPLE-BFC-VOF method using non-orthogonal grids in the body-fitted coordinate system, respectively. The simulated filling sequences are almost similar to each other in this geometry. However, there is a significant difference in the filling patterns. In case (a), there is a secondary rotational flow occurring around the semi-circular core, which is due to the loss of momentum energy along the stair-like grids of the semi-circular core surface.

References

[1] Campbell J (1991) *Castings* (Oxford: Butterworth Heinemann)
[2] Hwang W S and Stoehr R A 1983 *J. of Metals* Oct 22
[3] Hirt C W 1984 *Proc. Modeling of Casting Welding and Advanced Solidification Processes—II* ed J A Dantzig *et al* (Warrendale, PA: TMS) p 67
[4] Stoehr R A and Hwang W S 1984 *Proc. Modeling of Casting Welding and Advanced Solidification Processes—II* ed J A Dantzig *et al* (Warrendale, PA: TMS) p 47
[5] *A Series of Proceedings of Modeling of Casting Welding and Advanced Solidification Processes (MCWASP)—II–X* (Warrendale, PA: TMS)
[6] Cross M and Campbell J (1995) *Modeling of Casting Welding and Advanced Solidification Processes—VII* (Warrendale, PA: TMS)
[7] Usamani A S, Cross J I and Lewis R W (1992) *Int. J. Numer. Meth. Eng.* **35** 787
[8] Lee J, Mok J and Hong C P (1999) *ISIJ International* **39** 1252
[9] Mok J, Hong C P and Lee J (2003) *ISIJ International* **43** 1212
[10] Hong C P, Lee S Y and Song K (2001) *ISIJ International* **41** **9** 999
[11] Bennon W D and Incropera F P (1987) *Int. J. Heat Mass Transfer* **30** **10** 2161
[12] Beckermann C and Viskanta R (1988) *Int. J. Heat Mass Transfer* **31** **1** 35
[13] Mampaey F A and Zhu Z A (1995) *Modeling of Casting Welding and Advanced Solidification Processes—VII* ed M Cross *et al* (Warrendale, PA: TMS) p 4

Chapter 13

Modelling of microstructure evolution

Over the past decade, computational modelling, which enables extensive use of mathematics for solving complicated problems, has emerged as a powerful and important tool to predict the microstructure evolution in various casting and solidification processes. The significant advances of numerical modelling techniques have made it possible to analyze transport phenomena, including heat, mass and momentum transfer in the mushy zone during solidification to a high level of detail, leading to the development of various models of microstructure evolution in solidification of alloys. Recently, cellular automaton models have generated significant interest among researchers because of its potential to be applied in practical casting and solidification problems. In this chapter, we will describe the numerical algorithms of cellular automaton models and also show some examples how to put CA models into practical usage.

13.1 Introduction

Control of solidification microstructures is of prime importance for the control of the properties and quality of final casting products in modern casting technology. Prediction of microstructure evolution in solidification of alloys is a key factor in controlling solidification microstructures. When we consider the purpose of modelling the microstructure evolution in alloy solidification, there are two perspectives: if one puts emphasis on *physics*, quantitative understanding of physical phenomena is an important aspect. On the other hand, if one emphasizes the importance of *practice*, it is very important to put microstructure models into practical applications. Models of microstructure evolution in alloy solidification can be classified into two groups: deterministic and stochastic models.

Deterministic models, based on the solution of the continuum equations over some volume elements, were developed for the description of the nucleation, growth and impingement of equiaxed grains in solidification processes. As reviewed recently [1], deterministic models have made great progress in coupling with macroscopic heat flow calculations to model solidification processes and average microstructure features, such as grain size, secondary dendrite arm spacing and segregation pattern, at the scale of the whole process. Among the deterministic models phase field models, which are based on thermodynamics, offer the opportunity for precise and quantitative understanding the dynamics of pattern selection and side branching in dendritic growth. They have recently emerged as a viable computational tool for simulating the formation of complex interfacial patterns in solidification [2]. However, phase field models are presently limited to a very small computational domain due to the large computational capacity needed.

Compared with deterministic models, stochastic models, Monte Carlo (MC) and cellular automaton techniques are generally considered to provide some advantages, such as high efficiency in computation, a large computational domain, and easy applications to practical problems. The MC procedure, which accounts for bulk diffusion, attachment and detachment kinetics and uses a lowest free energy change algorithm, has been applied to predict recrystallization and grain growth in the solid-state phase transformation [3], as well as to simulate solidification grain structures in casting [4]. However, there are some drawbacks of the MC methods in simulating solidification structures. For example, they do not explicitly integrate the growth kinetics of the solid/liquid interface and ignore the specifics of macro- and micro-transport. In addition, the MC time step used in the calculation is not correlated with real time. Consequently, they are unable to realistically show the time-evolution solidification structures.

Another stochastic method, a cellular automaton (CA) technique [5], was applied to predict solidification grain structures. This method is based upon the consideration of physical mechanisms on nucleation, growth kinetics of a dendrite tip and crystallographic orientations. Furthermore, the mechanisms of competitive dendrite growth are directly embedded in the CA algorithm. So that the CA model can quantitatively carry out the time-dependent simulation for microstructure evolution, in which the individual grains are identified and their shapes and sizes can be shown graphically. However, it is well known that the classical CA models mentioned above have the limitations that they are unable to describe the detailed microstructure evolution including dendritic side branching, solute segregation and eutectic phase formation. Recently, a modified CA model [6] which can quantitatively describe the evolution of dendritic growth features, including the growing and coarsening of the primary trunks, the branching of the secondary and tertiary dendrite arms, as well as the solute segregation patterns, has been developed.

Figure 13.1 indicates the length scales of the typical microstructure models [7]. Classical cellular automaton models are suitable for simulating macroscopic-scaled structures such as solidification grains, ranging from mm to cm. On the other hand, phase field models are suitable for

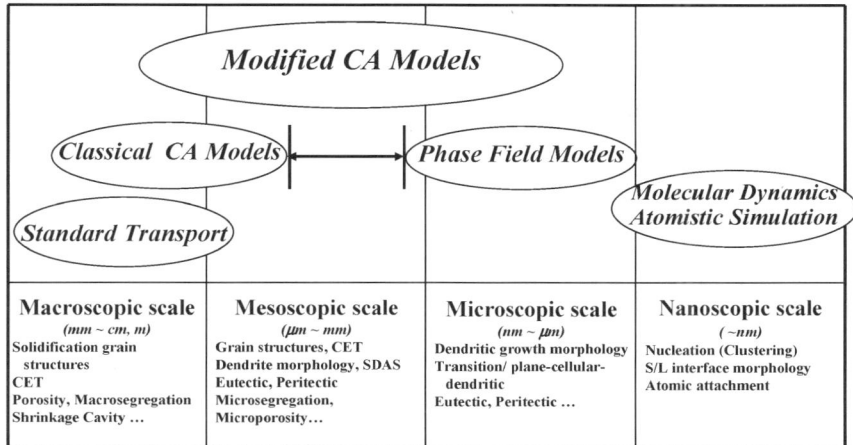

Figure 13.1. Length scales of numerical models for microstructure evolution [7].

microscopic-scaled structures such as dendritic growth morphology, ranging from nm to μm. It is to be noted that there is a wide gap between phase field and classical CA models. A modified CA model was found to bridge this gap [8]. Molecular dynamics and atomistic simulation models are considered to be suitable for modelling nanoscopic-scaled structures such as nucleation behavior, atomic attachment and solid/liquid interface morphology.

13.2 Nucleation and growth kinetics

13.2.1 Nucleation

The conditions of nucleation in solidification of liquid metals are very important in determining the characteristics of microstructure evolution. The number, shape and size of grains vary according to process variables in solidification. It is generally considered that the solidification of commercial alloys begins with heterogeneous nucleation. Heterogeneous nucleation is assumed to occur on nucleation sites randomly chosen on the mould surface and in the bulk liquid. There have been two significant models developed for treating heterogeneous nucleation: one is the continuous nucleation model in which a continuous dependency of nucleation density on temperature or undercooling is assumed, and the other is the instantaneous nucleation model in which all nuclei are assumed to be generated at a certain nucleation undercooling. However, because of the limited understanding of heterogeneous nucleation, fundamental calculation of available nucleation sites and undercoolings needed for the nuclei becoming active

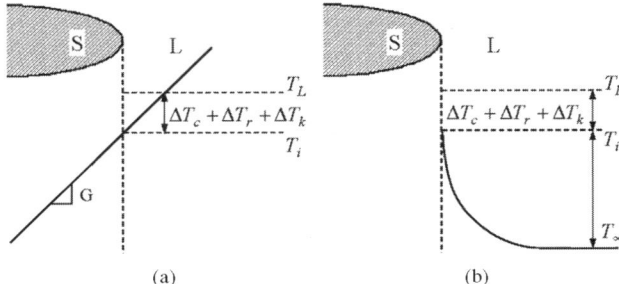

Figure 13.2. Schematics of temperature fields around a dendrite tip: (a) for the KGT model [11] and (b) for the LKT model [12].

during solidification is impossible. Therefore, in both cases, an empirical relationship between the resultant number of nuclei and the amount of undercooling must be deduced from a set of experiments. A schematic comparison of instantaneous and continuous nucleation models is shown in figure 13.2 [1], both for nucleation on the mould surface and in the bulk liquid. The basic equations and related parameters for various nucleation models that must be experimentally evaluated or assumed are reviewed in table 13.1 [1].

When the temperature range in which nucleation occurs is very small, the continuous model will be similar to the instantaneous model. According

Table 13.1. Summary of nucleation models [1].

Model	Type	Basic equation	Fitting parameters
Oldfield	Continuous	$\dfrac{\partial N}{\partial t} = -nK_1(\Delta T)^{n-1}\dfrac{\partial T}{\partial t}$	n, K_1
Maxwell and Hellawell	Continuous	$\dfrac{\partial N}{\partial t} = (N_s - N_i)\mu_2 \exp\left[-\dfrac{f(\theta)}{\Delta T^2(T_p - \Delta T)}\right]$	N_s, θ
Thervoz et al	Continuous (statistical)	$\dfrac{\partial N}{\partial(\Delta T)} = \dfrac{n}{\sqrt{2\pi}\,\Delta T_\sigma} \exp\left[-\dfrac{(\Delta T - \Delta T_{max})^2}{2(\Delta T_\sigma)^2}\right]$	$n, \Delta T_{max}, \Delta T_\sigma$
Goettsch and Dantzig	Continuous (statistical)	$N(r) = \dfrac{3N_s}{(R_{max} - R_{min})}(R_{max} - r)^2$	N_s, R_{max}, R_{min}
Stefanescu et al based on Hunt	Continuous	$\dfrac{\partial N}{\partial t} = (N_s - N_i)\mu_2 \exp\left[-\dfrac{\mu_3}{\Delta T^2}\right]$	$N_s(dT/dt)$

to the calculation for the peritectic Al–Ti system when nucleation was completed the fraction of solid was a mere 10^{-4} [9]. Thus, in cases of solidification of commercial alloys there will be no particular difference whether a continuous or an instantaneous model is adopted.

In this chapter we will adopt the continuous nucleation model in which two different Gaussian distributions were considered for heterogeneous nucleation at the mould surface and in the bulk liquid. A continuous nucleation distribution, $dn/d(\Delta T)$, was used to describe the grain density increase, dn, which is induced by an increase in the undercooling, $d(\Delta T)$, as in the following Gaussian distribution [10].

$$\frac{dn}{d(\Delta T)} = \frac{n_{max}}{\sqrt{2\pi}\Delta T_\sigma} \exp\left[-\frac{1}{2}\left(\frac{\Delta T - \Delta T_{nuc}}{\Delta T_\sigma}\right)^2\right] \qquad (13.2.1)$$

where ΔT_{nuc} is the mean nucleation undercooling, ΔT_σ is the standard deviation, and n_{max} is the maximum density of nuclei given by the integral of this distribution from 0 to ∞.

Thus, the density of grains, $n(\Delta T)$, formed at any undercooling ΔT, is given by

$$n(\Delta T) = \int_{t_n}^{t} \frac{dn}{d(\Delta T)} d\Delta T. \qquad (13.2.2)$$

The nucleation parameters in equation (13.2.1), n_{max}, ΔT_{nuc} and ΔT_σ, are dependent upon the alloy composition and the process variables. Generally, it is considered that the most important factor is the pouring temperature of a melt. It is apparent that lower pouring temperatures increase the number of free crystals that are able to survive and grow.

At the beginning of simulation, the nucleation sites both on the mould surface and in the bulk liquid are randomly set according to equation (13.2.1) and are identified with a reference number that relates to a corresponding undercooling for nucleation. At a time step interval, the change of a cell state from 'liquid' to 'solid' is initiated either by nucleation or growth of a solid cell. If a given cell is a predetermined nucleation site and the local undercooling is larger than that which is necessary for nucleation, this cell changes its state from liquid to solid and its corresponding index is randomly defined to be an integer among 48 classes, which represents its crystallographic orientation.

13.2.2 Growth kinetics

The characteristics of dendritic growth can be represented by two analytical models, the KGT (Kurz–Giovanola–Trivedi) model [11] and the LKT (Lipton–Kurz–Trivedi) model [12]. The KGT model describes a constrained dendritic growth with a positive temperature gradient in liquid at the

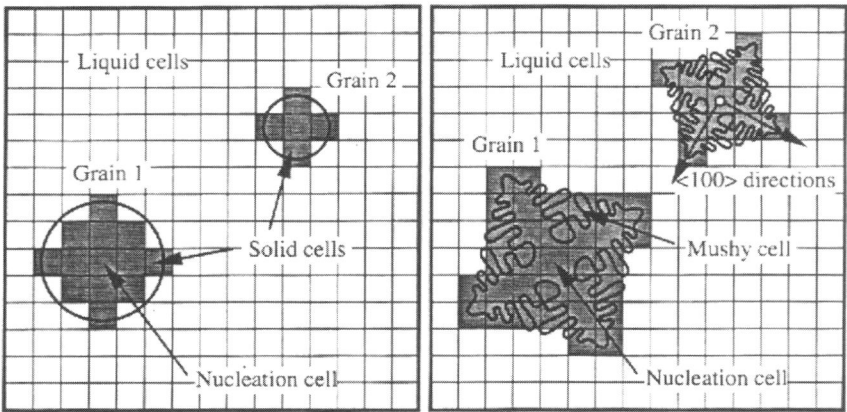

Figure 13.3. Computational space of a two-dimensional CA for dendrite envelop grain growth [5].

solid/liquid interface. On the other hand, the LKT model describes a free dendritic growth into an undercooled melt in which the temperature gradient in liquid ahead of the solid/liquid interface is assumed negative. Figure 13.3 indicates the schematics of temperature fields around a dendrite tip assumed in the KGT and the LKT models [5]. In rapid solidification processes, such as splat cooling on a substrate, melt spinning and atomization, the existence of undercooling in liquid ahead of the solid/liquid interface has been demonstrated both by experiments and by heat transfer analyses. Generally, the relationship between dendrite growth velocity and undercooling in liquid is evaluated with the aid of the KGT model for normal solidification conditions and the LKT model for rapid solidification conditions. Except for rapid solidification with higher undercoolings, the two models provide the same relationship between the dendrite growth velocity and the local undercooling in liquid [13]. Table 13.2 indicates the calculated growth kinetics parameters for the KGT model for Al–Cu and Al–Si alloys.

Table 13.2. Growth kinetics parameters for Al–Cu and Al–Si alloys systems [13].

Alloy system	Growth velocity (m/s) $v(\text{mass\% C}, \Delta T) = k_1 \Delta T^2 + k_2 \Delta T^3$
Al–Cu	$k_1 = 0.14 \exp(-\text{mass\% Cu}/0.513) - 0.157(-\text{mass\% Cu}/0.516)$
	$k_2 = 0.0046 \exp(-\text{mass\% Cu}/0.4) + 0.00014 \exp(-\text{mass\% Cu}/1.58)$
Al–Si	$k_1 = -0.03292 \exp(-\text{mass\% Si}/0.19) - 0.0016 \exp(-\text{mass\% Si}/0.51)$
	$k_2 = 0.0064 \exp(-\text{mass\% Si}/0.25) + 0.00016 \exp(-\text{mass\% Si}/1.17)$

13.3 Classical cellular automaton models

13.3.1 Model description

According to Wolfram's definition [14], a cellular automaton model is a mathematical idealization of physical systems in which space and time are discrete, and physical quantities take on a finite set of discrete values. The cellular automaton model for microstructure simulation consists of (1) the geometry of a cell, (2) the state of a cell, (3) the neighborhood configuration, and (4) several transition rules that determine the state of a given cell during one time step. The computational domain is divided into uniform cells, squares in two dimensions and cubes in three dimensions. Furthermore, each cell is characterized by different variables (such as temperature and crystallographic orientation) and states (solid or liquid). The definition of neighborhood includes the first four nearest neighbors or the whole eight surrounding neighbor cells of the first layer for two-dimensional models.

Figure 13.4 shows the schematics of a two-dimensional cellular automaton for dendrite envelop grain growth [5]. The cellular automaton evolves in discrete time steps, and the state of a cell at a particular time is calculated from the local rule, such as the nucleation and growth kinetics. In order to simulate dendritic grain structures, nucleation and growth kinetics should be implemented into the CA model.

It is to be noted that the stochastic aspect in CA models is only related to nucleation, i.e., randomly choosing the locations and crystallographic orientations of nuclei, whereas the growth is usually treated in a deterministic way. The governing equations and numerical algorithms for calculating

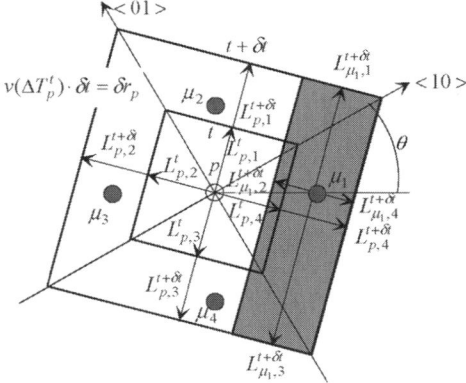

Figure 13.4. A schematic diagram illustrating the growth algorithm for a dendrite grain whose ⟨10⟩ direction is misoriented by an angle with respect to the horizontal axis of the CA network [5].

nucleation and growth, and coupling CA structure simulation with macroscopic heat transfer calculation are described in detail below.

13.3.2 Nucleation and growth algorithm implemented into CA

The heterogeneous nucleation kinetics can be implemented into CA models as follows. At the beginning of simulation, the nucleation sites both on the mould surface and in the bulk liquid are randomly set according to equation (13.2.1) and are identified with a reference number that relates to a corresponding undercooling for nucleation. At a time step interval, the change of a cell state from 'liquid' to 'solid' is initiated by either nucleation or growth of a solid cell. If a given cell is the predetermined nucleation site and the local undercooling is larger than that which is necessary for nucleation, this cell changes its state from liquid to solid. The preferential growth orientation of a grain would be determined randomly in the stage of nucleation. In cases of two dimensions, the orientation of the primary $\langle 10 \rangle$ dendrite arms of a nucleated cell is randomly determined among 48 classes in the range of $[0, \pi/2\,]$. In three dimensions, the orientation of the primary $\langle 100 \rangle$ dendrite arms can be determined similarly [15].

Once a cell has nucleated, it will grow with a preferential direction corresponding to its crystallographic orientation having a growth velocity determined by the local undercooling. Now, we will introduce the growth algorithm, which describes the two-dimensional growth of a randomly oriented grain in any thermal field [16]. Let us consider a *nucleation cell*, p, for which the formation of a new grain had occurred at a time t_n. As shown in figure 13.4, the main $\langle 10 \rangle$ and $\langle 01 \rangle$ crystallographic directions make an angle θ and $\theta + \pi/2$, respectively, with respect to the horizontal axis. This 'nucleation cell' p is surrounded by four nearest-neighbor cells labeled μ_1, μ_2, μ_3 and μ_4. It is assumed that the dendritic network within a cell p develops as a square envelop. The four half-diagonals ($j = 1, 4$) of this square correspond to the dendrite tip directions, and their extension, $^*L_{p,j}^t$, is therefore given by

$$^*L_{p,j}^t = \int_{t_n}^{t} v(\Delta T_p^t)\,dt \qquad (13.3.1)$$

where $v(\Delta T_p^t)$ is the growth velocity of the dendrite tip as deduced by the KGT model or the LKT model, T_p^t is the local temperature being given by the thermal field calculation.

The growth of the $\langle 11 \rangle$ planes can also be monitored. Their position, $L_{p,j}^t$, from the center of the cell, p, is given by

$$L_{p,j}^t = \frac{^*L_{p,j}^t}{\sqrt{2}} = \frac{1}{\sqrt{2}} \int_{t_n}^{t} v(\Delta T_p^t)\,dt. \qquad (13.3.2)$$

At a critical time, t_c, the square will catch the four cell centers μ_i, which will occur when

$$L_{p,j}^{t_c} = \frac{l}{\sqrt{2}}[\cos\theta + |\sin\theta|] \qquad (13.3.3)$$

where l is the spacing between the cells. At that time, the four neighboring cells will have their state index numbers set to the same value as that of the parent cell p, providing they are still liquid. Rectangles are grown from each μ_i center, as shown in figure 13.4. The four $\langle 11 \rangle$ directions ($j = 1, 4$) at each cell center, μ_i ($i = 1, 4$), grow with the same growth kinetics, $v(\Delta T_{\mu_i}^t)/\sqrt{2}$, as defined in equation (13.3.2), but with the local undercooling corresponding to each cell location. At the time of capture, t_c, these rectangles degenerate into segments whose size corresponds to the actual extension of the dendritic front which captured these sites. In figure 13.4, for example, the initial extension of the four $\langle 11 \rangle$ directions at site μ_1 will be such that: $L_{\mu_{1,1}}^0 + L_{\mu_{1,3}}^0 = \sqrt{2} \cdot l \cdot [\cos\theta + |\sin\theta|]$ and $L_{\mu_{1,2}}^0 = L_{\mu_{1,4}}^0 = 0$. In the time-stepping calculation, each of these extensions of the rectangle is updated as follow.

$$L_{\mu_{i,j}}^t = L_{\mu_{i,j}}^0 + \frac{1}{\sqrt{2}}\int_{t_c}^{t} v(\Delta T_{\mu_i}^t)\,dt \cong L_{\mu_{i,j}}^{t+\delta t} + \frac{v(\Delta T_{\mu_i}^{t+\delta t})\,\delta t}{\sqrt{2}} \qquad (i,j) = 1, 4.$$
$$(13.3.4)$$

The velocity, $v(\Delta T_{\mu_i}^t)$, can be computed as a function of the local undercooling, $\Delta T_{\mu_i}^t$, at the cell center, using the formula given in table 13.2. When all neighbors of a growing cell have solidified, its $\langle 11 \rangle$ directions are no longer incremented.

13.3.3 Coupling the macroscopic heat flow calculation with CA models

In general, the stochastic CA models consist of two schemes: the CA model to simulate the microstructural evolution and the finite volume method (FVM) or the finite element method (FEM) to calculate the macroscopic heat flow. In this section we will briefly describe the coupling of the CA with the FVM. The coupled model of CA with the FEM, called the CAFÉ model, are referred to the literature [5, 16].

13.3.3.1 Macroscopic heat flow calculation by the finite volume method

First, let us consider the macroscopic heat flow calculation using the finite volume method. The integral form of transient heat conduction equation, which will be used in the finite volume approach, is given by

$$\frac{\partial}{\partial t}\iiint_\Omega (\rho C_v T)\,d\Omega = -\iint_\Gamma (\mathbf{q}\cdot\mathbf{n})\,d\Gamma + \iiint_\Omega \dot{g}\,d\Omega. \qquad (13.3.5)$$

228   *Modelling of microstructure evolution*

The typical boundary conditions being used in heat flow calculation of casting solidification are those at the melt/mould interface and the air/mould interface as follows.

At the melt/mould interface,

$$q = h_{int}(T_{mould} - T_{melt}). \qquad (13.3.6)$$

At the air/mould interface,

$$q = h_{air}(T_{air} - T_{mould}) \qquad (13.3.7)$$

where q is the heat flux, T_{melt}, T_{mould} and T_{air} are the temperatures of the melt, the mould and the air. h_{int} and h_{air} are the interfacial heat transfer coefficients at the melt/mould and the air/mould interfaces, respectively.

The discretization of the governing equation, equation (13.3.5), on the computational domain can be made using the procedures mentioned in chapter 6.

13.3.3.2 Coupling of the CA model with heat transfer calculation

Figure 13.5 indicates a schematic diagram of the computational domain with control volumes (CV) which are used for macroscopic heat transfer calculation by the finite volume method. As shown in the figure, each CV is generally further divided into 4 × 4 (or 6 × 6) rectangular or square cellular automaton (CA) cells for microstructure simulation, in two dimensions. It is therefore simple to superimpose the CA cells on the FVM control volumes. Temperatures of the CV nodes, indicated by the solid circles, are calculated by the

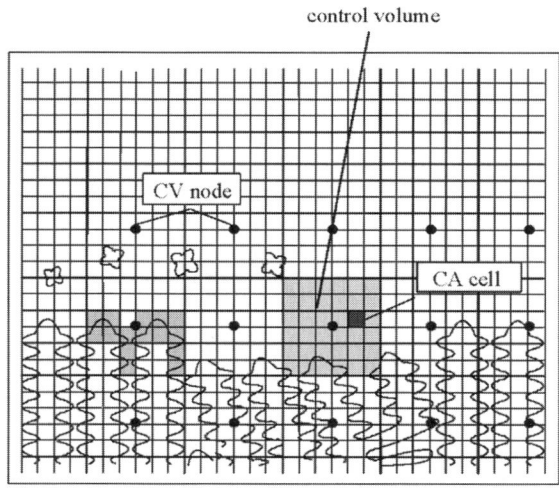

Figure 13.5. A schematic diagram of superimposing CA cells with control volume nodes for the finite volume method.

explicit finite difference scheme. Temperatures of the CA cells can then be linearly interpolated using the temperatures of the CV nodes. Whether a liquid cell nucleates or a solid cell grows with a certain growth velocity can be determined based on the calculated temperature profiles and the local undercoolings in the cells. If the undercooling of a liquid cell is greater than that which is necessary for nucleation, the liquid cell changes its state from liquid to solid. The growth velocity of a growing cell can be evaluated by the KGT or the LKT models using the parameters given in table 13.2.

The latent heat in the solidifying cells can be estimated by a modified temperature recovery method [17, 18]. The temperatures of the CV nodes are then recovered by interpolating the cell temperatures. Finally, using these updated temperatures, macroscopic heat transfer calculation can be continued. This series of calculation will be repeated until the end of solidification, as shown in the main flow diagram, figure 13.6.

In order to reduce the computational time, two different time steps for simulation can be used: one for the macroscopic heat transfer calculation

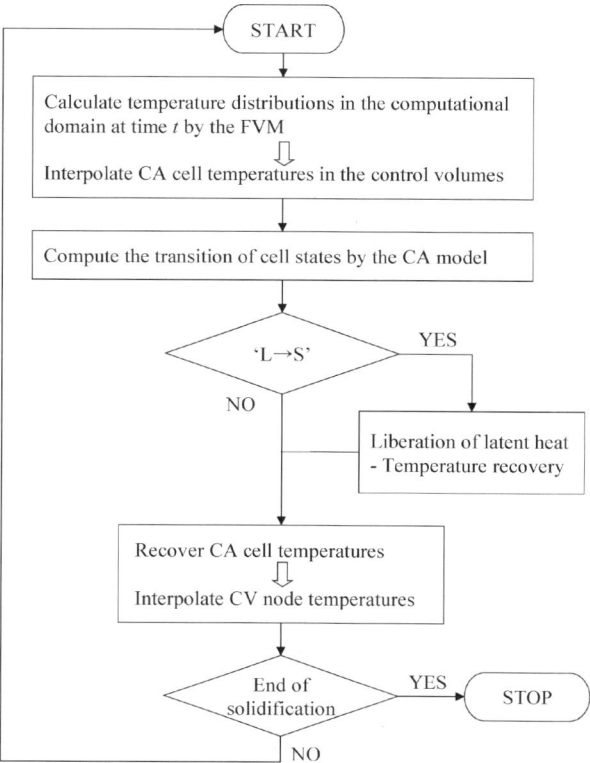

Figure 13.6. A flow diagram for the coupling of the CA model with the FVM.

230   *Modelling of microstructure evolution*

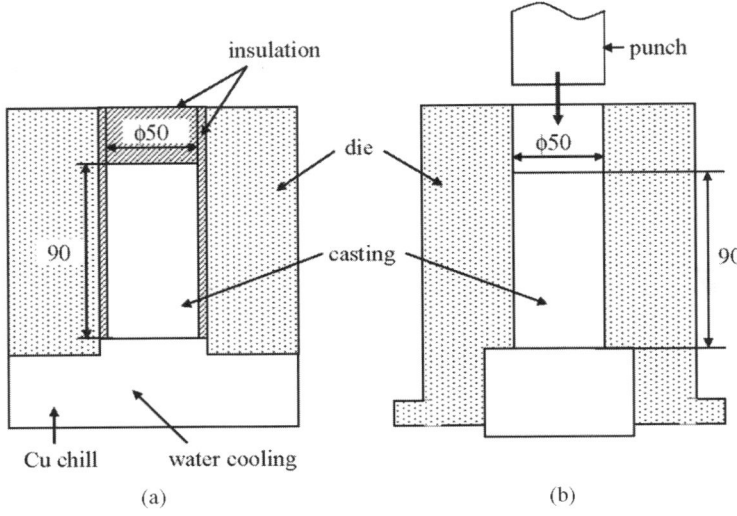

Figure 13.7. A schematic drawing of the die assembly: (a) directional casting and (b) squeeze casting [19].

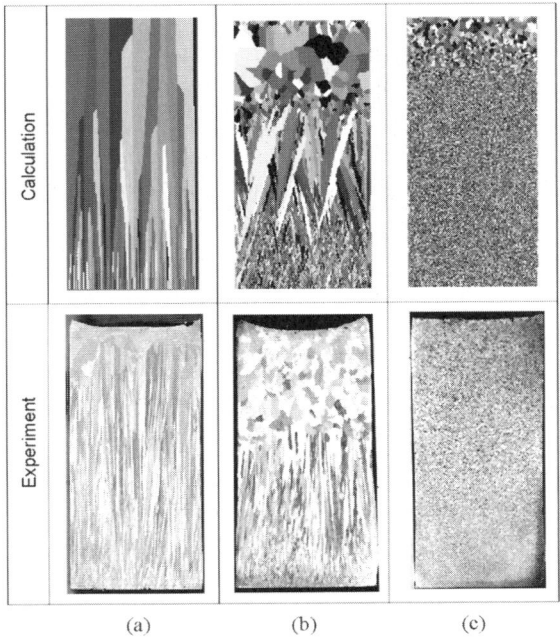

Figure 13.8. The simulated and experimental macrostructures of an Al–4.5 mass% Cu alloy directionally solidified upon a chill plate of 25 °C with various pouring temperatures: (a) 810 °C, (b) 760 °C and (c) 760 °C with inoculation [19].

based on CV, and the other for the microscopic CA calculation based on CA cells as follows.

Time step for macroscopic heat transfer calculation by the FVM is

$$\delta t_T = \frac{\Delta x_{CV}^2 \times \rho C_p}{5 \times \lambda} \quad (13.3.8)$$

where Δx_{CV} is the CV size, ρC_p is the volumetric specific heat, and λ is the thermal conductivity.

The time step for microscopic CA calculation is

$$\delta t_{CA} = \frac{\Delta x}{2 \times v_{max}} \quad (13.3.9)$$

where v_{max} is the maximum growth velocity obtained by scanning the growth velocities of all interface liquid cells during each time interval.

13.3.4 Examples of classical CA simulation

The coupled CA-FVM models have been applied to predict solidification grain structures formed in various casting processes [13, 15, 19]. Figure 13.7 illustrates a schematic drawing of die assemblies for (a) directional

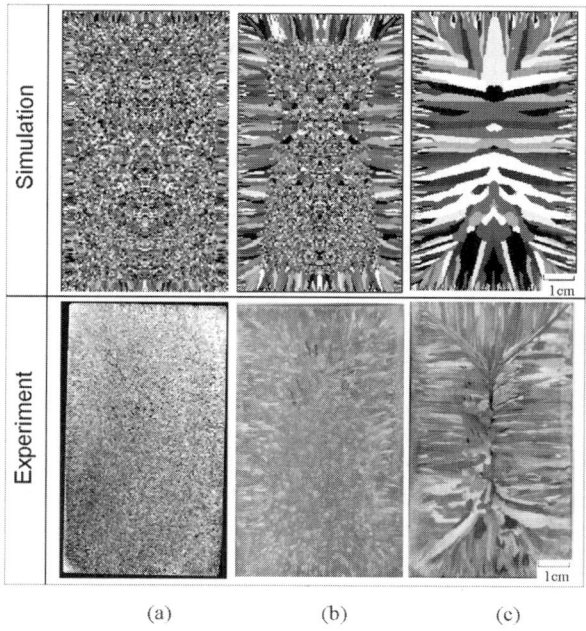

Figure 13.9. The simulated and experimental macrostructures of an Al–4.5 mass% Cu alloy in squeeze casting with a die temperature of 120 °C, an applied pressure of 50 MPa and various pouring temperatures: (a) 710 °C, (b) 760 °C and (c) 810 °C [19].

casting and (b) squeeze casting. Experimental results on directional casting and squeeze casting of an Al–4.5 mass% Cu alloy were compared with the CA simulations [19].

Figure 13.8 indicates the simulated and experimental grain structures of an Al–4.5 mass% Cu alloy directionally solidified upon a chill plate of 25 °C with various pouring temperatures of (a) 810 °C, (b) 760 °C and (c) 760 °C with inoculation using 0.01 mass% TiB_2. It can be noted that a high pouring temperature of 810 °C results in a fully columnar structure. The columnar to equiaxed transition (CET) occurs when the pouring temperature decreases to 760 °C, as shown in figure 13.8(b). However, fully equiaxed solidification grain structures can be formed with inoculation at a pouring temperature of 760 °C, as shown in figure 13.8(c).

Figure 13.9 shows the effect of pouring temperature on solidification grain structures in squeeze casting of an Al–4.5 mass% Cu alloy with a die temperature of 120 °C and an applied pressure of 50 MPa for various pouring temperatures of (a) 710 °C, (b) 760 °C and (c) 810 °C. It is obvious from the figure that, with an increase of pouring temperature, the grain structure changes gradually from fully equiaxed structures, to CET region, and finally to fully columnar structures.

13.4 Modified cellular automaton models

13.4.1 Model description

Classical CA models [5, 13, 15] mentioned in the previous section are based both on the probabilistic concept, such as the random determination of heterogeneous nucleation sites and the preferential growth orientations of nuclei, and on the deterministic concept for the growth kinetics of a dendrite tip. The growth velocity of dendritic grains is determined by the local undercooling based on a calculated thermal field. Therefore, the classical CA models can only provide grain structures, i.e., thermal dendritic grains, but dendritic growth features cannot be predicted, as shown in figures 13.8 and 13.9. The modified CA models [6, 7] retain both the probabilistic aspect and the deterministic concepts of the classical CA models mentioned above. However, different from the classical CA models in which only the temperature field was calculated, the modified CA models account not only for a thermal field, but also for solute concentration and fluid flow fields during solidification. In addition, they also account for the effect of curvature on the interface temperature. Thus, the modified CA models can provide dendritic growth features (solutal dendrites), microsegregation patterns, and other microstructural growth features such as eutectic growth and the formation of second phases.

For coupling the modified CA models for microstructural evolution with the FVM for macroscopic transport calculations, the same schemes used in the classical CA models described in the previous sections can be adopted.

13.4.2 Growth

As mentioned in the previous section, the growth velocity of a dendrite tip is calculated using the KGT or LKT models based on the local undercooling. The total undercooling in the liquid at the solid/liquid interface, $\Delta T(t_n)$, is considered to be the sum of four contributions.

$$\Delta T(t_n) = \Delta T_T + \Delta T_c + \Delta T_r + \Delta T_k \quad (13.4.1)$$

where ΔT_T, ΔT_c, ΔT_r and ΔT_k are the undercooling contributions associated with temperature, concentration, curvature and attachment kinetics, respectively. The last term on the right hand of equation (13.4.1), the kinetic undercooling, becomes significant only at very high solidification velocities such as rapid solidification and is negligible under normal solidification conditions. In general, three contributions (thermal, solutal and curvature effects) are taken into account in modelling of microstructure evolution.

Thus, the local undercooling $\Delta T(t_n)$, taken at the center of the cell i at time t_n is given by

$$\Delta T(t_n) = T_l - T_i(t_n) + m(\rho_{A,i}(t_n) - \rho_{A,0}) - \Gamma \bar{K}_i(t_n) \quad (13.4.2)$$

where T_l is the equilibrium liquidus temperature, m is the liquidus slope, $\rho_{A,0}$ is the initial concentration, and Γ the Gibbs–Thomson coefficient. $\bar{K}_i(t_n)$, $\rho_{A,i}(t_n)$ and $T_i(t_n)$ are the interface mean curvature, the concentration and the temperature of the cell i at the solid/liquid interface at time t_n, respectively.

There are several methods commonly used to calculate the interface mean curvature [20–22]. In this section we will briefly introduce a counting cell method proposed in the literature [20]. The mean curvature of an interface cell with the solid fraction f_s can be calculated by the following equation.

$$\bar{K}_i = \frac{1}{\Delta x} \left[1 - 2 \frac{f_s(i) + \sum_{j=1}^{n} f_s(j)}{n+1} \right] \quad (13.4.3)$$

where Δx is the cell size and n is the number of the neighboring cells. The values of the curvature calculated using equation (13.4.3) vary from $1/dx$ to 0 for convex surfaces and from 0 to $-1/dx$ for concave surfaces.

According to the KGT model and an iteration method, the relationship between the growth velocity, $v[\Delta T(t_n)]$, and the local undercooling, $\Delta T(t_n)$, can be expressed as

$$v[\Delta T(t_n)] = k_1 \Delta T(t_n)^2 + k_2 \Delta T(t_n)^3 \quad (13.4.4)$$

where k_1 and k_2 are the coefficients: $k_1 = 2.0 \times 10^{-6}$ m/s K^2 and $k_2 = 1.49 \times 10^{-6}$ m/s K^3 for an Al–7 mass% Si alloy, respectively. Detailed information can be found in the literature [6, 7].

Several growth algorithms for the modified CA models are introduced in the literature [23]. In this section we will introduce a growth algorithm of a modified CA model for the growth of non-faceted crystals [6].

Once a cell has nucleated, it will grow with a preferential growth direction corresponding to its crystallographic orientation having a growth velocity determined by the local undercooling. Let us consider a solidified cell labeled A which lies at the solid/liquid interface, as shown in figure 13.10. There must exist at least one liquid cell within its eight neighbors (including four nearest neighbors and four corner neighbors). Figure 13.10(a) describes the details of the growth algorithm between the solid cell A and its liquid neighbor cell i. The notation $l_A^i(t_n)$ indicates the growth length of the solid cell A with respect to its liquid neighbor cell i at time t_n, which can be calculated by

$$l_A^i(t_n) = \frac{\sum_{n=1}^{N} [v_n\{\Delta T(t_n)\} \times \Delta t_n]}{\{\cos\theta + |\sin\theta|\}} \quad (13.4.5)$$

where Δt_n is the time step and N indicates the iteration number. θ is the angle of the preferential growth direction of the cell A, $\vec{n}_A\langle 10 \rangle$, with respect to the linking line between the cell A and the cell i. Then, the solid fraction of the cell i at a certain time, $f_s^i(t_n)$, can be expressed by

$$f_s^i(t_n) = \frac{l_A^i(t_n)}{L} \quad (13.4.6)$$

where L is the length between the cell A and the cell i, as shown in figure 13.10(a): if i is one of the four nearest east, west, south or north neighbors, $L = dx$; and if i is the corner neighbor, $L = \sqrt{2}\,dx$. When $f_s^i(t_n) \geq 1$, which means the growth front of the solid cell A can touch the center of the liquid cell i, the cell i will then transform its state from liquid to solid and get the same orientation index as the cell A.

By means of the algorithm described above, the primary dendrite will grow and coarsen with the preferential $\langle 10 \rangle$ direction, as shown in figure 13.10(b). As the growth and the coarsening of a primary trunk proceed, the solute will be enriched in liquid near the solid/liquid interface due to the solute redistribution, which will destroy the interface stability and therefore cause the sidebranching into the secondary arms, as shown in figure 13.10(c).

13.4.3 Coupling the continuum model with a modified CA model

As mentioned earlier, the modified CA models, different from the classical CA models in which only the temperature field was calculated, account not

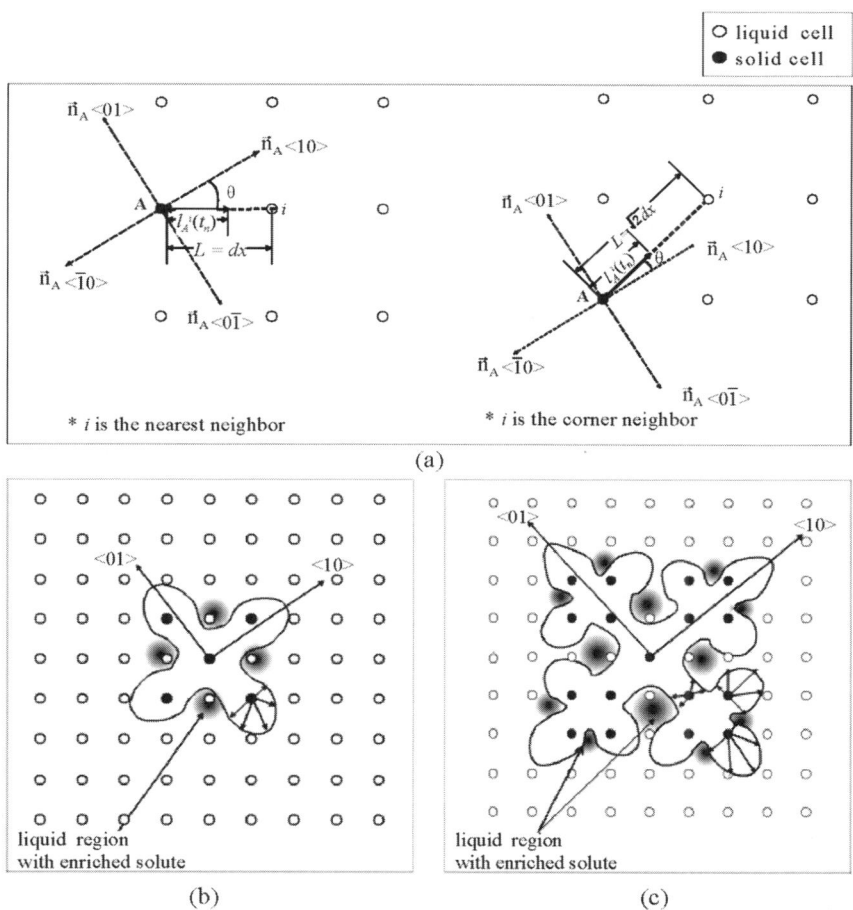

Figure 13.10. Schematic diagrams of the growth algorithm used in the modified CA model: (a) the details of the growth from the solid cell A to the liquid cell l, (b) the illustration of the growth and coarsening of a primary trunk and (c) the illustration of sidebranching [6].

only for a thermal field, but also for solute concentration and fluid flow fields during solidification. In this section we will briefly describe the coupling of the continuum model with a modified CA model.

13.4.3.1 Macroscopic transport

As described in chapter 3, the general form of transport equation is given by equation (3.3.7).

$$\frac{\partial}{\partial t} \iiint_\Omega \rho \Phi \, d\Omega = -\iint_\Gamma \rho \Phi \mathbf{u} \cdot \mathbf{n} \, d\Gamma + \iint_\Gamma K \nabla \Phi \cdot \mathbf{n} \, d\Gamma + \iiint_\Omega S \, d\Omega \quad (3.3.7)$$

where Φ is the mode of transfer (1 is the mass transfer, **u** is the momentum transfer, C_vT is the heat transfer, and ω_A is the species transfer), K is the diffusion coefficient, and S is the source term.

Considering that the system involves various states of liquid, solid and solid/liquid interface, the continuity and Navier–Stokes equations for incompressible fluid flow can be expressed as follows.

(Equation of continuity)

$$\iint_\Gamma (\xi\mathbf{u}) \cdot \mathbf{n}\,d\Gamma = 0. \tag{13.4.7}$$

(Navier–Stokes equation)

$$\frac{\partial}{\partial t}\iiint_\Omega \rho(\xi\mathbf{u})\,d\Omega + \iint_\Gamma \rho(\xi\mathbf{uu}) \cdot \mathbf{n}\,d\Gamma = \iint_\Gamma \mu\nabla(\xi\mathbf{u}) \cdot \mathbf{n}\,d\Gamma - \iiint_\Omega \nabla P\,d\Omega \tag{13.4.8}$$

where **u** is the velocity vector, ρ is the density which is considered to be identical and constant in liquid and solid, μ is the viscosity, and P is the hydrostatic pressure. ξ is a parameter which is dependent on the state of a cell. The no-slip boundary condition at the solid/liquid interface is applied by determining the parameter ξ as follows.

$$\xi = \begin{cases} 0 & (f_s = 1) \\ 1 & (f_s = 0) \end{cases}. \tag{13.4.9}$$

In the interface region, ξ varies smoothly from one to zero as the solid fraction of a cell goes from zero to one, which indicates that the flow velocity decreases gradually as the solid fraction increases in the interface region, and finally becomes zero in solid.

It is assumed that the solid/liquid interface behaves locally as if it were in a state of equilibrium, i.e., *local equilibrium*. When a cell transforms its state from liquid to solid by nucleation or growth, the solute partition between liquid and solid at the interface is then given by

$$\rho_s^* = k\rho_l^* \tag{13.4.10}$$

where k is the partition coefficient, and ρ_s^* and ρ_l^* are the interface equilibrium concentrations in the solid and liquid phases, respectively. Similarly, the governing equation for species transfer in liquid and solid is given by

$$\frac{\partial}{\partial t}\iiint_\Omega \rho_A\,d\Omega + \iint_\Gamma (\xi\mathbf{u}\rho_A) \cdot \mathbf{n}\,d\Gamma$$

$$= \iint_\Gamma (D_A\nabla\rho_A) \cdot \mathbf{n}\,d\Gamma + \iiint_\Omega \rho_A(1-k)\frac{\partial f_s}{\partial t}\,d\Omega \tag{13.4.11}$$

where D_A is the solute diffusion coefficient. The second term on the right-hand side of equation (13.4.11) indicates the amount of solute rejection or absorption at the solid/liquid interface. The species transfer in the solid region is purely controlled by diffusion ($\xi = 0$).

13.4.3.2 Coupling of the continuum model and a modified CA model

The coupling scheme of the continuum model with the MCA model is basically similar to the case of the classical CA model mentioned in the previous section.

The SIMPLE algorithm described in chapter 10 can be used to solve the transport equations. First, we evaluate the flow velocity and solute concentration fields in the computational domain by solving the transport equations, equations (13.4.7), (13.4.8) and (13.4.11). Based on the calculated solute distributions, the local undercooling and the growth velocity of an interface cell can be calculated using equations (13.4.2)–(13.4.5). The solid fraction increment, Δf_s, of an interface cell can be calculated by equation (13.4.6). The newly solidified cells in the computational domain are checked and considered as obstacle cells in the next step of flow simulation. The flow velocity distribution is then updated by the state parameter, ξ, according to the new solid fraction profile. Using these updated velocity and solute profiles, fluid flow and species transfer calculation can be continued. This series of calculations is repeated until the end of simulation.

As explained previously, fluid flow and species transfer calculations are implicitly carried out, whereas the CA simulation is simulated by an explicit scheme. Thus, the time step for iterative calculation is limited by the maximum change rate of solid fraction. In order to avoid numerical instability, it is considered that at least five time intervals are needed to complete the solidification of an interface cell. So, the stable time step for the simulation is determined by

$$\Delta t = \frac{1}{5}\left(\frac{\partial f_s}{\partial t}\right)^{-1}_{\max} \tag{13.4.12}$$

where $(\partial f_s/\partial t)_{\max}$ is the maximum rate of change of solid fraction obtained by scanning all interface cells during each interval of a time step. Detailed information can be found in the literature [22].

13.4.4 Examples of modified CA simulation

The MCA models have been applied to predict two-dimensional and three-dimensional free dendritic growth in an undercooled melt, multi-dendritic morphology in practical casting [6, 24], non-dendritic or globular microstructure evolution in semi-solid casting processes [7], eutectic phase formation in both regular non-faceted/non-faceted and irregular non-faceted/faceted eutectic alloy systems [25–27], and dendritic growth morphology with fluid flow [22].

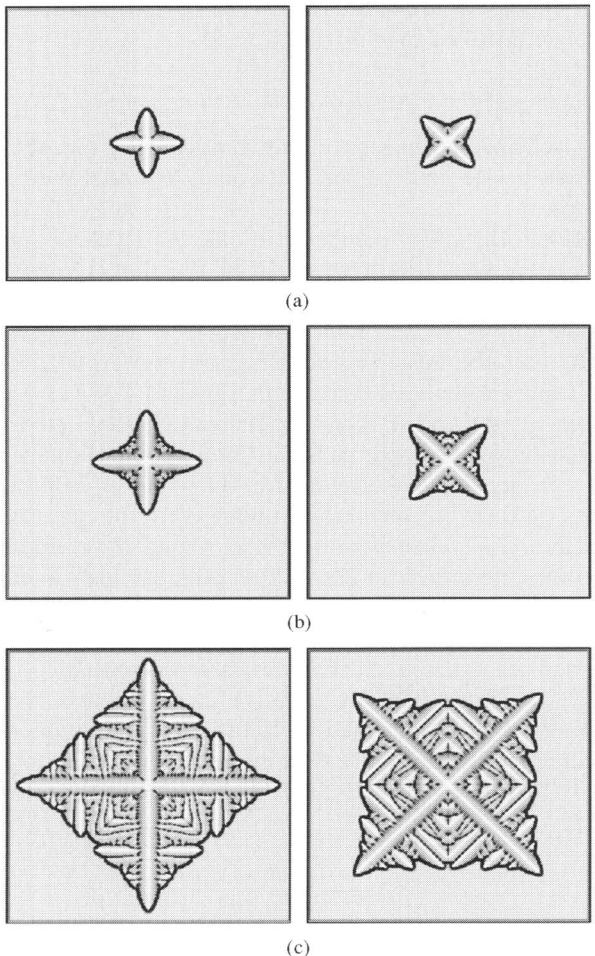

Figure 13.11. The simulated dendrite shapes during the isothermal growth of an Al–2.0 mass% Cu alloy at 888 K ($\Delta T = 40$ K): (a) the initial growth stage, (b) the initiation of sidebranching, and (c) the dendrite with well-developed sidebranches [6].

Figure 13.11 indicates an example of the simulated dendrite growth morphology of an Al–2.0 mass% Cu alloy solidified in an undercooled melt ($\Delta T = 40$ K) [6]. In order to simulate free dendritic growth into an undercooled melt, the computational domain was divided into 320×320 cells with a cell size of 0.2 μm, which is fine enough to resolve a dendrite tip radius. In the beginning of this simulation, one nucleus with the preferential growth orientation of $0°$ or $45°$ with respect to the horizontal

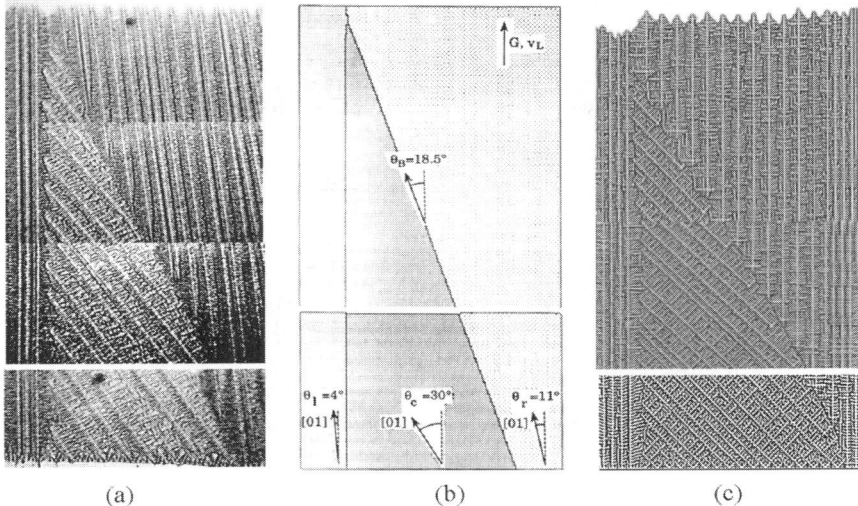

Figure 13.12. Competitive growth of columnar grains and dendrites in directional solidification: (a) the experimental result [28], (b) the CAFÉ simulation [16], and (c) the modified CA simulation [6].

direction was assigned in the center of the computational domain. The initial concentration was assumed to be identical to 2.0 mass% Cu.

The *in situ* observation of the dendritic solidification of transparent organic materials is a very powerful tool for the study of grain selection or competitive growth in the columnar zone [28]. Figure 13.12 shows the competition between three columnar grains of a succinonitrile–1.3 mass% acetone alloy solidified directionally with a constant thermal gradient, $G = 1900$ K/m, and a velocity of the liquidus isotherm, 86×10^{-6} m/s: (a) the experimental result [28], (b) the simulated grain structure by the CAFÉ model [16], and (c) the simulated dendritic grain structure by the modified CA model [6].

Recently, semi-solid forming (or casting) processes, which consist of casting and forging processes in a semi-solid state, have generated significant interest among casting manufacturers. The most important thing in semi-solid forming processes is how to obtain a semi-solid state metallic material, consisting of globular solid particles surrounded by liquid in an appropriate ratio, changing its shape easily by a small force due to its thixotropic properties, and being cast like a liquid due to its high fluidity [8]. Thus, the study on the formation of fine globular or non-dendritic microstructure becomes an important subject in this innovative metal technology. The modified CA model was applied to investigate the mechanism of evolution of globular microstructure in aluminum alloys [8]. In order to predict the primary

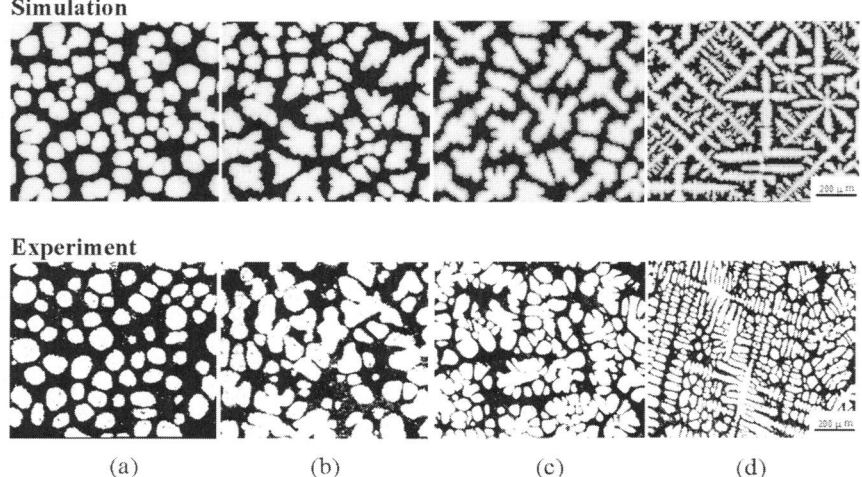

Figure 13.13. The simulated and experimental microstructures of an Al–7.0 mass% Si alloy with a cooling rate of 0.3 °C/s for various pouring temperatures: (a) 620 °C, (b) 630 °C, (c) 640 °C and (d) 660 °C [8].

phase evolution under different solidification conditions, the computational domain was divided into 400 × 330 cells with a cell size of 3 µm. The initial temperature field was assumed to be uniform and cooled down with a specified cooling rate. Figure 13.13 indicates the effect of pouring temperature on the evolution of primary phase in an Al–7.0 mass% Si alloy, indicating the predicted and experimental microstructures with a constant cooling rate of 0.3 °C/s for various pouring temperatures: (a) 620 °C, (b) 630 °C, (c) 640 °C and (d) 660 °C.

It can be noted that under a low pouring temperature of 620 °C with a low cooling rate, the primary α phase exhibits the globular morphology as shown in (a). However, with the increase of pouring temperature, the primary phase transits gradually to intermediate rosette-like morphology with short stout arms as shown in (b) and (c), and finally transits to dendritic structures as shown in (d). These phenomena are considered to be due to different nucleation conditions depending on the pouring temperature. It is well known that the evolution of primary phase in solidification of alloys is closely related not only to the pouring temperature, but also to the cooling rate. Figure 13.14 shows the predicted and experimental microstructures of an Al–7.0 mass% Si alloy with a pouring temperature of 620 °C for various cooling rates: (a) 0.3 °C/s, (b) 3 °C/s, (c) 10 °C/s and (d) 30 °C/s. It can be clearly seen from the figures that, with the increase of cooling rate, the primary phase morphology changes gradually from globular to rosette-like, and finally to dendritic structure.

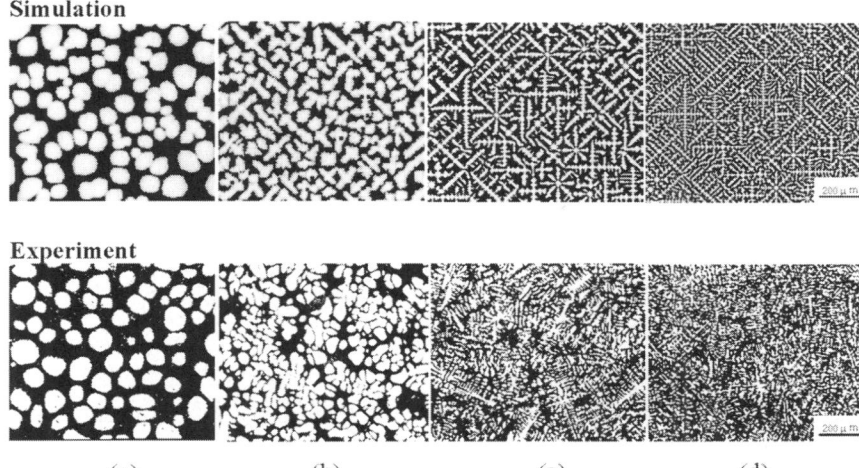

Figure 13.14. The simulated and experimental microstructures of an Al–7.0 mass% Si alloy with a pouring temperature of 620 °C for various cooling rates: (a) 0.3 °C/s, (b) 3 °C/s, (c) 10 °C/s and (d) 30 °C/s [8].

Flow behavior of molten metal in casting processes has been regarded as an unavoidable phenomenon during solidification. It has been known that forced convection in a melt affects dendrite growth morphology. Therefore, quantitative understanding of combined flow behavior of molten metal during solidification is necessary for controlling microstructures. The modified CA model coupled with a continuum model for calculating solute transfer by both convection and diffusion during solidification was used to investigate the dendrite growth morphology solidified in a flowing melt. Figure 13.15 indicates the simulated dendrite growth morphology, solute profiles and flow velocity vectors solidified from an undercooled melt ($\Delta T = 12$ K) in a flowing melt for various solute contents: (a) 1.5 mass% Cu, (b) 3.0 mass% Cu and (c) 5.0 mass% Cu. For this simulation, the computational domain is divided into 201×201 cells with a cell size of 0.4 μm. The figures on the left column indicate the dendrites growing in a static melt and the ones on the right column the dendrites growing in a flowing melt with a flow velocity of 0.01 m/s. It is apparent that with the increase of solute concentration the dendrites become finer with the enhanced sidebranches, as observed experimentally [29]. It is also to be noted as shown in the figure that the dendrite arms in the upstream direction grow faster, while the growth of the dendrite in the downstream direction is much delayed. The effect of flow on the asymmetry of dendrite growth is closely related to the solute concentration. As the solute concentration increases, the asymmetric growth of dendrites increases.

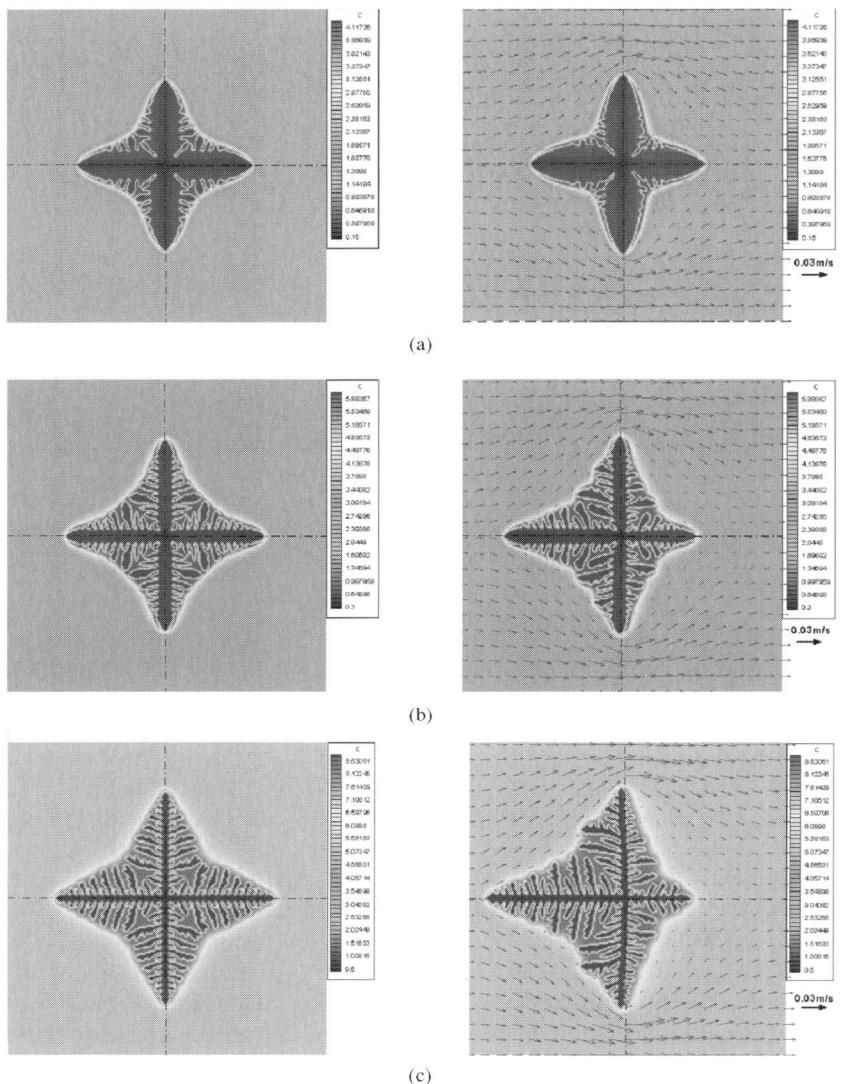

Figure 13.15. The simulated dendrite growth morphology solidified from an undercooled melt ($\Delta T = 12$ K) in a flowing melt for various solute contents: (a) 1.5 mass% Cu, (b) 3.0 mass% Cu and (c) 5.0 mass% Cu.

Figure 13.16 indicates the simulated dendrite growth morphology, solute profiles and flow velocity vectors of an Al–3 mass% Cu alloy solidified from an undercooled melt ($\Delta T = 12$ K) in a flowing melt for various flow velocities and directions: (a) without flow, (b) horizontal flow with a flow velocity of 0.03 m/s, and (c) flow from 45° with a flow velocity of 0.02 m/s. For this

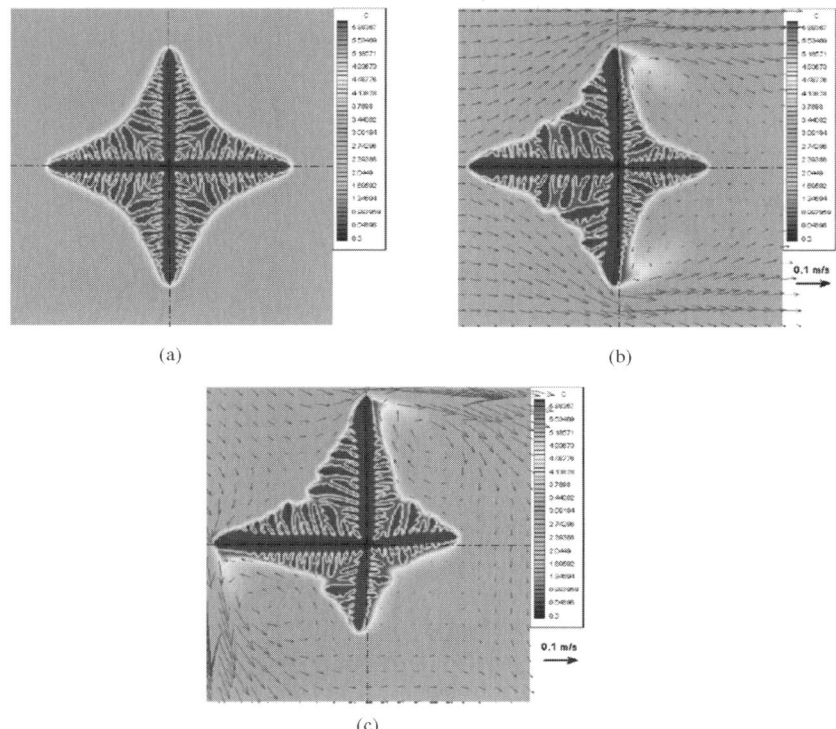

Figure 13.16. The simulated dendrite growth morphology solidified from an undercooled melt ($\Delta T = 12$ K) in a flowing melt for various flow velocities and directions: (a) without flow, (b) horizontal flow with a flow velocity of 0.03 m/s, and (c) flow from 45° with a flow velocity of 0.02 m/s.

simulation, the computational domain is divided into 201 × 201 cells with a cell size of 0.4 μm. In the beginning of simulation, one nucleus with the preferential growth orientation of 0° with respect to the horizontal direction was assigned in the center of the computational domain. The top and bottom surfaces of the domain are assumed to be the symmetrical boundaries. As shown in the figure, the dendrite arms in the upstream direction grow faster, while the growth of the dendrite in the downstream direction is much delayed.

As the dendrite grows, solute atoms are rejected in liquid ahead of the solid/liquid interface, which are washed away from the upstream to the downstream direction by fluid flow, resulting in the asymmetrical solute profile in liquid, i.e., the concentration in the left region is lower than that in the right region. According to equation (13.4.2), under a uniform temperature field, the lower the concentration, the larger the local undercooling. Therefore, the dendrite growth velocity in the upstream region is faster than that in the downstream region, resulting in asymmetrical dendritic

244   *Modelling of microstructure evolution*

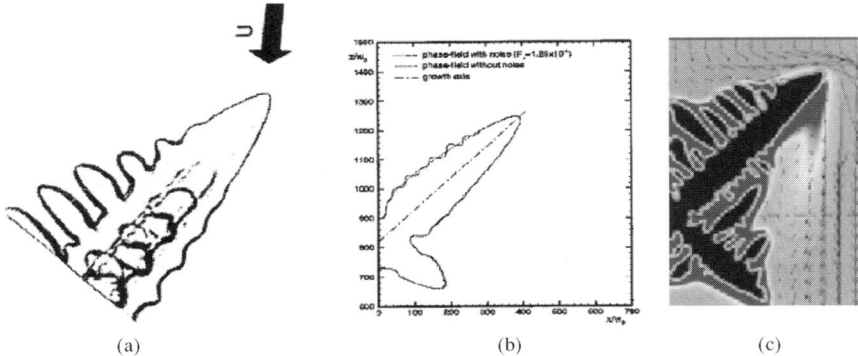

Figure 13.17. A single dendritic growth with fluid flow from 45°, indicating sidebranching in the upstream direction: (a) the experimental result [30], (b) a phase field simulation [31] and (c) a modified CA simulation [7].

growth morphology, such as the deflection behavior of dendrite arms and the sidebranching in the upstream direction.

Figure 13.17 indicates that when single dendritic growth occurs with fluid flow from 45° degrees, side branches appear in the upstream direction: (a) the experimental result [30], (b) a phase field simulation [31], and (c) a modified CA simulation. In case of a phase field simulation, if there is no artificial noise as indicated by dotted lines, no differences in dendrite morphology between the upstream and downstream directions are found. However, according to the modified CA simulation, asymmetrical growth and sidebranching for a dendrite in the upstream direction can be precisely predicted.

13.5   Case studies

13.5.1   Simulation of solidification grain structures by classical cellular automaton models

13.5.1.1   Description of the problem

Generally, solidification grain structures consist of three parts: the competitive growth zone including the chill zone, the columnar zone and the equiaxed zone. In the competitive growth zone near the chill surface, high cooling rates cause relatively high growth velocities. In the columnar zone, the growth velocity is relatively low due to the latent heat of freezing released ahead of the solidification front. However, as the solidification takes place in the columnar zone, the growth velocity steadily increases because of the increased local undercooling, which is caused by the decreased temperature gradient in the bulk liquid. Finally, when the local undercooling of the bulk liquid is greater than that which is necessary for heterogeneous

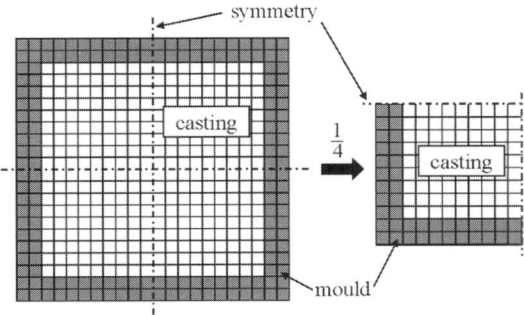

Figure 13.18. Illustration of the computational domain.

nucleation, equiaxed grains can nucleate and grow in the bulk liquid. This results in the columnar-to-equiaxed transition.

In this example, we will simulate the evolution of grain structures of Al alloys solidified in a metal mould and study the columnar-to-equiaxed transition. Figure 13.18 indicates a schematic drawing of the computational domain with control volumes, which is used for the simulation of solidification of Al alloys in a metal mould by the finite volume method. Each control volume consists of 10 × 10 square CA cells for the cellular automaton model. The thickness of a metal mould is 4 mm and the dimension of the casting is 20 mm × 20 mm. The metal mould is exposed to air.

13.5.1.2 Thermo-physical properties and nucleation parameters

Table 13.3 indicates the thermal and physical properties. The nucleation parameters used in the present simulation is shown in table 13.4.

Table 13.3. Thermal and physical properties.

Symbol	Meaning	Casting (Al–2.5 mass% Cu)	Mould (SKD61)
T_L	Liquidus temperature	928	
T_E	Eutectic temperature	822	
T_m	Melting temperature	933	
k	Partition coefficient	0.17	
C_p	Specific heat (J/Kg · K)	1086	580
ρ	Density (kg/m^3)	2780	7890
λ	Thermal conductivity (W/m · K)	192.5	14.4
ΔH_V	Volumetric latent heat (J/m^3)	1.107×10^9	
H_{int}	Heat transfer coefficient at the mould/casting interface (W/m^2 · K)	400	
H_{air}	Heat transfer coefficient at the mould/air interface (W/m^2 · K)	15	

246 *Modelling of microstructure evolution*

Table 13.4. The nucleation parameters.

T_{pour} (K)	$n_{max,s}$ (m^{-2})	$\Delta T_{nuc,s}$ (K)	$\Delta T_{\sigma,s}$ (K)	$n_{max,b}$ (m^{-2})	$\Delta T_{nuc,b}$ (K)	$\Delta T_{\sigma,b}$ (K)
1093	2.5×10^4	0.5	0.1	2.5×10^4	5.0	0.7
983	1.2×10^6	0.5	0.1	3.0×10^7	5.0	0.7
933	1.5×10^6	0.5	0.1	3.0×10^9	5.0	0.7

Note: The subscripts s and b correspond to the nucleation parameters on the mould surface and in the bulk liquid, respectively.

13.5.1.3 Simulation

Let us simulate solidification grain structures with various pouring temperatures using the execution file [ca.exe].

Data input
```
> Input Tpour(K): 1093
> Input dTnuc,s and dTnuc,b: 0.5 5.0
> Input dTd,s and dTd,b: 0.1 0.7
> Input Nmax,s(1-939) and Nmax,b(1-219961): 25 25
Maximum dt: 0.0001568
> Input dt(time increment): 0.00015
```

Results
Figure 13.19 indicates the simulated grain structures of an Al–2.5 mass% Cu alloy solidified in a metal mould with various pouring temperatures. As shown in the figure, a fully columnar structure was obtained when the pouring temperature is 1093 K. As the pouring temperature decreases, the length of the columnar zone decreases, and finally when the pouring temperature is 933 K, a fully equiaxed grain structure except for the chill zone is obtained.

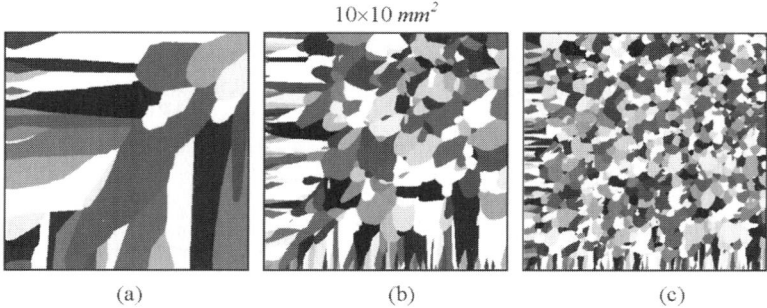

Figure 13.19. The simulated grain structures of an Al–2.5 mass% Cu alloy: (a) a fully columnar structure with $T_{pour} = 1093$ K, (b) a CET region with $T_{pour} = 983$ K and (c) a fully equiaxed structure with $T_{pour} = 933$ K.

```
---- Start of Iterative Loop ----
Iter( 2000) Time( 0.3) CAST:Tmax(1091.2) Tmin(1080.9) fs( 0.0)
.....................omitted...................
Iter( 94000) Time(14.1) CAST:Tmax(926.9) Tmin( 919.4) fs( 46.8)
Iter( 96000) Time(14.4) CAST:Tmax(925.7) Tmin( 918.1) fs( 66.5)
Iter( 98000) Time(14.7) CAST:Tmax(924.1) Tmin( 916.8) fs( 87.4)
Iter(100047) Time(15.0) CAST:Tmax(922.2) Tmin( 915.2) fs(100.0)
---- End of Calculation ----
```

13.5.2 Simulation of dendritic growth by modified cellular automaton models

13.5.2.1 Description of the problem

As the dendrite grows, solute atoms are rejected in liquid ahead of the solid/liquid interface, which are washed away from the upstream to the downstream direction by fluid flow, resulting in the asymmetrical solute profile in liquid, i.e., the concentration in the upstream region is lower than that in the downstream region. According to equation (13.4.2), under a uniform temperature field, the lower the concentration, the larger the local undercooling. Therefore, the dendrite growth velocity in the upstream region is faster than that in the downstream region, resulting in asymmetrical dendritic growth, such as the deflection behavior of dendrite arms and the sidebranching in the upstream direction.

Figure 13.20 illustrates the computational domain used in the present simulation. The computational domain is divided into 201 × 201 cells with a cell size of 0.4 μm and a radius of the seed of 4.43 μm. It is assumed that melt flows into the domain with a bulk flow velocity of U_{in} from left to

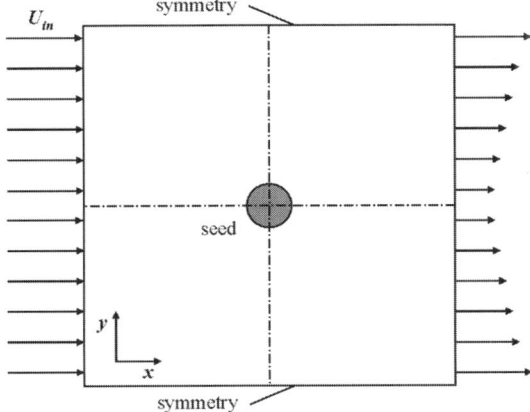

Figure 13.20. Illustration of the computational domain used for the simulation of a free dendritic growth from an undercooled melt with melt convection.

248   *Modelling of microstructure evolution*

right. The top and bottom surfaces of the domain are assumed to be the symmetrical boundaries.

13.5.2.2 Simulation

Let us simulate dendritic growth with and without fluid flow using the execution file [mca.exe]. The effect of the preferential growth orientation are examined for three cases: 0°, 30° and 45° with two kinds of Cu content.

(i) Without flow

Data input
```
> Input Cu concentration (0-8 mass%): 3
> Input preferential growth orientation (0-90 deg.): 0
> Input undercooling (K): 12
> Input final solid fraction (%): 15
> Input flow velocity (m/s): 0
```

Results
```
* Flow velocity     = 0
* Reynolds number   = 0
* Fluid density     = 2475
* Laminar viscosity = 0.00014

---- Start of Iterative Loop ----
time(1.289254e-004) iter( 20) fs(0.0011)
.................omitted.............
time(4.165477e-003) iter(420) fs(0.1230)
time(4.372631e-003) iter(440) fs(0.1349)
time(4.572554e-003) iter(460) fs(0.1471)
---- End of Calculation ----
```

(ii) With flow

Data input
```
> Input Cu concentration (0-8 mass%): 3
> Input preferential growth orientation (0-90 deg.) : 0
> Input undercooling (K) : 12
> Input final solid fraction (%) : 15
> Input flow velocity (m/s) : 0.03
```

Results
Figures 13.21 and 13.22 indicate the simulated dendritic growth morphology of Al–Cu alloys with and without flow in the bulk liquid for various preferential growth orientations: (a) 0°, (b) 30° and (c) 45°. The figures show that the deflection of the primary dendrite arms occurs in the upstream direction. As the concentration of solute increases from 1.5mass%Cu to 3.0mass%Cu, the sidebranching of dendrites increases.

Case studies 249

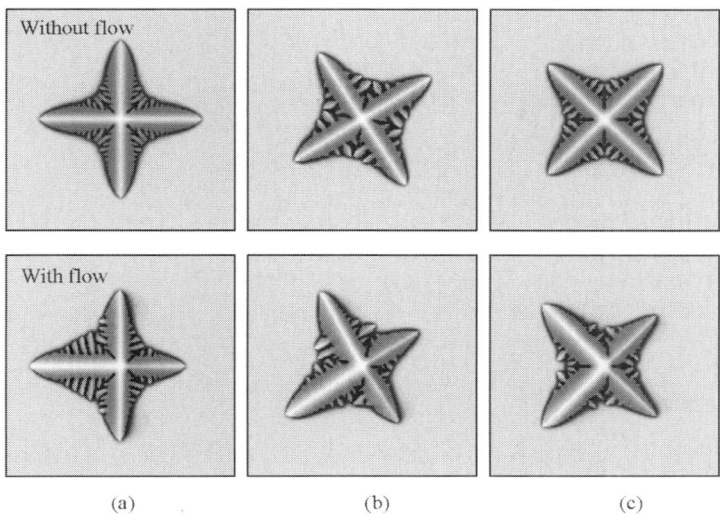

Figure 13.21. The simulated dendritic growth morphology of an Al–1.5 mass% Cu alloy solidified from an undercooled melt ($\Delta T = 12\,\mathrm{K}$) for various preferential growth orientations: (a) 0°, (b) 30° and (c) 45°. The bulk flow velocity is 0.03 m/s.

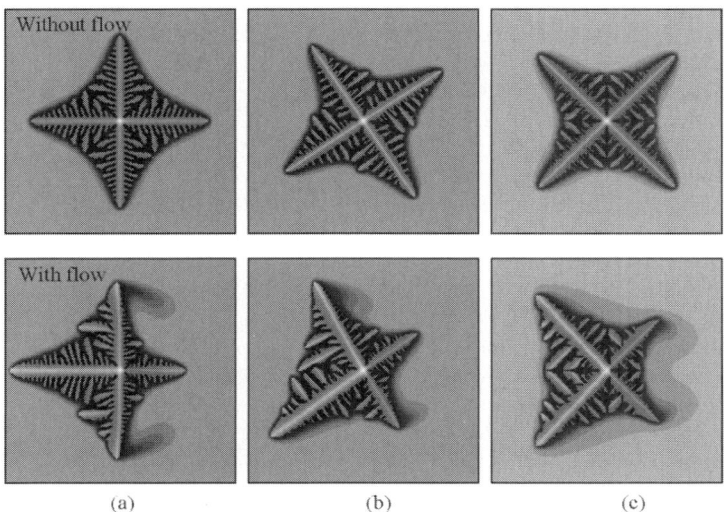

Figure 13.22. The simulated dendritic growth morphology of an Al–3.0 mass% Cu alloy solidified from an undercooled melt ($\Delta T = 12\,\mathrm{K}$) for various preferential growth orientations: (a) 0°, (b) 30° and (c) 45°. The bulk flow velocity is 0.03 m/s.

```
* Flow velocity      = 0.03
* Reynolds number    = 530357
* Fluid density      = 2475
* Laminar viscosity  = 0.00014

---- Start of Iterative Loop ----
time(1.157297e-004) iter( 20) fs(0.0011)
................omitted................
time(3.447532e-003) iter(520) fs(0.1218)
time(3.567567e-003) iter(540) fs(0.1311)
time(3.686503e-003) iter(560) fs(0.1410)
---- End of Calculation ----
```

References

[1] Stefanescu D M 1995 *ISIJ Int.* **35** 637
[2] Kobayashi R 1993 *Physica D* **63** 410
[3] Brown S G R and Spittle J A 1989 *Mater. Sci. Tech.* **5** 362
[4] Zhu P and Smith R W 1992 *Acta Metall. Mater.* **40** 683
[5] Rappaz M and Gandin Ch-A 1993 *Acta Metall. Mater.* **41** 345
[6] Zhu M F and Hong C P 2001 *ISIJ Int.* **41** 436
[7] Hong C P and Zhu M F 2003 *Proceedings of 8th Asian Foundry Congress* Bangkok, Thailand, p 2
[8] Zhu M F and Hong C P 2001 *ISIJ Int.* **41** 992
[9] Maxwell I and Hellawell A 1975 *Acta Metall.* **23** 229
[10] Thevoz P, Desbioles J L and Rappaz M 1989 *Metall. Trans.* **20A** 311
[11] Kurz W, Giovanola B and Trivedi R 1986 *Acta Metall.* **34** 823
[12] Lipton J, Kurz W and Trivedi R 1987 *Acta Metall.* **35** 957
[13] Lee K Y and Hong C P 1997 *ISIJ Int.* **37** 38
[14] Wolfram S 1983 *Rev. Mod. Phys.* **55** 601
[15] Chang Y Y, Lee S M, Lee K Y and Hong C P 1998 *ISIJ Int.* **38** 63
[16] Gandin Ch-A and Rappaz M 1994 *Acta Metall. Mater.* **42** 2233
[17] Hong C P, Umeda T and Kimura Y 1984 *Metall. Trans.* **15B** 91&101
[18] Ahn J Y, Lee K Y, Lee S M and Hong C P 1995 *Modeling of Casting Welding and Advanced Solidification Processes—VII* ed M Cross *et al* (Warrendale, PA: TMS Publications) p 687
[19] Cho I S and Hong C P 1997 *ISIJ Int.* **37** 1098
[20] Sasikumar R and Sreenivasan R 1994 *Acta Metall. Mater.* **42** 2381.
[21] Nastac L 2000 *Proc. Modeling of Casting and Solidification Processes—IV* ed C P Hong *et al* (Seoul: Hanrimwon) p 31
[22] Shin Y H and Hong C P 2002 *ISIJ Int.* **42** 359
[23] Beltran-Sanchez L and Stefanescu D M 2003 *Metall. and Mater. Trans.* **34A** 367
[24] Zhu M F and Hong C P 2002 *ISIJ Int.* **42** 520
[25] Zhu M F and Hong C P 2002 *Phys. Rev. B* **66** 155428
[26] Hong C P and Zhu M F 2003 *Proc. Modeling of Casting Welding and Advanced Solidification Processes—X* ed D M Stefanescu *et al* (Warrendale, PA: TMS) p 63

[27] Zhu M F and Hong C P 2004 *Metall. and Mater. Trans.* to be published in 2004
[28] Esaka H 1986 PhD Thesis No. 615, Ecole Polytechnique Fédérale de Lausanne
[29] Hutt J E C, St John D H, Hogan L and Dahle A K 1999 *Mater. Sci. and Tech.* **15** 495
[30] Bouissou P, Perrin B and Tabeling P 1989 *Phys. Rev. A* **40** 509
[31] Tong X, Beckermann C, Karma A and Li Q 2001 *Phys. Rev. E* **63** 061601

Index

acceptor cells, 171
adiabatic, 26
analytical solution, 124–126
atomistic simulation, 221
asymmetric growth, 241
asymmetrical dendritic growth, 247

back pressure, 198
backward difference, 39–40
balance equations, 14
bilinear interpolation, 166
body fitted coordinate (BFC) system, 180
   discretization method, 184–195
   surface cell, 196–200
   transformation, 182–184
   VOF method, 195–196
body force, 19
boundary cell, 167
boundary conditions, 22, 25–26, 27–28, 48–50, 54–55, 171–172, 196
   BFC, 196
   flow, 171–172
   heat conduction, 48–50, 54–55, 73
   Γ_0-type boundary, 54
   Γ_I-type boundary, 54
   Γ_{II}-type boundary, 54
   Γ_{III}-type boundary, 54
   Γ_{IV}-type boundary, 55
   Γ_V-type boundary, 55
boundary element method (BEM), 38
boundary layer, 3–13
   mass transfer, 9
   momentum, 12–13
   thermal, 3–5
Brody–Flemings model, 107

bulk fluid, 2–3, 8–9, 12, 24, 28
buoyancy, 3
buoyancy force, 198

CA-FVM model, 227–232
Cartesian coordinate system, *see* rectangular coordinate system
Cartesian velocity, 185, 190–191
cellular automaton (CA), 220, 225–244
   classical, 220, 225–232
   modified, 220, 232–244
cells, 228
central difference, 39–41
central difference scheme, 127–129
columnar to equiaxed transition (CET), 232
chain rule, 182
checker-board pressure field, 151
classical cellular automaton, 220, 225–232
computational domain, 49, 50, 182
computational grids, 50–51
conduction, 1–2
conductivity, 1
conservation laws, *see* balance equations
continuity equation, 16–19, 127
   integrated, 127
continuum model, 234, 235, 237
contravariant velocity, 184, 192–193, 199
   surface cell, 199
control volume, 14, 44–47, 50–51, 184, 228
   BFC, 184
   classical CA, 228

253

254   *Index*

control-volume form, 16
control-volume nodes, 228
control-volume surface, 14, 50, 184
　　BFC, 184
convection, 2–4
　　forced, 3
　　free, 3
convective heat transfer coefficient, 4–5
convective mass transfer, 8–9
convective mass transfer coefficient, 28
convective momentum transfer, 11–13
convective transfer, 33
correction formulae, 144
coupling, 227–231
　　classical CA, 227–231
　　modified CA, 234–237
Crank–Nicolson method, 79
cross derivatives, 186
cross diffusion term, 186
crystallographic orientation, 223
curvature, 233
curved gating system, 212
curved shaped surface, 179
CV nodes, *see* control volume nodes
cylindrical coordinate system, 16–17

damping effect, 146, 210
deflection behavior, 244
dendrite arm spacing, 107
deterministic model, 219–220
deviation, *see* standard deviation
diffusion, 6–7
diffusion coefficient, 27
diffusivity, 11, 27, 31
　　mass, 27
　　momentum, 11, 31
　　thermal, 31
discretization method, 38–47, 51–53,
　　74–75, 152–160, 184–191
　　BFC, 184–191
discretization schemes, 127–134
　　comparison, 133–134
discretized equations, 46, 53, 58–59, 75,
　　85, 155–156, 158–159, 188–189,
　　190–191
　　BFC, 188–189, 190–191
donor and acceptor flux approximation
　　(DAFA), 195

donor cells, 171
double diffusive convection, 146, 210
downstream direction, 241–247

elliptic equation, 136–137
emmissivity, 5
empty cell, 167
enthalpy method, 111–112
equilibrium solidification, 106
equivalent specific heat method,
　　109–111
Eulerian approach, 166
Eulerian cells, 166
Eulerian mesh, 166
evolution, *see* microstructure evolution
explicit method, 76–78, 139

F value, 168–169, 209
Fick's law of diffusion, 7, 31–32
finite difference methods (FDM),
　　37–47
finite element method (FEM), 38
finite volume method (FVM), 38, 44–47,
　　50–55, 227
first law of thermodynamics, 14, 23
fixed grid methods, 103–104
forced convection, 3, 241
forward difference, 39–40
forward-time and centered-space,
　　76–77
Fourier number, 77
Fourier's law of conduction, 1–2,
　　30–31
free convection, 3
free slip condition, 22
free stream, 12
free surface, 165–171, 207–211
　　tracking, 165–171, 207–211
FTCS, 76
full cell, 167

gate, 167, 207
Gauss divergence theorem, 16
Gaussian distribution, 223
generalized coordinate system, 182
geometric coefficient, 184
globular, 239–241
globular morphology, 240

governing equations, 14–29, 33–35, 105, 124, 148–150
  for energy transfer, 23–26
  involving latent heat, 105
  for mass transfer, 14–19
  for momentum transfer, 19–22
  for species transfer, 26–29
  for steady convection and diffusion, 124
gravity casting, 205
grid point, 38
growth algorithm, 226–227, 234
  classical CA, 226–227
  modified CA, 234
growth kinetics, 223–224

heat conduction, 48–100
  steady state, 48–72
  transient, 73–100
heat flux, 1
heat transfer, 1–5
heat transfer coefficient, 4
heterogeneous nucleation, 221, 223, 244–245
homogeneous nucleation, 221, 223, 244–245
hybrid difference scheme, 130–131

implicit method, 78–79, 139
incompressible flow, 16, 20
initial conditions, 25–26, 27–28, 48–50, 54–55, 73
  heat conduction, 48–50, 54–55, 73
inlet condition, 22
integral form, see control-volume form
integral method, 41–44

Jacobian of the transformation, 183

KGT model, 223
kinetic energy, 23
kinetics, 221–224

Lagrangian approach, 166
Lagrangian mesh, 166
laminar flow, 10
length scale, 221
lever rule, 106

linear interpolation, 127, 150
LKT model, 223
local equilibrium, 236

marker and cell, 140–141, 166–168, 208–209
  free surface, 166–168, 208–209
MAC method, see marker and cell
marker, 166–168, 208–209
marker particle, 166–168, 208–209
mass transfer, 6–9
  convective 8–9
maximum density of nuclei, 223
MCA, see modified cellular automaton
mesh, 83
microstructure evolution, 219–250
  with flow, 241–244
modified cellular automaton, 220, 232–244
molecular dynamics, 221
momentum interpolation method, 152, 184, 199
momentum transfer, 9–13
  convective, 11–13
  viscous, 10–11
momentum transfer coefficient, 12
Monte Carlo (MC) model, 220
mould cavity, 205
mould filling, 205–215
mushy zone, 105

Navier–Stokes equation, 20–21, 135–136
neighboring nodes, 50
neighbors, 159, 234
Newton's law, 11–13, 31
  of cooling, 13
  of viscosity, 11, 31
Newtonian fluids, 11
non-dendritic, 239–241
non staggered grids, 150, 184
non-orthogonal grids, 179
normal stress condition, 171, 196
no-slip condition, 10, 22, 236
nucleation cell, 226
nucleation model, 221–223, 226
  continuous, 221
  heterogeneous, 221
  instantaneous, 221

nucleation parameters, 223

one-dimensional
  convection–diffusion, 40–41, 44, 124
  diffusion, 6
  heat conduction, 1, 60, 86
outlet condition, 22

parabolic equation, 136–137
partial differential form, 16
partition coefficient, 236
Peclet number, 125
phase change problems, 101–114
  alloy solidification, 105–114
  fixed grid methods, 103–104
  transformed grid methods, 104
  variable grid methods, 104
physical domain, 84, 104, 180, 182
Poisson equation, 137
potential energy, 23
potential flow, 48
power-law scheme, 131–132
preferential growth orientation, 234
pressure correction, 138, 144–145, 160–163, 192–195
pressure gradient, 146, 158, 189
primary diffusion term, 186
primitive variable , 138

radiation, 5
radiation heat transfer coefficient, 26, 49
rate equations, 14
rectangular coordinate system, 16–17
relaxation factor, 191
Reynolds number, 10
Reynolds stress, 173
Reynolds stress equation, 172–173
rosette-like morphology, 239–241
runner, 207

Scheil model, 107
secondary diffusion term, *see* cross diffusion term
semi-circular core, 174, 200, 215
semi-solid, 239–240
sidebranching, 244, 248
similarities, 30–36

SIMPLE method, 143–145, 160–165, 191–195, 210–211
  algorithm, 160–165
  BFC, 191–195
  surface tracking, 210–211
SIMPLE-BFC-VOF method, 212–214
SIMPLE-VOF method, 206, 211
SMAC method, 141–142
SOLA method, 142–143
SOLA-VOF method, *see* VOF method
solid fraction, 105
solution scheme of fluid flow, 163–165, 194–195
specific heat, 3
spherical coordinate system, 16–17
sprue, 207
stability criterion, 78, 80–84
staggered grids, 150
standard deviation, 223
steady state, 10, 48
  heat conduction, 48–55
Stefan–Boltzmann law, 5
stochastic model, 219–220
stream function, 137
surface cell, 168, 196
surface phenomena, 24
surface tension, 171
surface tension pressure, 172

tangential stress condition, 171, 196
Taylor-series formulation, 38–39
TDMA, 55, 146
temperature gradient, 1
temperature recovery method, 107–109
thermal diffusivity, 31
thermal energy, 23
Thomas algorithm, 55
three-dimensional
  coordinate system, 17
  heat conduction, 59–60, 84
thixotropic, 239
time step, 229–231, 237
transformed grid methods, 104
transient, 41, 73
  heat conduction, 73–79
transport phenomena, 1, 14, 30
tri-diagonal matrix algorithm, 55
truncation error, 39

turbulent flow, 10. 172–173
turbulent stress, *see* Reynolds stress
two-dimensional
  BFC, 186–191
  CA, 225–231, 232–237
  flow, 136–138, 153–160
  heat conduction, 58–59, 84

undercooling, 223, 233
  local, 233
  nucleation, 223
under-relaxation factor, 191
unsteady condition, 23
upstream direction, 243–247
upwind difference scheme, 129–130

variable grid methods, 104

velocity correction, 144–145, 160–161, 192–195
viscosity, 11
  kinematic, 11
volume of fluid, 168–171, 195–196, 209–211
  BFC, 195–196
  mould filling, 209–211
VOF, *see* volume of fluid
volumetric specific heat, 231
vorticity, 137
vorticity transport equation, 137
vorticity-stream function approach, 137–138

wall boundary condition, 171
wave number, 82